The Release of
Genetically Modified
Microorganisms—REGEM 2

FEDERATION OF EUROPEAN MICROBIOLOGICAL SOCIETIES SYMPOSIUM SERIES

Recent FEMS Symposium volumes published by Plenum Press

A Continuation Order Plan is available for this series. A continuation order will bring delivery of each new volume immediately upon publication. Volumes are billed only upon actual shipment. For further information please contact the publisher.

The Release of Genetically Modified Microorganisms — REGEM 2

Edited by

Duncan E. S. Stewart-Tull

University of Glasgow
Glasgow, Scotland
United Kingdom

and

Max Sussman

Medical School
University of Newcastle upon Tyne
Newcastle upon Tyne
United Kingdom

PLENUM PRESS • NEW YORK AND LONDON

Library of Congress Cataloging-in-Publication Data

The Release of genetically modified microorganisms--REGEM 2 / edited
by Duncan E.S. Stewart-Tull, Max Sussman.
       p.   cm. -- (FEMS symposium ; no. 63)
    "Proceedings of a symposium held under the auspices of the
Federation of European Microbiological Societies, August 29-31,
1991, in Nottingham, United Kingdom"--T.p. verso.
    Includes bibliographical references and index.
    ISBN 0-306-44302-3
    1. Recombinant microorganisms--Environmental aspects--Congresses.
I. Stewart-Tull, D. E. S. (Duncan E. S.) II. Sussman, Max.
III. Federation of European Microbiological Societies. IV. Series.
QR100.R45 1992
660'.65--dc20   92-30222
                                                              CIP

Proceedings of a symposium held under the auspices of the
Federation of European Microbiological Societies,
August 29–31, 1991, in Nottingham, United Kingdom

ISBN 0-306-44302-3

© 1992 Plenum Press, New York
A Division of Plenum Publishing Corporation
233 Spring Street, New York, N.Y. 10013

Printed in the United States of America

THE LORD IRONSIDE

FOREWORD

If ripple effect is a measure of greatness in scientific discovery then GEMMOs have a lot going for them and this book dramatically illustrates the risks associated with advances being made by researchers to mobilize and control the power of the microorganism in the world's fight to perfect nature and find remedies for its imperfections.

In the field of genetic science it is abundantly clear that so much more can be achieved through prevention rather than cure and that the indirect kill, by reason of its logic is a much more powerful weapon for winning results. Nevertheless the dilemma facing politicians arises over whether man should tamper with something which is God-given such as Radioactivity and Genetic endowment.

The Roman Catholic church finds difficulty in accepting the proposition that what is God-given can be treated as a product under human control and maybe that is why recently half a century of genetic research on a strain of bees resistant to a devastating parasite at the Buckfastleigh Benedictine Monastery has inexplicably ceased whilst verging on scientific success.[1] The Anglican Community on the other hand does not see the sacrosanctity of Radioactivity and Genetic material as a bar to man-manipulation with appropriate safeguards.

In seeking appropriate safeguards we should consult widely with countries having similar cultural heritage to our own to ensure that we take account of the through life costs and logistics of genetic modification, without constraining the researcher unnecessarily in his quest for knowledge. Regulations drafted more out of fear of the unknown than confidence in the outcome will bring no rewards for research or society.

The advantages to be obtained from the release of GEMMOs are exciting and regulatory constraints must reflect the future benefits likely to accrue to Society. Pressure Groups who have battled over the nuclear industry are now setting their sights on biotechnology and the ripple effects must generate their own harmonic to keep out of trouble. This book helps all readers to understand the implications of the scientific advances being made.

> The Lord Ironside
> Member of the Parliamentary
> and Scientific Committee
> House of Lords
> London
> SW1 0PW, UK

(1) The Sunday Times, London, 1 March 1992.

PREFACE

The potential benefits which could be derived from the use of genetic modification techniques are apparent. However, scientists are aware of their obligations to society and are striving to ensure that the release of modified microorganisms will be safe. The first international conference on release, REGEM 1, was held in Cardiff, UK in 1988. It was a recognized success and there was agreement that 'release' should be discussed regularly . Consequently, under the sponsorship of the Federation of European Microbiological Societies, the Society for Applied Bacteriology and the Society for General Microbiology, REGEM 2 was held in Nottingham, UK in the autumn of 1991. In his message of welcome Sir Hans Kornberg said "I greatly welcome the proposal to hold a second Conference on the release of genetically-modified microorganisms. Not only has much scientific information and experience been gained since REGEM 1 in 1988, but this year sees the implementation in national law of the two European Directives on Biotechnology. The provisions of the 'Deliberate Release' Directive will put in place a Community-wide, harmonized, approach to the regulatory oversight of both small-scale trials and of commercial-scale releases of GEMMOs. It will clearly be of benefit to all who use and intend to release GEMMOs into the environment to familiarize themselves with these important advances, and to attend REGEM 2."

This book forms the permanent record of the Plenary sessions, Workshops and Poster contributions. The text of the camera-ready manuscript was prepared from computer-disk presentations with conversion programmes to Apple Mac. Mistakenly, it was believed this would be straightforward but imagine the reaction to thirty-four pages of the following converted hieroglyphics:

˘ÉM˘ icrobiol. 57, 366-373.ÆÆ˘ÉM˘ cDermott, ˘ÉJ˘ .B., ˘ÉU˘ nterman, R., ˘ÉB˘ rennan, ˘ÉM˘ .J., ˘ÉB˘ rocks, ˘ÉR˘ .E., ˘ÉM˘ obley, ˘ÉD˘ .P., ˘ÉS˘ chwartz, ˘ÉC˘ .kcC. and ˘ÉD˘ ietrich, ˘ÉD˘ .K. (1989). Two strategies for ˘ÉP˘ CB soil ˘Ér˘ emediation:biodegradation and ˘És˘ urfactant extraction. Environ. ˘ÉP˘ rog. 8, 46-51.ÆÆ˘ÉM˘ ileski, ˘ÉG˘ .J., ˘ÉB˘ umpus, ˘ÉJ˘ .A., ˘ÉJ˘ urek, ˘ÉM˘ .A. and ˘ÉA˘ ust, ˘ÉS˘ .D. (1988). Biodegradation of ˘ÉP˘ entachlorophenols by the White Rot fungus ˘ÅP˘ hanerochoraete ˘Ac˘ hrysosporium˘ . ˘ÉA˘ ppl. Environ. ˘ÉM˘ icrobiol. 54, 2885-2889.ÆÆ˘ÉO˘ livieri, R., ˘ÉB˘ acchin, P., ˘ÉR˘ obertiello, A., ˘ÉO˘ ddo, N., ˘ÉD˘ egen, L. and

Sanity was maintained with the constant computer expertise provided in Glasgow by Mulu Gedle — the extent of our gratitude is considerable. Without her contribution this book would have ended up as one of those symposium volumes with different fonts and typefaces of varying shades of grey.

The help provided by Janie Curtis, Nicola Clark and Gregory Safford of Plenum Publishing Corporation is gratefully acknowledged.

Duncan E.S. Stewart-Tull
Max Sussman

May 1992

ACRONYM

You will notice throughout the book that we have changed the acronym for genetically-modified microorganisms, because we were cautioned that GEMS conflicted with the Global Environment Monitoring Service used by WHO since 1975.

During the conference various alternatives were proposed to many of the delegates:

GMO	-	Gēē mō	(a little horse named MO ?)
GAMO	-	Gă mō	(GA, genetically-altered but slightly gaga or a general accident)
GROMs	-	Gr̄r̄ ŏ ms	(may be reorganized but sounds like a gremlin problem)
GEMOMO	-	Gĕ Mō Mō	(too near to Geronimo)

After two days of deliberation there was consensus agreement for:

GEMMO - Gĕ m mō (pl **GEMMOs**) - quite close to the original GEMs and formed from GEnetically-Modified Micro Organism. It even sounds a gem when translated literally into Dutch "Edelsteen maaien".

CONTENTS

PLENARY LECTURES

WORKSHOPS

POSTERS (UNREFEREED)

SUMMARY

ENVIRONMENTAL PRESSURE IMPOSED ON GEMMOS IN SOIL

J.D. van Elsas

Institute for Soil Fertility Research
P.O.Box 48
6700AA Wageningen
The Netherlands

This chapter will examine the different types of environmental stress imposed on GEMMOs in soil, and how these stress factors may affect GEMMO establishment and survival. In addition, putative selective advantages or disadvantages to which GEMMOs are subjected in relation to their heterologous DNA are considered. The focus is exclusively on genetically modified bacteria, because these organisms are currently mostly studied with respect to biosafety and effectivity. Evidence is presented which points to an ecological disadvantage of certain GEMMOs compared with their respective parent strains. On the other hand, selective pressure that favours certain GEMMOs, mainly biodegradative bacteria, may enhance their competitiveness in soil. It is concluded that because of the complexity of the soil system, it is virtually impossible to obtain a picture of all potential selective pressures that favour or disfavour introduced GEMMOs.

INTRODUCTION

Genetically modified microorganisms (GEMMOs), in particular bacteria, will certainly be released on a large scale into open-field situations, if regulatory constraints permit. Likely candidates for release world-wide are rhizobia with improved nitrogen-fixing efficiency, genetically modified avirulent derivatives of plant pathogens, for example *Pseudomonas solanacearum* , *Agrobacterium* spp. or *Erwinia carotovora* (Lacy and Stromberg, 1991) and ice-*Pseudomonas syringae* (Panopoulos, 1986). Other organisms developed for biological control purposes, such as pseudomonads and rhizobia that carry *Bacillus thuringiensis* insect toxin genes, have also been proposed for field releases (van Elsas *et al.*, 1991b; Waalwijk *et al.*, 1991; Skot *et al*, 1990; Watrud *et al.*, 1985). In addition, organisms modified to contain functional genes for the degradation of environmental pollutants, such as chlorinated organic compounds, are being developed and tested; these might be used for remediation of polluted soil (Short *et al.*, 1990; Ramos *et al.*,1991).

Release of unmanipulated bacteria has a long history of unpredictable and often disappointing results (van Elsas and Heijnen, 1990). One of the major problems has been the often poor establishment and survival of the introduced bacteria. Of the many attempts to establish certain inoculants in soils, in particular those carried out in the Soviet Union in the 1950s and early 1960s (Mishustin and Naumova, 1962), probably the only organism which was introduced with some degree of success, was *Rhizobium*. Nevertheless, the now firmly established *Rhizobium* inoculant industry continues to be plagued by lack of success of inoculation, which is often due to competition for nodulation by the less effectively nitrogen-fixing rhizobia indigenous to the soil (Thies *et al.*, 1991).

Given this recognized difficulty to establish an inoculum in a soil system and the generally poor performance of introduced populations, it is attractive to speculate about the potential

The Release of Genetically Modified Microorganisms
Edited by D.E.S. Stewart-Tull and M. Sussman, Plenum Press, New York, 1992

1

behaviour of GEMMOs derived from the wild-type organisms. Unless the extra genetic element can provide some predictable or unanticipated ecological advantage for the GEMMO, the general expectation is that the GEMMO will be at an ecological disadvantage as compared with the parent because of the extra metabolic load that leads to decreased fitness (Tiedje *et al.*,1989). However, this should not be regarded as certain, since previous work in chemostat systems on the effects of transposon Tn5 on bacterial fitness has shown unexpected growth advantages for Tn5-carrying strains as opposed to the parent strain (Biel and Hartl, 1983). Also, the presence of the bacteriophage *cos* site on a plasmid in an *Escherichia coli* host increased bacterial fitness (Edlin *et al.*, 1984). Finally, the presence of Tn10 in certain hosts also leads to increased competitiveness (Chao *et al.*, 1983). These examples serve to show that when novel genetic sequences are inserted, unexpected effects on fitness and survival of the host may result. Such effects may work out differently in soil as compared with a laboratory system, or they may be preferentially expressed in a stressed soil environment.

This chapter will focus on the types of stress imposed on GEMMOs in soil and on their differential effects on GEMMOs and parent strains. The soil environment is chosen, because most releases to be dealt with are into this ecosystem. After briefly discussing the dominant stress factors that occur in the soil environment and the overall behavioral response of introduced bacterial populations, the focus will be on specific population effects due to the extra genetic and physiological load carried by the GEMMO. Questions to be addressed will be if and to what extent the extra gene(s) affect bacterial establishment and persistence, and whether environmental stress acting positively on the introduced phenotype, possibly providing selective force, is able to promote its ecological persistence.

STRESS FACTORS IN SOIL

Soil represents an environment dominated by the solid phase, although liquid and gaseous compounds also play a role. Thorough descriptions of the intricacies of the soil environment and of its influences on soil microbial life have been provided elsewhere (Stotzky, 1986, 1989; Stotzky *et al.*, 1991; Oades, 1988) and are not within the scope of this chapter. Perhaps the most important characteristic of the soil environment is its heterogeneity, that is every site in soil may inherently be different from any other site, even at the scale of individual microbial cells (μm). This has serious implications for the life of bacteria present in that system, as well as for any interpretation of experimental data derived from it.

Many different factors in soil can affect the establishment and population dynamics ("fate") of introduced bacteria. These factors can be subdivided into biotic and abiotic factors. Biotic factors are, for example, antagonism, antibiosis, competition and predation or parasitism, brought about by the mixed community present in the soil. In addition, the presence of plant roots may be considered as a biotic factor. Abiotic factors of importance in soil are fluctuations in moisture content, nutrient supply, soil textural type, pH, temperature, presence of charged surfaces (clays) or plant surfaces such as roots. All of these factors may act differentially on various introduced organisms and, in addition, they may constitute a web-like interaction pattern. Stotzky *et al.* (1991) gave some examples of different soil factors acting upon incoming microbes. From this and other information it may be concluded that the effects of such factors on microbial fate can only be described in broad generalizations. This makes any prediction of the fate of incoming GEMMOs inherently difficult. In the following, however, a brief account will be given of well-established dominant soil factors that affect bacterial life in soil and their consequences for establishment and survival of incoming microbes.

Perhaps the most important abiotic factor that dominates the life of bacteria in soil is the presence of water, expressed as the soil water activity. With the exception of soils flooded with water, the most commonly found water condition of soil is one in which only certain pores of the soil void are completely water-filled, while other soil pores are merely covered by a water film or are dry. As outlined by Stotzky *et al.* (1991), this limits the possibilities for biological activity, since the many possible modes of bacterial development and activity, including movement and colonization of additional spaces in soil, are limited by spatial constraints. In addition, microbes in such water-depleted soil conditions must cope with water potential stress, that is they need mechanisms to maintain their cellular water content against the forces of soil water suction.

A second factor of importance is the gross lack of available substrate, in particular carbon, in most soils. It is widely accepted that soils are not inherently substrate-poor, but that readily-available bacterial nutrients are generally in short supply. Physical (spatial separation) and/or chemical (lack of degradability) constraints prevent bacteria from freely utilizing these nutrients. In addition, most carbonaceous compounds may already have been utilized by the indigenous microflora. An obvious implication for invading GEMMOs is that these will commonly enter a state of nutrient deprivation (starvation) in soil.

Soil textural type is another factor that affects bacterial establishment. Bacteria introduced into soils with a higher clay and/or silt content (finer-textured soils) showed better survival than those introduced into coarser-textured soils (Van Veen and van Elsas, 1986). This has been attributed to the capacity of the former soil to preserve a greater biomass, and this in turn has been linked to a higher number of protective microniches for bacteria (refuge sites) because of the larger number of pores with small orifices (Heijnen and van Veen, 1991).

Soil pH and soil temperature may also drastically affect bacterial establishment and subsequent survival in soil (Van Elsas and Trevors, 1991). Studies on the effects of pH on rhizobia in soil showed that organism survival tended to decrease at lower pH (Lowendorf *et al.*, 1981; Thornton and Davey, 1984). However, strains tolerant to lower pH have been selected and these were shown better to withstand soil acidity (Thornton and Davey, 1984). Temperature may act either directly on the introduced GEMMOs, or do so indirectly, via effects on the indigenous microflora. Van Elsas *et al.* (1991a) showed that survival of transposon Tn5 containing *Pseudomonas fluorescens* introduced into soil was higher at lower temperatures (4 and 15°C) than at the higher temperature used (27°C). Similarly, Wessendorf and Lingens (1989) found that survival of *P. fluorescens* in soil microcosms is better at 4 ^{o}C than at 25°C. Bolton *et al.* (1991), comparing survival of a psychrophilic *Pseudomonas* strain in soil microcosms, a climate chamber and the field, found survival in the field and climate chamber to be greater than in the soil microcosms. They related this observation to the lower average temperatures in the former two systems. It thus appears that lower temperatures in soil may be favorable for survival of invading microorganisms, possibly due to a reduction of biotic processes which tend to reduce their population size.

Finally, another abiotic factor of potential influence on bacterial establishment and survival in soil is the presence of compounds that exert selective pressure which directly favors the heterologous genes present in the introduced GEMMO. An example of such selection is soil polluted with heavy metals in which a virtual monoculture of *Alcaligenes eutrophus* carrying a heavy metal resistance plasmid developed (Mergeay, M, personal communication).

Biotic factors that exert pressure on introduced GEMMOs are related to biological interactions between organisms of the soil environment and the incoming GEMMO. Such interactions have been grossly classified by Strauss *et al.* (1986) and Stotzky *et al.* (1991), as 'neutral' where there is no advantage to any partner, 'parasitic' where there is exclusive advantage to one partner or 'beneficial' to both partners. The complex web of different biological interactions in the soil is probably responsible for the observed homeostasis of the soil system, that is the tendency of the system to counterbalance disturbances (Strauss *et al.*, 1986). In addition, natural soil has also been termed microbiostatic, that is soil generally does not permit growth of added microbes unless some disturbance to the system is brought about (Ho and Ko, 1985). Soil microbiostasis is probably also largely caused by biological factors, since it is relieved by soil sterilization (Ho and Ko, 1985). Amongst the many microbial interactions which potentially govern bacterial establishment and survival, predation by protozoa and competition for available substrate and space are probably predominant.

Predation by protozoa is an ubiquitous stress factor in soil and is probably most frequently responsible for the decline of introduced bacterial populations (Habte and Alexander, 1975, 1977; Heijnen *et al.*, 1988). Predators, however, do not usually completely eliminate their prey but do so only to a certain level. This has been attributed to an energy-expenditure phenomenon, on the assumption that at low prey densities it becomes too energy-expensive for predators to graze (Alexander, 1981). Competition between microbes for the sparse nutrients or available space -"ecological niches"- in the soil is a second biotic factor of utmost importance. That competition plays a large role has become apparent from experience with *Rhizobium* inoculants, which are often outcompeted by indigenous, less effectively nitrogen-fixing, strains (Thies *et al.*, 1991). For biosafety evaluation, it is equally important to obtain information about the competitive ability of the GEMMO *versus* the correponding wild-type or parent organism, since this determines possible hazards of GEMMOs outcompeting wild-type bacteria in soils.

It is evident that both the abiotic and the biotic factors mentioned, and their interactions, affect the fate (establishment, survival) of introduced bacteria, regardless of whether these are genetically modified or not. For a given soil it is inherently difficult to predict which factors will predominantly control the persistence of an inoculant. However, it seems safe to suggest that the homeostatic and microbiostatic nature of the soil environment represents a natural mechanism that controls the size of bacterial populations, including those released into the environment. In addition, released microbial populations will often be at an ecological disadvantage as compared with indigenous ones, since they tend to occupy less protected niches in the soil environment.

PATTERNS OF BEHAVIOR OF INTRODUCED BACTERIAL POPULATIONS

The population dynamics of an introduced GEMMO depends on its initial establishment and subsequent growth and/or survival, which in turn are heavily affected by the aforementioned stress factors. Measurements of bacterial dynamics in the heterogeneous soil environment provide overall pictures, which represent the net results of all the different forces acting on the population.

The gross patterns of behavior of microbial populations upon introduction into the environment have been classified into the three following categories Gillett *et al.*, 1984):

1. Rapid disappearance, to levels below the limit of detection, in a short time, for example 3 days or less.
2. Exponential decline for about 7 to 10 days, followed by persistence of a surviving population at a low population density,for example 10 to 100 cells per g of soil or ml liquid.
3. Relatively rapid exponential decline of 1 to 4 logs in 1 to 14 days, followed by a slow progressive decline to extinction.

The response of each introduced population will obviously also depend on the ecological hardiness of the organism, for example *Escherichia coli* introduced into soil, in which it is not a native organism, may react according to the first pattern (Henschke and Schmidt, 1990), *Bacillus* added to soil according to pattern 2 (van Elsas *et al.*, 1986), and *Rhizobium* spp. according to pattern 3 (Postma *et al.*, 1988). Fluorescent pseudomonads, genetically modified or unmodified, are widely developed for application to soils (Van Elsas *et al.*, 1991b; Waalwijk *et al.*, 1991; Kluepfel and Tonkyn, 1991). A survey of the behavioral patterns of different fluorescent pseudomonads upon introduction into soils revealed that, without exception, all decayed (Table 1). Decay rates differed widely, ranging from roughly 0.2 to 1 (Log decrease per 10 days), and probably depended on the type of organism and soil. The behavior was difficult to attribute to any of the patterns described above, and the major problem associated with this was the question of detection and eventual resuscitation of low numbers of surviving bacteria. However, the general response was inevitably a decline to low numbers, confirming the consensus statement about the fate of introduced bacteria in soil.

EFFECT OF THE GENETIC ALTERATION ON THE BEHAVIORAL PATTERN OF THE GEMMOS

Probably the most urgent question pertaining to the biosafety issue is the putative effect of the GEMMO on the ecosystem due to its genetic alteration, that is any effect different from the effect brought about by the wild-type organism. The establishment and survival of the GEMMO as opposed to that of the wild-type, affect the extent of these effects and are therefore of interest. In particular, the possibility of a GEMMO displacing its wild-type counterpart from a natural system, thereby effectively taking over the niche occupied by the wild-type and possibly establishing its novel trait permanently, has been of concern. At the same time, in such a situation, the effectiveness of the application would obviously be greatly increased.

To find a suitable "niche" and establish itself in a soil system, a GEMMO will probably have to compete with oligotrophic strains in a nutrient-deprived environment. It will thus rely on the availability of niches occupied by indigenous strains similar to itself, which use the same space and nutrients. The possible outcome of the competition of invading GEMMOs with both oligotrophic organisms and organisms with similar trophic behavior, and thus the likelihood of inoculant establishment, may be estimated by examining the available information on the behavior of the GEMMO *versus* the wild-type in natural (nonsterile) soil. These studies have been performed according to two different strategies, one in which the parent strain and

Table 1.   Decay rates of different fluorescent pseudomonads introduced into soils in soil microcosms or in field microplots[a]

Soil	Experimental system	Introduced strain/marker	Decay rate[b]	period (days)	Reference
Loamy sand	Field microplot	P. fluorescens (chr::Tn5)	0.5-0.6	60	Van Elsas et al., 1986
Silt loam	Field microplot	P. fluorescens (chr::Tn5)	0.3	60	Van Elsas et al., 1986
Loamy sand (wheat rhizosphere)	Field microplot	P. fluorescens (chr::Tn5)	0.8	60	Van Elsas et al., 1986
Silt loam (wheat rhizosphere)	Field microplot	P. fluorescens (chr::Tn5)	0.2	60	Van Elsas et al., 1986
Loamy sand	Microcosm	P. fluorescens (chr::Tn5)	0.8	55	Heijnen et al., in prep
L. sand + bentonite	Microcosm	P. fluorescens (chr::Tn5)	0.2	55	Heijnen et al., in prep
Loamy sand	Microcosm	P. fluorescens (RP4)	0.9	60	Van Elsas & Trevors, 1990
L. sand + bentonite	Microcosm	P. fluorescens (RP4)	0.4	60	Van Elsas & Trevors, 1990
Sandy loam	Microcosm	P. fluorescens Pf1-2 Rp^{R}	0.2-0.4	36	Compeau et al., 1988
Sandy loam	Microcosm	P. fluorescens Pf1-8 Rp^{R}	0.8	30	Compeau et al., 1988
Silt loam	Microcosm	P. fluorescens R1 Rp^{R}	1.1	29	Wessendorf & Lingens, 1989
Sandy loam	Microcosm	P. putida Pp1-2	0.7	30	Compeau et al., 1988
Clay loam	Field microplot	P. putida N-1R Rp^{R}	0.5	60	Dupler & Baker, 1984
Sandy loam	Field microplot	P. putida N-1R Rp^{R}	0.5	60	Dupler & Baker, 1984
Silty clay loam	Microcosm	P. aeruginosa	0.8	49	Zechman & Casida, 1982

[a] From: Van Elsas et al. (1991c). Initial cell numbers added were in the order of 10^7/g soil in most cases. Cells were added from washed fresh cultures without using carrier materials.

[b] Defined as the overall $\log_{10}$ decline in cfu counts per 10 days (calculated over the experimental period).
chr::Tn5: chromosomal insertion of transposon Tn5 encoding kanamycin resistance;
RP4: plasmid encoding kanamycin, tetracycline and ampicillin resistance.
Rp^{R}: resistant to rifampicin.

GEMMO were added to different parallel soil portions and the other in which both were added to the same soil portion. Whereas the first approach provided information on the survival and competition patterns of each strain *versus* the wild flora, the second approach directly examined the competitive ability of the GEMMO versus the parent in a background of indigenous organisms affecting both partners. By the first approach, Pillai and Pepper (1990) did not detect any differences in the survival patterns of *Rhizobium leguminosarum* biovar *phaseoli*, whether or not labelled with transposon Tn5, in desert soils. In addition, drying of the soil did not result in any detectable difference in the survival patterns of the two strains. Devanas *et al.* (1986), studying the survival in natural soil of *E. coli* strains carrying different antibiotic resistance plasmids *versus* their parent strains without plasmids, found that, for reasons unknown, the persistence of the plasmid-bearing strains was actually enhanced as compared with the parent strain. However, the persistence of these strains was still limited.

Short *et al.* (1990) compared the dynamics of a constitutively 2,4-dichlorophenoxyacetate (TFD)- degrading *Pseudomonas putida* (plasmid-encoded), a TFD-inducible comparable strain and the corresponding parent strain in soils without TFD, and found no difference in survival patterns between the three strains. Population densities of all three strains declined slowly over 20 days. Ramos *et al.* (1991) also observed no difference in behavior in sterile soil for three isogenic *P. putida* strains which differed only in their plasmid content (i.e. the parent strain, a plasmid pWWO-loaded derivative and a pWWO-EB62 loaded derivative). In addition, a modified *Pseudomonas* strain (chr:: *lac* ZY) introduced onto wheat in the field (Drahos *et al.*, 1988) showed survival kinetics on wheat roots similar to that of the parent strain. Finally, Scanferlato *et al.* (1989) found that the survival patterns after introduction into aquatic microcosms were similar between a modified *Erwinia carotovora* ssp. *carotovora* strain, containing the Tn903 kanamycin resistance gene on its chromosome, and its parent, both in the water and the sediment column phases of the system. Total introduced populations tended to decline from the established levels of 10^6 or 10^9 cfu / ml of water to extinction in 16 days. As a general conclusion from all these data, it might be stated that any differences in ecological behavior due to extra genetic and/or metabolic load were probably too small to be detectable in these studies in which conventional microbiological techniques were used. The difference in behavior of the plasmid-bearing *E. coli* strains as opposed to the parent strains might be attributable to a selective advantage provided by the presence of the plasmids. It is tempting to speculate about the possibility of antibiotics playing a role. The next section will deal with this aspect.

The second approach seems to be more precise in detecting any subtle effects of the additional genetic moieties, since it directly detects competitive differences in the same system.

Thus, Orvos *et al.* (1990) added mixtures of an *E. carotovora* ssp. *carotovora* parent strain and its GEMMO derivative carrying the Tn903 kanamycin resistance gene on its chromosome to soil, and observed a decline of the GEMMO derivative in concert with the observed decline of the total introduced population, which suggests that the GEMMO derivative behaved like the parent strain. In contrast, van Elsas *et al.* (1991b) found two transposon Tn5-based GEMMO derivatives of *P. fluorescens* strain R2f to be less fit than the parent strain in a loamy sand soil, both in terms of survival in an introduced mixed population and in the degree of colonization of the rhizosphere of developing wheat plants. Surprisingly, this slight competitive disadvantage was not detectable in the same soil amended with bentonite clay, suggesting that the soil type into which the organism is introduced may affect the balance between incoming GEMMO and wild-type organisms. In more recent experiments, performed with the strain R2f GEMMOs produced by recombinational insertion into the chromosome Ar-1, which contains the functional kanamycin resistance gene *npt* II, and Art-3, which contains an *npt* II-*cry* IVB hybrid in the same position, a competitive disadvantage was noted for both GEMMOs tested against the parent (Fig. 1). In addition, colonization of the rhizoplane was also severely impaired. The similarity in behavior between strains Ar-1 and Art-3 led to the conclusion that both were equally impaired in their fitness in soil as compared with the parent strain, either because of the effect of the *npt* II gene product on cellular physiology or to an undetected impairment of functioning of genes adjacent to the chromosomal insertion site.

The overall conclusion to be drawn from these still fragmentary data, might be that as yet there is no evidence of a drastic increase in fitness in soil of a GEMMO *versus* its parent strain. It is obvious that this contention cannot be generalized, since the available information pertains to a limited number of strains, genes and soil systems. In addition, the observation that

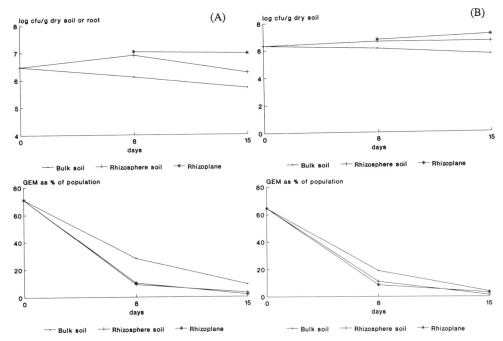

Fig. 1 Decrease of proportion of GEMMO in GEMMO/parent mix in soil microcosms planted to wheat. (A) *Pseudomonas fluorescens* R2f Rp^r /strain Art-3 mixture,(B) strain R2f Rp^r /strain Ar-1 mixture. Strains Art-3 and Ar-1 are R2f Rp^r derivatives which carry inserts of, respectively, *npt* II-*cry* IVB and *npt* II on their chromosomes.

the presence of antibiotic resistance genes did not result in enhanced survival of the hosts, is interesting in the light of possible selective pressure in soil favoring the occurrence of these genes (van Elsas, 1991).

PUTATIVE SELECTIVE ADVANTAGE OF GEMMO PHENOTYPE IN SOIL

Selective forces favouring GEMMOs in soil as opposed to the indigenous microflora can be diverse. For instance, the presence of a suitable host plant could select for a 'super-nodulating' *Rhizobium* , the presence of a suitable target insect for *Bacillus thuringiensis* endotoxin-carrying bacteria, and the presence of chemical selective force(s) in the form of xenobiotic compounds, antibiotics or heavy metals, could favour organisms with the capacity to degrade or resist these compounds.

Information on the selective pressure acting on biodegradative organisms introduced into soil and, in addition, on the possibility of antibiotic pressure acting on introduced bacteria carrying antibiotic resistance genes will be discussed below. The latter has relevance for practical purposes, since antibiotic resistance genes such as the *npt* II gene which confers resistance to kanamycin and neomycin, have frequently been used as genomic markers for cloning and ecological studies (Van Elsas *et al.*, 1986, 1991b; Fredrickson *et al.*, 1988).To insert the *cry*IVB gene into the chromosome of target bacteria, it was first inserted into the unique *Bam* H1 site of transposon Tn5, a procedure which leaves the kanamycin resistance gene *npt* II and the streptomycin resistance gene intact. Alternatively, *cry* IVB was linked to the selectable marker *npt*II. Secondly, *cry* IVB was co-inserted with the respective selectable marker genes into the chromosomes of the target bacteria. Any selective pressure favouring *npt* II or the streptomycin resistance gene in soil, for example due to the presence of biologically functional concentrations of kanamycin, neomycin or streptomycin, may therefore also select the adjacent *cry* IVB gene.

Biodegradative Genes

The classical example of a selective chemical pressure positively affecting an organism introduced in soil is that of *P. cepacia* whose biodegradative capacity was resuscitated by the addition of 2,4,5-trichlorophenoxyacetic acid (2,4,5-T) to soil (Chatterjee *et al* ., 1982 ; Kilbane *et al* ., 1983). This served to show that organisms which had declined in numbers to undetectable levels, 'resurfaced' after the application of the selective pressure which provided an advantage to its phenotype. Unfortunately, the possible occurrence of viable-nonculturable organisms (Colwell *et al*., 1985) was not tested, so that it is unclear whether the revival was due to a few cells growing out into a large population or considerably more cells in a non-culturable state again becoming viable. Alternatively, temporary loss of the functional genotype in a large part of the population due to the absence of 2,4,5-T, for example due to a deletion, may also have played a role.

Removal of the xenobiotic from the system by biodegradation, and therefore diminution of the selective force, resulted in a decline of the biodegradative population to levels below detection (Chatterjee *et al*., 1982; Kilbane *et al*., 1983). Golovleva *et al.* (1988) also reported that the persistence in soil of an introduced population of *P. aeruginosa* capable of degrading kelthane was greatly enhanced by the presence of kelthane, and the population tended to disappear after the degradation of this compound. Other more recent data on the selective force provided by recalcitrant xenobiotic compounds degradable by modified bacteria, were provided by Ramos *et al.* (1991). Similar to the previous findings, they showed that a phenylbenzoate-degrading *P. putida* strain survived significantly better in soil in the presence of the compound than in its absence; yet, its population decreased slowly. In contrast, Short *et al.* (1990) found that TFD added to soil did not enhance the persistence of a *P. putida* strain capable of degrading this compound. This lack of stimulation of inoculant survival was related to its inability to obtain energy from the (partial) degradation of the compound (Short *et al.*, 1990). On the other hand, a population of *P. putida* cells which possessed the ability to derive energy from TFD degradation, was shown to be resuscitated upon the addition of this compound to soil (Short *et al.*, 1990). Brokamp and Schmidt (1991), studying the biodegradation of 2,2-dichloropropionate (DCPA) by *Alcaligenes xylosoxidans* containing a plasmid-encoded halidohydrolase responsible for DCPA degradation, found that the presence of DCPA stimulated the development of the introduced biodegradative population, even though cells were still detectable in the absence of the selective agent. Together, these examples clearly show the forces of selection operate on different incoming biodegradative GEMMOs in a mixed bacterial community in soil, where the introduced organisms possess a unique property, biodegradation, which provides them with an extra energy source. At the same time, it also became evident that with the disappearance of this selective force, the relative predominance of these GEMMOs tended to diminish, often as far as to extinction. Unfortunately, data on how the population at low cfu density is present in the soil, that is whether as low, undetectable, cell numbers, or as higher, non-culturable cell numbers, and on the fate of the biodegradative genes, that is the putative occurrence of deletions, gene transfer and/or plasmid loss, are largely lacking.

Antibiotic Resistance Genes

Antibiotic resistance genes are other genetic moieties which may potentially be selected in soil, since soil is a reservoir of organisms with the capacity to produce antibiotics. Nevertheless, it has long remained obscure whether antibiotics are actually produced in soil in ecologically significant quantities, given the inherent difficulties of detecting antibiotics in soil (Van Elsas, 1991). First, all attempts to show the occurrence of antibiotics in soil have long remained elusive (Gottlieb, 1976). Secondly, antibiotics may interact with clay particles and other surfaces in soil and become inactivated. Also, the gross conditions that favor antibiotic production are not commonly found in bulk soil, because most antibiotics are secondary metabolites produced under conditions of nutrient excess. Nevertheless, recent evidence, both indirect and direct, has pointed to a role for antibiotics in soil microbial interactions. Indirect evidence comes from the observation that rhizospheric populations of *Rhizobium* contained an enhanced proportion of cells with resistance against antibiotics, such as streptomycin, after liming of acid soils (Scotti *et al.*, 1982; Ramos *et al.*, 1987). This was attributed to an enhanced presence and activity in the rhizosphere of antibiotic-producing streptomycetes. In addition, Li and Alexander (1990) recently showed that co-inoculation of alfalfa with a streptomycin-resistant *R. meliloti* inoculant and the streptomycin-producer *Streptomyces*

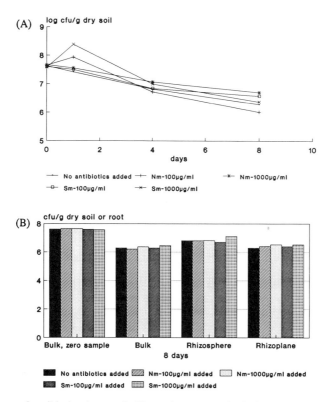

Fig. 2 Influence of antibiotics (neomycin-Nm and streptomycin-Sm) on survival and wheat root colonization of *Pseudomonas fluorescens* R2f (chr::Tn5) in Ede loamy sand soil microcosms. A. Inoculant survival in unplanted soil. B. Dynamics of inoculant cfu numbers in bulk, rhizosphere and rhizoplane of wheat (planted soil) after 8 days. Bulk and rhizosphere cfu counts expressed per g dry soil, rhizoplane counts expressed per g root fresh weight.

griseus, resulted in an increase in root colonization and nodulation. Since controls in which antibiotic non-producing *S. griseus* were used showed no such increase, the effect was attributed to antibiotic production by the co-inoculant. In addition, a fluorescent pseudomonad has recently been shown detectably to produce the antibiotic phenazine-1-carboxylic acid in wheat rhizosphere soil (Thomashow *et al.*, 1990). Crucial to this finding was the detection, by extraction techniques, of the antibiotic in soil. In addition, antibiotic production was recently implied in the suppression of radish damping-off by introduced *P. cepacia* (Homma and Suzui,1989). Thus, although both indirect and direct evidence seems to leave no doubt as to the production and role for antibiotics in soil, there is still a general lack of knowledge about the extent of the selective pressure for different antibiotics in soil and the soil conditions which promote such selective force. It is likely that the classical difficulties of detection of antibiotics in soil have been related to their occurrence in soil microsites where conditions were favorable for the adequate metabolism of the producers. Since the number of these microsites may be limited in soil and the transport of the organic compounds limited due to adsorption processes, the significance of antibiotics may also be constrained when the soil as a whole is considered, yet be very significant for the microbial interactions, e.g. competition, at the microsite level (Van Elsas, 1991).

In addition, the human input of antibiotics into the environment, via releases into sewage and manure, has been increasing over the years (Van Elsas, 1991). The extent of this additional selective antibiotic pressure in soil is largely unknown. However, 'worst-case' calculations of the release of kanamycin in Dutch soils suggested that on average, the annual kanamycin input may be as high as 0.13 µg/ g of top soil (Bijvoet,J., pers. comm.). Since the input is probably not evenly spread in the soil, local kanamycin concentrations may be substantially higher. Experiments aimed at detecting the putative ecological advantage to hosts due to the presence of antibiotic-resistance genes can be executed regardless of whether knowledge about the actual occurrence of antibiotics in soil is available. Studies in which the behavior of isogenic strains with or without extra antibiotic-resistance genes are compared could provide indirect evidence of the occurrence of antibiotic pressure. As outlined above, such studies, performed with the antibiotic resistance genes of transposon Tn5 or Tn903, have not provided evidence of selective forces positively influencing the survival of the antibiotic-resistance-containing microorganisms. In addition, experiments performed in antibiotic-amended soils, as opposed to unamended soils, provide evidence about the likelihood of selection of an introduced GEMMO which carries antibiotic-resistance genes *versus* the wild flora. Bashan (1986) showed a slight enhancement of wheat root-colonization by antibiotic resistant *Azospirillum brasilense* on repeated amendment of the soil with four bacterium- and fungi-inhibiting compounds, one of which was streptomycin. The total rhizospheric bacterial population was temporarily depressed as a result of the amendment. More recent experiments with added antibiotics (Recorbet, G., pers.comm.; Henschke *et al.*, 1989; our own work) have failed to reveal a drastic increase in fitness of transposon Tn5 loaded organisms in kanamycin-, neomycin- and/or streptomycin-amended soils. However, slight positive effects on fitness have been noted by Henschke *et al.* (1989) and, recently, in our laboratory. The effect of antibiotics on the dynamics of introduced *P.fluorescens* (chr::Tn5) populations in soil is shown in Fig. 2A. There was no significant effect of added neomycin at 100 or 1000 μg/ ml soil water (equivalent to about 20 or 200 μg/ g dry soil) or of streptomycin at 100 μg/ml. However, an initial effect of streptomycin at 1000 μg/ml was noted, which disappeared in the course of the incubation (Fig. 2A). The effects of antibiotics on wheat root colonization by the inoculant were similar in that only streptomycin at high concentrations induced higher numbers of introduced pseudomonads in the wheat rhizosphere (Fig. 2B). One reason for this apparent lack of large effects may be the known adsorption of antibiotics, in particular kanamycin, to soil particles, thereby lowering the locally effective concentrations. This adsorption would obviously have similar consequences for naturally-produced antibiotics. Another explanation might be the presence in soil of substantial kanamycin-resistant populations, sometimes as high as 10^4 to 10^5 cfu / g, leading to a 'dilution' of the advantage provided by kanamycin. Recent work with heavy metal resistance genes, however, has presented clear evidence of heavy metal selection acting on introduced microbial populations (De Rore *et al.*, 1990). In studies aimed at comparing the persistence of *Alcaligenes eutrophus* carrying a plasmid with resistance determinants to Co, Zn and Cd (*czc* genotype) and that of the parent in soil, a significantly higher population level of the plasmid-carrying strain developed in sterilized heavy metal polluted soil as opposed to the parent strain, whereas in unpolluted soil the response of both populations, in terms of increase in population size, was similar. However, in non-sterile soil such a differential response was not noted, leading to the contention that biological factors counteracted the positive selection of the *czc* phenotype.

CONCLUSIONS

As previously found for unmodified organisms for release, the establishment of GEMMOs in soil depends on a myriad of interacting soil factors, the net result of which is often homeostatic or bacteriostatic. Permanent establishment of a GEMMO population at ecologically significant levels is therefore inherently difficult, and optimization of the efficiency of each introduction will meet difficulties dictated by the strain and by intricacies of the soil. Of the biotic soil factors, predation by protozoa and competition against the wild flora are probably predominant in controlling the size of introduced populations. Since these are factors both unmodified and modified organisms must cope with when introduced, it becomes important to know whether the extra genetic load will provide any ecological advantage or disadvantage as compared with corresponding wild strains. The still fragmentary evidence presented here clearly showed that the extra genetic load present in the different GEMMOs examined provided an ecological disadvantage in some cases and was ecologically neutral in others. However, in cases where selective pressure was applied to soil, populations were either stimulated, as in the case of most xenobiotic-degrading organisms, or could be considered as ecologically neutral, as often in the case of antibiotic-resistant organisms. However, this may have been due to too low a selective force acting on the introduced GEMMOs in their microsites in soil, given the probable adsorption of added antibiotics to soil particles, which may render them unavailable.

Finally, it seems evident that when considering the effects of introduced GEMMOs on the ecosystem, these should be compared with the effects of parent strains. Only effects significantly different from those of parent strains should be taken into account when the biosafety of released GEMMOs is to be judged.

ACKNOWLEDGEMENTS

I thank Anneke Wolters for excellent technical assistance. Eric Smit, Hans van Veen and Caroline Heijnen are acknowledged for critically reading the manuscript.

REFERENCES

Alexander, M. (1981). Why microbial predators and parasites do not eliminate their prey and host. Ann. Rev. Microbiol. 35, 113-133.

Bashan, Y. (1986). Enhancement of wheat root colonization and plant development by *Azospirillum brasilense* Cd. following temporary depression of rhizosphere microflora. Appl. Environ. Microbiol. 1,1067-1071.

Biel, S.W. and Hartl, D.L. (1983). Evolution of transposons: natural selection for Tn5 in *Escherichia coli* K12. Genet. 103, 581-592.

Bolton, H. jr., Fredrickson, J.K., Bentjen, S.A., Workman, D.J., Li, S.W. and Thomas, J.M. (1991). Field calibration of soil-core microcosms: Fate of a genetically altered rhizobacterium. Microb. Ecol. 21, 163-173.

Brokamp, A. and Schmidt, F.R.J. (1991). Survival of *Alcaligenes xylosoxidans* degrading 2,2-dichloropropionate and horizontal transfer of its halidohydrolase gene in a soil microcosm. Curr. Microbiol. 22, 299-306.

Chao, L., Vargas, C., Spear, B.B. and Cox, E.C. (1983). Transposable elements as mutator genes in evolution. Nature 303, 633-635.

Chatterjee, D.K., Kilbane, J.J. and Chakrabarty, A.M. (1982). Biodegradation of 2,4,5-trichlorophenoxyacetic acid in soil by a pure culture of *Pseudomonas cepacia*. Appl. Environ. Microbiol. 44, 514-516.

Colwell, R.R., Brayton, P.R., Grimes, D.J., Roszak, D.B., Huq, S.A. and Palmer, L.M. (1985). Viable but non-culturable *Vibro cholerae* and related pathogens in the environment: implications for release of genetically-engineered microorganisms. Biotechnol. 3, 817-820.

Compeau, G., Al-Achi, B.J., Platsouka, E. and Levy, S.B. (1988). Survival of rifampin-resistant mutants of *Pseudomonas fluorescens* and *Pseudomonas putida* in soil systems. Appl. Environ. Microbiol. 54, 2432-2438.

De Rore, H., Top, E., Höfte, M., Mergeay, M. and Verstraete, W. (1990). Intergeneric transfer of heavy metal resistance genes in non-polluted and heavy metal-polluted soil microcosms. Abstr. Symposium "Environmental Impact of GEMS", University of Warwick, UK.

Devanas, M.A., Rafaeli-Eshkol, D. and Stotzky, G. (1986). Survival of plasmid-containing strains of *Escherichia coli* in soil: effect of plasmid size and nutrients on survival of hosts and maintenance of plasmids. Curr. Microbiol. 13, 269-277.

Drahos, D.J., Barry, G.F., Hemming, B.C., Brandt, E.J., Skipper, H.D., Kline, E.L., Kluepfel, D.A., Hughes, T.A. and Gooden, D.T. (1988). Pre-release testing procedures: US field test of a *lac* ZY-engineered soil bacterium, in "The Release of Genetically-Engineered Microorganisms" (Eds. Sussman, M., Collins, C.H., Skinner, F.A. and Stewart-Tull, D.E.). pp. 181-191. Acad. Press, London.

Dupler, M. and Baker, R. (1984). Survival of *Pseudomonas putida*, a biological control agent, in soil. Phytopathol. 74, 195-200.

Edlin, G., Tait, R.C. and Rodriguez, R.L. (1984). A bacteriophage λ cohesive ends (*cos*) DNA fragment enhances the fitness of plasmid-containing bacteria growing in energy-limited chemostats. Biotechnol. 2, 251-254.

Fredrickson, J.K., Bezdicek, D.F., Brockman, F.J. and Li, S.W. (1988). Enumeration of Tn5 mutant bacteria in soil by using a most-probable-number-DNA hybridization procedure and antibiotic resistance. Appl. Environ. Microbiol. 54, 446-453.

Gillett, J.W., Levin, S.A., Harwell, M.A., Alexander, M., Andow, D.A. and Stern, A.M. (1984). Potential impacts of environmental release of biotechnology products: Assessment, Regulation, and Research Needs, Ecosystems Research Center, ERC-075.

Golovleva, L.A., Pertsova, R.N., Boronin, A.M., Travkin, V.M. and Kozlovsky,S.A. (1988). Kelthane degradation by genetically engineered *Pseudomonas aeruginosa* BS827 in a soil ecosystem. Appl. Environ. Microbiol. 54, 1587-1590.

Gottlieb, D. (1976) . The production and role of antibiotics in soil. J. Antibiot. 29, 987-1000.

Habte, M. and Alexander, M. (1975). Protozoa as agents responsible for the decline of *Xanthomonas campestris* in soil. Appl. Microbiol. 29, 159-164.

Habte, M. and Alexander, M. (1977). Further evidence for the regulation of bacterial populations in the soil by protozoa. Archiv. Microbiol. 113, 181-183.

Heijnen, C.E. and van Veen, J.A. (1991). A determination of protective microhabitats for bacteria introduced into soil. FEMS Microbiol. Ecol.85, 73-80.

Heijnen, C.E., van Elsas, J.D., Kuikman, P.J. and van Veen, J.A. (1988). Dynamics of *Rhizobium leguminosarum* biovar *trifolii* introduced into soil; the effect of bentonite clay on predation by protozoa. Soil Biol. Biochem. 20, 483-488.

Henschke, R.B., Nucken, E. and Schmidt, F.R.J. (1989). Fate and dispersal of recombinant bacteria in a soil microcosm containing the earthworm *Lumbricus terrestris*. Biol. Fertil. Soils 7, 374-376.

Henschke, R.B. and Schmidt, F.R.J. (1990). Plasmid mobilization from genetically engineered bacteria to members of the indigenous soil microflora in situ. Curr. Microbiol. 20, 105-110.

Ho, W.C. and Ko, W.H. (1985). Soil microbiostasis: effects of environmental and edaphic factors. Soil Biol. Biochem. 17, 167-170.

Homma, Y. and Suzui, T. (1989). Role of antibiotic production in suppression of radish damping-off by seed bacterization with *Pseudomonas cepacia*. Ann. Phytopathol. Soc. Jpn. 55, 643-652.

Kilbane, J.J., Chatterjee, D.K. and Chakrabarty, A.M. (1983). Detoxification of 2,4,5-trichlorophenoxyacetic acid from contaminated soil by *Pseudomonas cepacia*. Appl. Environ. Microbiol. 45, 1697-1700.

Kluepfel, D.A. and Tonkyn, D.W. (1991). Release of soil-borne genetically modified bacteria: biosafety implications from contained experiments, in:Biological Monitoring of Genetically Engineered Plants and Microbes" (Eds. MacKenzie, D.R. and Henry, S.C.). pp.55-65. Agricult. Res. Inst. Bethesda, Maryland.

Lacy, G.H. and Stromberg, V.K. (1991). Pre-release microcosm tests with a genetically engineered plant pathogen, in: "Biological Monitoring of Genetically Engineered Plants and Microbes" (Eds. Mackenzie, D.R. and Henry, S.C.). pp.81-98. Agricult. Res. Inst. Bethesda, Maryland.

Li, D.-M. and Alexander, M. (1990). Factors affecting co-inoculation with antibiotic-producing bacteria to enhance rhizobial colonization and nodulation. Plant and Soil 129, 195-201.

Lowendorf, H.S., Baya, A.M. and Alexander, M. (1981). Survival of *Rhizobium* in acid soils. Appl. Environ. Microbiol. 42, 951-957.

Oades, J.M. (1988). The retention of organic matter in soils. Biogeochemistry 5, 35-70.

Mishustin, E.N. and Naumova, A.N. (1962). Bacterial fertilizers, their effectiveness and mode of action. Microbiologyia 31, 543-555.

Orvos, D.R, Lacy, G.H. and Cairns, J. jr. (1990). Genetically engineered *Erwinia carotovora* : survival, intraspecific competition, and effects upon selected bacterial genera. Appl. Environ. Microbiol. 56, 1689-1694.

Panopoulos, N.J. (1986). Tactics and feasibility of genetic engineering of biocontrol agents, in "Microbiology of the Phyllosphere" (Ed. Fokkema,.N.J. and van den Heuvel, J.). pp. 312-332. Cambridge Univ. Press, London.

Pillai, S.D. and Pepper, I.L. (1990). Survival of Tn5 mutant bean rhizobia in desert soils: phenotypic expression of Tn5 under moisture stress. Soil Biol. Biochem. 22, 265-270.

Postma, J., van Elsas, J.D., Govaert, J.M. and van Veen, J.A. (1988). The dynamics of *Rhizobium leguminosarum* biovar *trifolii* introduced into soil as determined by immunofluorescence and selective plating techniques. FEMS Microbiol. Ecol. 53, 251-260.

Ramos, J.L., Duque, E. and Ramos-Gonzalez, M.-I. (1991). Survival in soils of an herbicide-resistant *Pseudomonas putida* strain bearing a recombinant TOL plasmid. Appl. Environ. Microbiol. 57, 260-266.

Ramos, M.L.G., Magelhaes, N.F.M. and Boddey, R.M. (1987). Native and inoculated rhizobia isolated from field-grown *Phaseolus vulgaris* : effects of liming an acid soil on antibiotic resistance. Soil Biol. Biochem. 19, 179-185.

Scanferlato, V.S., Orvos, D.R., Cairns, J. jr. and Lacy, G.H. (1989). Genetically engineered *Erwinia carotovora* in aquatic microcosms: survival and effects on functional groups of indigenous bacteria. Appl. Environ. Microbiol. 55, 1477-1482.

Scotti, M.R.M.M.L., Sa, N.M.H., Vargas, M.A.T. and Dobereiner, J. (1982). Streptomycin resistance of *Rhizobium* isolates from Brazilian cerrados. Anales da Academia Brasileira de Ciencias 54, 733-738.

Short, K.A., Seidler, R.A. and Olsen, R.H. (1990). Survival and degradative capacity of *Pseudomonas putida* induced or constitutively expressing plasmid-mediated degradation of 2,4-dichlorophenoxyacetate (TFD) in soil. Can. J. Microbiol. 36, 821-826.

Skøt, L., Harrison, S.P., Nath, A., Mytton, L.R. and Clifford, B.C. (1990). Expression of insecticidal activity in *Rhizobium* containing the δ–endotoxin gene cloned from *Bacillus thuringiensis* subsp.*tenebrionis*. Plant and Soil 127, 285-295.

Stotzky, G. (1986). Influence of soil mineral colloids on metabolic processes, growth, adhesion, and ecology of microbes and viruses, in: "Interactions of Soil Minerals with Natural Organics and Microbes".p. 305-428. SSSA Spec. Publ. 17, Madison, Wisconsin.

Stotzky, G. (1989). Gene transfer among bacteria in soil, in "Gene transfer in the Environment" (Eds. Levy, S.B. and Miller, R.V.). pp. 165-222. McGraw-Hill, New York.

Stotzky, G., Zeph, L.R. and Devanas, M.A. (1991). Factors affecting the transfer of genetic information among microorganisms in soil, in "Assessing Ecological Risks of Biotechnology" (Ed. Ginzburg, L.R.), pp.95-122. Butterworth-Heinemann, Boston.

Strauss, H.S., Hattis, D., Page, G., Harrison, K., Vogel, S. and Caldart, C.(1986). Genetically-engineered microorganisms: II. Survival, multiplication and genetic transfer. Recomb. DNA Techn. Bull. 9, 69-88.

Thies, J.E., Singleton, P.W. and Bohlool, B.B. (1991). Influence of the size of indigenous rhizobial populations on establishment and symbiotic performance of introduced rhizobia on field-grown legumes. Appl. Environ.Microbiol. 57, 19-28.

Thomashow, L.S., Weller, D.M., Bonsall, R.F. and Pierson III, L.S. (1990). Production of the antibiotic phenazine-1-carboxylic acid by fluorescent *Pseudomonas* species in the rhizosphere of wheat. Appl. Environ. Microbiol. 6, 908-912.

Thornton, F.C. and Davey, C.B. (1984). Saprophytic competence of acid-tolerant strains of *Rhizobium trifolii* in acid soil. Plant and Soil 80, 337-344.

Tiedje, J.M., Colwell, R.K., Grossman, Y.L., Hodson, R.E., Lenski, R.E., Mack, R.N. and Regal, R.J. (1989). The planned introduction of genetically-engineered organisms: ecological considerations and recommendations. Ecol. 70, 298-315.

Van Elsas, J.D. (1991). Antibiotic resistance gene transfer in the environment: an overview, in "Genetic Interactions between Microorganisms in the Natural Environment" (Eds. Wellington, E.H.M. and Van Elsas, J.D.) Manchester Univers. Press (in press).

Van Elsas, J.D. and Heijnen, C.E. (1990). Methods for the introduction of bacteria into soil: a review. Biol. Fertil. Soils 10, 127-133.

Van Elsas, J.D. and Trevors, J.T. (1990). Plasmid transfer to indigenous bacteria in soil and rhizosphere: problems and perspectives, in "Bacterial Genetics in Natural Environments" (Eds. Fry, J.C. and Day, M.J.), pp. 188-199. Chapman and Hall, London.

Van Elsas, J.D. and Trevors, J.T. (1991). Environmental risks and fate of genetically engineered microorganisms in soil. J. Environ. Sci. and Health A26, 981-1001.

Van Elsas, J.D., Dijkstra, A.F., Govaert, J.M. and van Veen, J.A. (1986). Survival of *Pseudomonas fluorescens* and *Bacillus subtilis* introduced into two soils of different texture in field microplots. FEMS Microbiol. Ecol.8, 151-160.

Van Elsas, J.D., Trevors, J.T. and van Overbeek, L.S. (1991a). Influence of soil properties on the vertical movement of genetically-marked *Pseudomonas fluorescens* through large soil microcosms. Biol. Fertil. Soils 10, 249-255.

Van Elsas, J.D., van Overbeek, L.S., Feldmann, A.M., Dullemans, A.M. and De Leeuw, O. (1991b). Survival of genetically engineered *Pseudomonas fluorescens* in soil in competition with the parent strain. FEMS Microbiol.Ecol. 85, 53-64.

Van Elsas, J.D., Heijnen, C.E. and van Veen, J.A. (1991c). The fate of introduced genetically engineered microorganisms (GEMs) in soil, in microcosm and the field: impact of soil textural aspects, in: "Biological Monitoring of Genetically Engineered Plants and Microbes" (Eds. MacKenzie, D.R. and Henry, S.C.), pp.67-79. Agricult. Res.Instit., Bethesda, Maryland.

Van Veen, J.A. and van Elsas, J.D. (1986). Impact of soil structure and texture on the activity and dynamics of the soil microbial population, in "Perspectives in Microbial Ecology" (Eds. Megusar, F. and Gantar, M.).pp.481-488. Slovena Soc. Microbiol., Ljubljana.

Waalwijk, C., Dullemans, A., and Maat, C. (1991). Construction of a bio-insecticidal rhizosphere isolate of *Pseudomonas fluorescens* . FEMS Microbiol. Letters 77, 257-264.

Watrud, L.S., Perlak, F.J., Tran, M.-T., Kusano, K., Mayer, E.J., Mille-Wideman, M.A., Obukowicz, M.G., Nelson, D.R., Kreitinger, J.P. and Kaufman, R.J. (1985). Cloning of the *Bacillus thuringiensis* subsp. *kurstaki* delta-endotoxin gene into *Pseudomonas fluorescens*: molecular biology and ecology of an engineered microbial pesticide, in "Engineered Organisms in the Environment: Scientific Issues" (Eds.Halvorson, H.O., Pramer, D. and Rogul, M.). pp. 40-46. Am. Soc.Microbiol., Washington, D.C.

Wessendorf, J. and Lingens, F. (1989). Effect of culture and soil conditions on survival of *Pseudomonas fluorescens* R1 in soil. Appl. Microbiol. Biotechnol. 31, 97-102.

Zechman, J.M. and Casida, L.E. jr. (1982). Death of *Pseudomonas aeruginosa* in soil. Can. J. Microbiol. 28, 788-794.

INTERACTIONS IN COMMUNITIES OF MICROORGANISMS

Mark J Bale[1], Mike Hinton[2] and John E Beringer[1]

Department of Botany[1]
University of Bristol
Woodland Road
Bristol BS8 1UG, UK

Division of Veterinary Public Health and Food Safety [2]
University of Bristol
Langford House
Langford
Bristol, BS18 7DU, UK.

INTRODUCTION

The survival and persistence of microorganisms in the environment depends upon a number of factors, among these is the requirement that the organisms concerned are able to compete with indigenous microorganisms that have similar growth requirements. Van Elsas and others in this volume describe many of the problems faced by released microbes. In this chapter we will concentrate on genetic interactions that occur among members of the same species which, almost by definition, are uniquely able to interact with introduced strains because they have essentially the same requirements for growth. There is fairly extensive experience in practical terms of interactions between members of the same species obtained from studies of the natural colonization of animals by pathogenic strains of *Enterobacteriaceae*, and from the inoculation of crops with symbiotic microorganisms, such as *Rhizobium*.

GENETIC INTERACTIONS BETWEEN COMMUNITIES

There is ample evidence for genetic interactions between bacterial communities. This has come from the demonstration of gene transfer in model habitats with introduced donor and recipient bacteria, and from the observation that identical or related plasmids, or resistance genes, occur in a variety of different bacterial strains or species. Because of its clinical and commercial importance much of the work has concerned the acquisition of antibiotic resistance by bacteria infecting humans and other animals (Bale and Hinton, 1991). In many respects, the existence of genetic interactions involving highly mobile plasmids and strong selective pressures upon the genes, is a 'worst case scenario' for the dissemination of cloned genes from introduced GEMMO's, because it is inconceivable that strongly selectable genes on plasmid vectors will ever be licensed for release.

The evidence for genetic interactions between communities involving chromosomal gene exchange can best be examined using data acquired by population geneticists. Fortunately there has been some extremely good work of this nature during the last few years.

The Release of Genetically Modified Microorganisms
Edited by D.E.S. Stewart-Tull and M. Sussman, Plenum Press, New York, 1992

Direct Evidence for Genetic Interactions

Detailed studies have been made to examine the transfer of plasmid-borne genes in aquatic habitats, soil and animal intestinal tracts. There are several reviews of these results (see Fry and Day, 1990; Bale and Hinton, 1991) and our discussion will be limited to salient points that have emerged.

The physical characteristics of the environment, such as temperature, pH and moisture content, appear to affect conjugation in a manner entirely consistent with their effects on the physiology of the donor and recipient bacteria. There is evidence of differing optima of temperature, pH and nutrient status for different combinations of host and plasmid (Rochelle *et al.*, 1989).

An important feature of many direct experiments of gene transfer is the observation that indigenous microorganisms almost invariably interfere with conjugation. This has been observed for aquatic habitats (Bale *et al.*, 1987; O' Morchoe *et al.*, 1988), for soil systems (van Elsas *et al.*, 1990) and in the intestinal tracts of animals (Freter *et al.*, 1983). In the latter example, mathematical models have indicated that the interference is due to a lack of available adhesion sites and to the subsequent 'wash-out' and reduced conjugation between bacteria suspended in the gut lumen.

The availability of niches, or empty adhesion sites, may also influence the survival of bacteria in a variety of habitats (Alexander, 1981; Kuikman *et al.*, 1990). It has been shown that bacteria adhering to surfaces commonly occur as slowly growing microcolonies with a few emigrating daughter cells (Lawrence & Caldwell, 1987; Lawrence *et al.*, 1987; Fletcher, 1991). This does not appear to allow many opportunities for genetic exchange between communities, because bacteria established in a microcolony, under favorable conditions, are only capable of conjugation with equally favored bacterial microcolonies, which are likely to be spatially separated. Thus, only in artificial experiments involving the suppression of competing microorganisms and an ample supply of nutrients is conjugation readily detectable.

The very low potential for gene exchange between microcolonies means that the temporal separation of bacterial communities, that is succession, and its effects on genetic interactions must be considered.

Extracellular DNA is apparently abundant in marine habitats, in both water and sediments (Paul *et al.*, 1987; Steffan *et al.*, 1988). It also appears that extensive binding of such DNA to sand grains occurs, whereupon it becomes well-protected against extracellular nucleases (Romanowski *et al.*, 1991). Such DNA is also capable of transforming naturally transformable bacteria, such as *Bacillus* spp. (Lorenz *et al.*, 1988) and *Vibrio* spp. (Paul *et al.*, 1991). Similarly there is evidence for the packaging of bacterial DNA in bacteriophage (Morrison *et al.*, 1978; Saye *et al.*, 1987, 1990; Germida and Khachatourians, 1988; Zeph *et al.*, 1988) and its survival in the environment within bacteriophage before transduction of a susceptible host. Recent studies have indicated that bacteriophages are more abundant and active in aquatic systems (Bergh *et al.*, 1989; Ogunseitan *et al.*, 1990; Proctor and Fuhrman, 1990; Kokjohn *et al.*, 1991) and soil (O' Sullivan *et al.*, 1990) than was previously thought.

The potential for the temporal separation of donor and recipient implies that a detection system for released microorganisms and, in particular, the cloned genes they carry, should be based upon the detection of DNA, rather than the detection of the introduced bacteria.

The overall conclusion from the direct studies is that genetic interactions are unlikely to occur between bacterial communities in natural conditions, with possibly low nutrient status, sub-optimum physical conditions and in the presence of competing microorganisms. However, the existence of sophisticated conjugation, transduction and transformation systems in a wide range of naturally-occurring bacterial species argues that genetic interactions do occur in nature (Reanney *et al.*, 1982); evidence for this is provided by studies on the distribution of antibiotic-resistant phenotypes.

Indirect Evidence of Genetic Interactions

The increase in antibiotic resistance amongst clinically important bacteria provides evidence that plasmid transfer is widespread in hospitals and on farms. The examination of bacterial isolates from the pre-antibiotic era has shown that plasmids existed in many strains but that they were not normally associated with antibiotic resistance (Hughes and Datta, 1983). After the introduction of antibiotic therapy, similar groups of plasmids, as judged by incompatibility groups, acquired antibiotic resistance genes and transferred them to a variety of different bacterial species.

In the related veterinary field, the widespread use of antibiotics for growth promotion, prophylaxis, and for therapeutic purposes led to the appearance and transfer of antibiotic resistance genes. For example, Campbell *et al.,* (1986) and Campbell and Mee, (1987) showed that isolates of *Escherichia coli* from humans and pigs carried a related IncFIV plasmid expressing trimethoprim resistance. The distribution of plasmid types between *E. coli* biovars suggested that initially an ancestral plasmid spread into porcine isolates, acquired the resistance determinant and then transferred it into *E. coli* colonizing the human population. Similarly, resistance to trimethoprim in *E. coli* from broiler chicks was associated with a 65kb plasmid, which over a seven-week period was transferred into an unusual *E. coli* isolate (Chaslus-Dancla *et al.,* 1987).

Current UK legislation (Zoonoses Order 1989), requires that salmonella isolates from food animals be reported to the Ministry of Agriculture, Fisheries and Food, and a proportion of these isolates are examined further by the Central Veterinary Laboratory, Weybridge or the Central Public Health Laboratory, Colindale. This has allowed the evolution of various clones of *Salmonella* to be followed. For example, in 1974 the predominant phage-type of *Salmonella enterica* serotype Typhimurium isolated from calves was type 49 (Threlfall and Rowe, 1984). At this time a new phage type (DT204) was isolated which had evolved from the DT49 strain by the acquisition of a plasmid encoding tetracycline-resistance and a "typing phage restriction" phenotype which altered the observed phage type. Additional drug resistance markers, carried on a conjugative IncH2 plasmid, were subsequently acquired. After 1-2 years a new phage-type was recognized (DT204a) which had acquired a lysogenic bacteriophage which altered the phage type. Subtle shifts involving the loss of the "typing phage restriction" phenotype from the small Tet^r plasmid, the acquisition of a transposon encoding trimethoprim-resistance on the IncH2 plasmid and a different lysogenic phage, led to the appearance of DT204c, which became the predominant phage type that infected calves for several years (Threlfall and Rowe, 1984; Wray *et al.,* 1987). Bacteriophage-mediated genetic conversions in the genus *Salmonella* have been observed in other, direct, studies and appear to be common (Smith and Lovell, 1985; Barrow, 1986).

More recently *S. enterica* serotype Enteritidis phage-type 4, associated with an epidemic of food poisoning in humans, was shown to have converted to a multi-resistant strain of phage type 24 by the acquisition of a 37.5 kb IncN plasmid (Frost *et al.,* 1989). Phage-type 24 is now being isolated from humans and animals but it has yet to displace type 4 as the most common *Salmonella* Enteritidis phage-type.

These examples indicate that bacterial communities exchange genetic material and that such events can be recognized over a period of weeks or years where a selective pressure exists. Thus, as might be expected, the persistence of a genetic characteristic within a population can be shown to be strongly favored by selective pressure. In the absence of strong selective pressure, is there any evidence that genes which are exchanged are maintained in populations?

GENETIC INTERACTIONS INVOLVING CHROMOSOMAL GENES

Direct Evidence Involving Naturally Competent Bacteria

Although there are many methods for introducing and recombining chromosomal genes in the laboratory, there is less direct evidence for this type of interaction occuring in natural communities. Probably the earliest demonstration that this might occur was provided by Griffith (1928) who injected mice with killed virulent and live avirulent strains of *Streptococcus pneumoniae*. DNA released from the virulent cells was subsequently taken up by the avirulent bacteria which then became virulent. Studies on recombination between *Bacillus* spp. in soil (Graham and Istock, 1978; Duncan *et al.,* 1989) showed that many different recombinant types could be recovered. The presumed mechanism in both these cases was transformation but it was not proven. More recently naturally-competent *Pseudomonas stutzeri* strains were transformed in sediment (Stewart and Sinigalliano, 1990) and when the transforming DNA was adsorbed to sand-grains (Lorenz and Wackernagel, 1990). This type of interaction is made extremely likely by the widespread existence of exogenous DNA in aquatic environments (Paul *et al.,* 1987) and its stabilization on sediment particles (Lorenz *et al.,* 1988), whilst still being available for transformation of naturally competent bacteria.

It would appear, however, that this recombination of chromosomal genes is confined to closely-related naturally competent bacteria (Stewart and Carlson, 1986). This is because different DNA codon preferences lead to sequence divergence between even closely-related genera and species (Wilkins, 1988), which reduces the opportunity for homologous recombination. Even if exogenous and relatively homologous DNA enters the cell, it may be unable to recombine with the chromosome, this inability appears to be associated with mismatch repair systems. For example, mutations in the *mutL, mutS* and *mutH* genes of *S. enterica* st. Typhimurium enabled recombination to occur with *E. coli*, even though there was up to 20% divergence in DNA sequence (Rayssiguier *et al.*, 1989). However, despite these provisos there is evidence of chromosomal recombination between natural bacterial communities.

Indirect Evidence of Chromosomal Gene Transfer

The population structure of bacterial species was examined by exploiting the natural variations in the electrophoretic mobilities of certain enzymes from bacteria isolated from a variety of sources (reviewed by Young, 1989). It appears that of all of the possible combinations of enzyme polymorphism, only a small fraction are represented in natural populations. This means that many bacterial species exist as a relatively small number of largely clonal populations. Some species, such as *E. coli*, have a relatively low genetic diversity (Young, 1989), whilst others such as *Pseudomonas cepacia* (McArthur *et al.*, 1988) and *Rhizobium meliloti* (Eardly *et al.*, 1990) have a greater genetic diversity. This may in part be due to the differing variablility of the ecological niches occupied by different species (McArthur *et al.*, 1988; Zavarzin *et al.*, 1991).

Within a bacterial species the same clonal types are widespread, with little variation in different countries, suggesting that migration is rapid (Young, 1989). Different clones of a species often have specialized niches or properties, for example some clones of *E. coli* are frequently associated with urinary tract infections, whilst rarely causing other infections (Orskov and Orskov, 1983; Whittam *et al.*, 1989). The same principle applies to species such as *Haemophilus influenzae* (Musser *et al.*, 1985) and different clonal types of *S. enterica* serotype Derby and serotype Newport (Beltran *et al.*, 1988).

The existence of discrete clonal types argues that recombination within a bacterial species is sufficiently rare to maintain the observed linkage disequilibrium. This is not, however, incompatible with recombination of small regions of DNA. For example, there is ample indirect evidence for chromosomal rearrangements in certain genes of the naturally competent *Strep. pneumoniae* and *Neisseria* spp. (reviewed by Maynard Smith *et al.*, 1990). These genes tend to code for strongly selected functions, such as the penicillin-binding proteins (PBP; Spratt *et al.*, 1989) or adhesive fimbriae (Seifert *et al.*, 1988). The sequence, for example, of a central 275bp region of the PBP gene from penicillin-resistant and -sensitive strains of *Strep. pneumoniae* differed by 21%, whilst the sequences for other genes in these strains differed by only 0.5-1% (Maynard-Smith *et al.*, 1990).

This 'mosaic' structure of certain genes indicates that naturally competent populations can be transformed with short sequences of DNA under natural conditions, allowing the existence of considerable variability whilst still preserving the clonal population commonly observed in naturally competant genera such as *Haemophilus* and *Neisseria* (Musser *et al.*, 1985; Caugent *et al.*, 1987).

There is also evidence for horizontal gene transfer in species which are not naturally competent. Within the 1871 bp alkaline phosphatase gene (*phoA*) in eight natural and one laboratory strain of *E. coli* there were 87 polymorphic nucleotides. It was possible to deduce that the gene had different origins in each of the isolates, which suggests that gene transfer has occurred, probably by transduction (DuBose *et al.*, 1988). Similarly, analysis of three *E. coli* K1 serotypes each with different O serotypes (Achtman *et al.*, 1986) showed identical electrophoretic protein mobilities in all but one of the 15 enzymes examined. The enzyme which differed, 6-phosphogluconate dehydrogenase, is encoded by the *gnd* gene, which is adjacent to the *rfb* gene responsible for the different O chain of lipopolysaccharide. These two adjacent genes were probably introduced by transduction and recombination as a consequence of selection for the *rfb* gene. Sequence divergences within the *gnd* gene of nine natural *E. coli* isolates provide further evidence for recombination, both within *E. coli* and between *E. coli* and *S. enterica* serotype Typhimurium (Bisercic *et al.*, 1991). There is similar evidence for this in other virulence-associated genes such as the P-associated fimbria region (*pap*) of *E. coli* (Plos *et al.*, 1989).

Similar evidence for chromosomal gene recombination by transduction exists within the *Salmonella* genus. There are many different subspecies and serotypes in the

Salmonella enterica complex (Le Minor and Popoff, 1987). The structure of the species is essentially clonal (Beltran *et al.*, 1988); many of the serotypes are members of a single clone, whilst other serotypes are found in several different clonal types, some of which differ in pathogenicity (Beltran *et al.*, 1988). Sequence data of the flagellar genes from serotypes Heidelburg, Rubislaw and Typhimurium indicate that some serotypes (Heidelburg and Typhimurium) are closely related clones, but have large divergences (19%) in flagellar gene sequence. Conversely, other serotypes (Heidelburg and Rubislaw) are distantly related clones, but show very little sequence divergence in their flagellar genes (Smith *et al.*, 1990).

TABLE 1. Correlation between plasmid and chromosomal polymorphism (data derived from Young and Wexler, 1988)

Chromosome types	Plasmid types Site 1	Site 2
FMP/O	A_1 6, C_1 3	A_1 1
MFF/B	E_2 2, E_5 1	A_1 2, E_1 1, E_2 1, E_3 1, E_4 2, G_1 1, O_1 1, O_2 1, Q_1 1, Q_2 1
MFF/P	A_1 2, A_2 1, D_1 1 D_2 2, N_1 1	-
SSQ/J	-	A_1 2, B_1 1
SSQ/M	-	A_1 1, E_4 1, E_5 1 F_1 1, F_2 2
SSQ/K	H_3 2	-
MSM/R	H_2 1	-
MSM/L	H_1 1	-
MSM/N	J_1 1, K_1 3, K_2 3, K_5 1	K_1 1
MFS/H	-	H_1 1
MFS/I	-	K_5 1
MFS/F	K_9 4	-
MFS/E	-	I_1 3, I_2 1, I_3 1
MSS/E	K_4 1, K_6 3, K_{12} 2 L_1 3, M_1 1	-
MSS/D	-	P_1 1
MFK/C	-	K_8 2
MFK/Q	-	K_3 3, K_7 1, K_8 1, K_{10} 1, K_{11} 1

Chromosome types were determined on the basis of the electrophoretic migration of three different isoenzymes (the first three letters) and the hybridisation pattern of total DNA, cut with a restriction enzyme, to a probe carrying chromosomal genes (final letter). Plasmid types are determined by hybridisation of total DNA, cut with a restriction enzyme, to DNA from a Sym plasmid. A_1 6, L_1 3 etc representing different binding patterns, and the number of isolates having this pattern. In total 85 isolates were tested, '-' indicates that no such chromosomal types were isolated from that site.

The suggested mechanism for gene transfer and recombination in *E. coli* and *Salmonella* is transduction. Bacteriophage lysogeny is common amongst natural *E. coli* strains (Dhillon and Dhillon, 1981) and in *Salmonella* (Threlfall & Rowe, 1984; Smith and Lovell, 1985; Barrow, 1986).

So far, the large variability and 'mosaic' gene structure appears to be localized to a few genes, especially those related to external antigenic structures, or those that are closely linked to such genes and have a very high probability of being co-inherited. Whilst it is undoubtably a selective advantage to alter the antigenic structure, the observed mosaic structure of such genes may partly be due to the ability of serological methods to detect small antigenic variations. It is possible that in other, less well studied, cellular components similar divergent mosaic genes will exist.

Interactions between rhizobia in soils Soon after the demonstration by Hellriegel and Wilfarth (1888) and others that leguminous plants fixed nitrogen because they contained rhizobia in root nodules, an inoculation industry developed which has led to the introduction of rhizobia into millions of hectares of agricultural land annually. Studies of inoculation experiments have produced useful information about survival and competition in soil (Beringer and Bale, 1988).

Perhaps the most useful information about genetic interactions between populations of *Rhizobium* has come from the work of Young and colleagues (Young and Wexler, 1988; Young, 1989; Harrison *et al.*, 1989). In studies of the genetic variability of existing populations of *Rhizobium* they compared the chromosomal type (determined by isoenzyme production and DNA homology) with the type of plasmid present in the strains (determined by DNA homology). Their results for two sites in Norfolk, UK, are summarized in Table 1.

Some strains such as MFF/B seem to be compatible with a range of plasmid types, whereas others, such as FMP/O are only associated with one or two plasmid types. There are several interpretations of these observations; the host and plasmid may be co-adapted and different combinations are not stable and are therefore not found. Alternatively, since all of the *Rhizobium* strains examined were isolated from clover nodules, it is possible that other, fully random, host/plasmid combinations existed but did not nodulate the genotypes present. Despite the apparent promiscuity of antibiotic-resistance plasmids, detailed earlier, there is some evidence for the widespread existence of co-adapted plasmids and hosts, particularly with small 'cryptic' plasmids observed in natural bacteria.

From the point of view of releasing microorganisms, or plasmids, to the environment, it would appear that for *R. leguminosarum*, a strain of chromosome type MFS/E (Table 1), would be intrinsically less likely to indulge in extensive plasmid exchanges. This is because the three isolates carried almost identical plasmids, which were distinctly different from the plasmids in the other strains (Young & Wexler, 1988). Taken together these observations suggest that the association occurred a long time ago and the strains have not participated in crosses with other strains since (P. Young, personal communication). Likewise, if recombinant genes needed to be plasmid-borne, plasmids such as the P1 N1 and M1 types in Table 1 would offer the least potential to be inherited and propagated within indigenous *R. leguminosarum* strains because each was found only once. More research is needed before specific predictions can be made, because we do not know what stabilizes the plasmid/chromosome combinations, whether such combinations are determined by the environment, or whether it is feasible to test for compatibility in the laboratory.

CONCLUSIONS

The importance of the work on *Rhizobium*, and on the medically-important bacteria is that the results are derived from organisms isolated from soil or animals, and thus indicate what happens in nature. The importance of such studies, as opposed to laboratory-based studies which show what might happen, cannot be overemphasized. An assessment of the potential risks arising from the release of genetically modified microorganisms to the environment is made difficult because gene transfer between very diverse groups of microorganisms is relatively simple to demonstrate in the laboratory (Heinemann & Sprague, 1989). There is ample evidence that genetic exchange between bacteria in natural communities has occurred and is undoubtably still occurring. However, for genes or operons engineered into the chromosome of released bacteria there is a low, but finite, chance that such a gene could be transferred to other bacteria. Such bacteria will probably be closely-related and already carry an allele of the gene, because of the constraints posed by, for example, homologous recombination, sequence divergence and bacteriophage host range. This will tend to simplify the prediction of the consequences of genetic exchange because smaller groups of organisms will have to be considered. While the work discussed here shows that gene transfer does occur in nature, it is clear that, as expected, the inheritance of such DNA and its expression is a rare event, and very much subject to the need for a selective advantage to the recipient organism.

REFERENCES

Achtman, M., Heuzenroeder, M., Kuseck, B., Ochman, H., Caugent, D., Selander, R.K., Vaisanen-Rhen, V., Korhonen, T.K., Stuart, S., Orskov, F. and Orskov, I. (1986) Clonal analysis of *Escherichia coli* O2:K1 isolated from diseased humans and animals. Infect. Immun. 51, 268-276.

Alexander, M. (1981) Why microbial predators and parasites do not eliminate their prey and hosts. Ann. Rev. Microbiol. 35, 113-133.

Bale, M.J., Fry, J.C. and Day, M.J. (1987) Plasmid transfer between strains of *Pseudomonas aeruginosa* on membrane filters attached to river stones. J. Gen. Microbiol. 133, 3099-3107.

Bale, M.J. and Hinton, M. (1991) Bacteria and agricultural animals - survival and gene transfer, in "Environmental release of genetically engineered and other microorganisms" (Fry, J.C. and Day, M.J. Eds.) (in press) University Press,Cambridge.

Barrow, P.A. (1986) Bacteriophages mediating somatic antigenic conversions in *Salmonella cholerae-suis* - isolation from sewage. J. Gen. Microbiol. 132, 835-837.

Beltran, P., Musser, J.M., Helmuth, R., Farmer, J.J., Frerichs, W.M., Wachmuth, I.K., Ferris, K., McWhorter, A.C., Wells, J.G., Cravioto, A. and Selander, R.K. (1988) Toward a population genetic analysis of *Salmonella*: Genetic diversity and relationships among strains of serotypes *S. choleraesuis, S. derby, S. dublin, S enteritidis, S. heidelberg, S. infantis, S newport* and *S. typhimurium*. Proc. Nat. Acad. Sci. USA 85, 7753-7757.

Bergh, O., Borsheim, K.Y., Bratbak, G. and Heldal, M. (1989) High abundance of viruses found in aquatic environments. Nature (London) 340, 467-468.

Beringer, J. E. and Bale, M. J. (1988). The survival and persistence of genetically-engineered micro-organisms, in "The Release of Genetically-Engineered Micro-Organisms" (Sussman, M., Collins, C. H., Skinner, F. A. and Stewart-Tull, D. E., Eds.), pp. 29-46. Academic Press, London.

Bisercic, M., Feutrier, J. Y. and Reeves, P. R. (1991) Nucleotide sequences of the *gnd* genes from nine natural isolates of *Escherichia coli*: evidence of intragenic recombination as a contributing factor in the evolution of the polymorphic *gnd* locus. J. Bacteriol. 173, 3894-3900.

Campbell, I.G., Mee, B.J. and Nikoletti, S.M. (1986) Evolution and spread of IncFIV plasmids conferring resistance to trimethoprim. Antimicrob. Agents Chemother. 29, 807-813.

Campbell, I.G. and Mee, B.J. (1987) Mapping of trimethoprim resistance genes from epidemiologically related plasmids. Antimicrob. Agents Chemother. 31, 1440-1441.

Chaslus-Dancla, E., Lagorce, M., Lafort, J.-P., Courvalin, P. and Gerbaud, G. (1987) Probable transmission between animals of a plasmid encoding aminoglycoside 3-N-acetyltransferase IV and dihydrofolate reductase I. Vet. Microbiol. 15, 97-104.

Caugant, D. A., Mocca, L. F., Frasch, C. E., Froholm, O., Zollinger, W. D. and Selander, R. K. (1987) Genetic structure of *Neisseria meningitidis* populations in relation to serogroup, serotype and outer membrane protein patterns. J. Bacteriol. 169, 2781-2792.

Dhillon, T.S. and Dhillon, E.K.S. (1981) Incidence of lysogeny, colicinogeny and drug resistance in enterobacteria isolated from sewage and rectums of humans and some domesticated species. Appl. Environ. Microbiol. 41, 894-902.

DuBose, R.F., Dykhuizen, D.E. and Hartl, D.L. (1988) Genetic exchange among natural isolates of bacteria: recombination within the *phoA* gene of *Escherichia coli*. Proc. Nat. Acad. Sci. USA 85, 7036-7040.

Duncan, K.E., Istock, C.A., Graham, J.B. and Ferguson, N. (1989) Genetic exchange between *Bacillus subtilis* and *Bacillus licheniformis* - variable hybrid stability and the nature of bacterial species. Evolution 43, 1585-1609.

Eardly, B.D., Materon, L.A., Smith, N.H., Johnson, D.A., Rumbaugh, M.D. and Selander, R.K. (1990) Genetic structure of natural populations of the nitrogen-fixing bacterium *Rhizobium meliloti*. Appl. Environ. Microbiol. 56, 187-194.

Fletcher, M. (1991) The physiological activity of bacteria attached to solid surfaces. Adv. Microb. Physiol. 32, 53-85.

Freter, R., Freter, R.R. and Brickner, H. (1983) Experimental and mathematical models of *Escherichia coli* plasmid transfer *in vitro* and *in vivo*. Infect. Immun. 39, 60-84.

Frost, J.A., Ward, L.R. and Rowe, B. (1989) Acquisition of a drug resistance plasmid converts *Salmonella enteritidis* phage type 4 to phage type 24. Epidemiol. Infect. 103, 243-248.

Fry, J.C. and Day, M.J. Eds (1990) Bacterial genetics in natural environments. London, Chapman and Hall, London.

Germida, J.J. and Khachatourians, G.G. (1988) Transduction of *Escherichia coli* in soil. Can. J. Microbiol. 34, 190-193.

Graham, J.B. and Istock, C.A. (1978) Genetic exchange in *Bacillus subtilis* in soil. Mol. Gen. Genet. 166, 287-290.

Griffith, F. (1928) The significance of pneumococcal types. J. Hyg. (Camb.) 27, 113-158.

Harrison, S.P., Jones, D.G. and Young, J.P.W. (1989) *Rhizobium* population genetics: Genetic variation within and between populations from diverse locations. J. Gen. Microbiol. 135, 1061-1069.

Heinemann, J. A. and Sprague, G. F. (1989) Bacterial conjugative plasmids mobilize DNA transfer between bacteria and yeast. Nature, (London) 340, 205-209.

Hellriegel, H. and Wilfarth, H. (1888) "Untersuchungen uber die Stickstoff-Nahrung der Gramineen und Leguminosum". Zeitschrift fur der verschiedige Rubenzucker Industrie des Deutsches Reichs (Beilageheft).

Hughes, V.M. and Datta, N. (1983) Conjugative plasmids in bacteria of the 'pre-antibiotic' era. Nature (London) 302, 725-726.

Kokjohn, T. A., Sayler, G. S. and Miller, R. V. (1991) Attachment and replication of *Pseudomonas aeruginosa* bacteriophages under conditions simulating aquatic environments. J. Gen. Microbiol. 137, 661-666.

Kuikman, P.J., van Elsas, J.D., Jansen, A.G., Burgers, S.L. and van Veen, J.A. (1990) Population dynamics and activity of bacteria and protozoa in relation to their spatial distribution in soil. Soil Biol. Biochem. 22, 1063-1073.

Lawrence, J. R. and Caldwell, D. E. (1987) Behavior of bacterial stream populations within the hydrodynamic boundary layers of surface microenvironments. Microb. Ecol. 14, 15-27.

Lawrence, J. R., Delaquis, P. J., Korber, D. R. and Caldwell, D. E. (1987) Behavior of *Pseudomonas fluorescens* within the hydrodynamic boundary layers of surface microenvironments. Microb. Ecol. 14, 1-14.

Le Minor, L. and Popoff, M. Y. (1987) Designation of *Salmonella enterica* sp. nov., nom. rev., as the type and only species of the genus *Salmonella*. Int. J. Syst. Bacteriol. 37, 465-468.

Lorenz, M.G., Aardema, B.W. and Wackernagel, W. (1988) Highly efficient genetic transformation of *Bacillus subtilis* attached to sand grains. J. Gen. Microbiol. 134, 107-112.

Lorenz, M.G. and Wackernagel, W. (1991) High frequency of natural genetic transformation of *Pseudomonas stutzeri* in soil extract supplemented with carbon/energy and phosphorous source. Appl. Environ. Microbiol. 57, 1246-1251.

Maynard Smith, J., Dowson, C.G. and Spratt, B.G. (1991) Localized sex in bacteria. Nature (London) 349, 29-31.

McArthur, J.V., Kovacic, D.A. and Smith, M.H. (1988) Genetic diversity in natural populations of a soil bacterium across a landscape gradient. Proc. Nat. Acad. Sci. USA 85, 9621-9624.

Morrison, W.D., Miller, R.V. and Sayler, G.S. (1978) Frequency of F116-mediated transduction of *Pseudomonas aeruginosa* in a freshwater environment. Appl. Environ. Microbiol. 36, 724-730.

Musser, J.M., Granoff, D.M., Pattison, P.E. and Selander, R.K. (1985) A population genetic framework for the study of invasive diseases caused by serotype b strains of *Haemophilus influenzae*. Proc. Nat. Acad. Sci. USA 82, 5078-5082.

Ogunseitan, O.A., Sayler, G.S. and Miller, R.V. (1990) Dynamic interactions of *Pseudomonas aeruginosa* and bacteriophages in lake water. Microb. Ecol. 19, 171-185.

O'Morchoe, S.B., Ogunseitan, O., Sayler, G.S. and Miller, R.V. (1988) Conjugal transfer of R68.45 and FP5 between *Pseudomonas aeruginosa* strains in a freshwater environment. Appl. Environ. Microbiol. 54, 1923-1929.

Orskov, F. and Orskov, I. (1983) Summary of a workshop on the clone concept in the epidemiology, taxonomy and evolution of the enterobacteriacae and other bacteria. J. Infect. Dis. 148, 346-357.

O'Sullivan, M., Stephens, P.M. and Ogara, F. (1990) Interactions between the soil-borne bacteriophage-Fo-1 and *Pseudomonas* spp. on sugarbeet roots. FEMS Microbiol. Lett. 68, 329-333.

Paul, J.H., Frischer, M.E. and Thurmond, J.M. (1991) Gene transfer in marine water column and sediment microcosms by natural plasmid transformation. Appl. Environ. Microbiol. 57, 1509-1515.

Paul, J.H., Jeffrey, W.H. and Deflaun, M.F. (1987) Dynamics of extracellular DNA in the marine environment. Appl. Environ. Microbiol. 53, 170-179.

Plos, K., Hull, S.I., Hull, R.A., Levin, B.R., Orskov, I., Orskov, F. and Svanborg-Eden, C. (1989) Distribution of the P-associated-pilus (*pap*) region among *Escherichia coli* from natural sources: Evidence for horizontal gene transfer. Infect. Immun., 57, 1604-1611.

Proctor, L.M. and Fuhrman, J.A. (1990) Viral mortality of marine bacteria and cyanobacteria. Nature (London) 343, 60-62.

Rayssiguier, C., Thaler, D.S. and Radman, M. (1989) The barrier to recombination between *Escherichia coli* and *Salmonella typhimurium* is disrupted in mismatch-repair mutants. Nature (London) 342, 396-400.

Reanney, D.C., Roberts, W.P. and Kelly, W.J. (1982) Genetic interactions among microbial communities, in. "Microbial Interactions and Communities vol.I" (Bull A.T. and Slater. J.H. Eds.), pp. 287-322. Academic Press, London.

Rochelle, P.A., Fry, J.C. and Day, M.J. (1989) Factors affecting conjugal transfer of plasmids encoding mercury resistance from pure cultures and mixed natural suspensions of epilithic bacteria. J. Gen. Microbiol., 135, 409-424.

Romanowski, G., Lorenz, M.G. and Wackernagel, W. (1991) Adsorption of plasmid DNA to mineral surfaces and protection against DNaseI. Appl. Environ. Microbiol. 57, 1057-1061.

Saye, D.J., Ogunseitan, O., Sayler, G.S. and Miller, R.V. (1987) Potential for transduction of plasmids in a natural freshwater environment: Effect of plasmid donor concentration and a natural microbial community on transduction in *Pseudomonas aeruginosa*. Appl. Environ. Microbiol. 53, 987-995.

Saye, D.J., Ogunseitan, O.A., Sayler, G.S. and Miller, R.V. (1990) Transduction of linked chromosomal genes between *Pseudomonas aeruginosa* strains during incubation *in situ* in a freshwater habitat. Appl. Environ. Microbiol. 56, 140-145.

Seifert, H.S., Ajioka, R.S., Marchal, C., Sparling, P. F. and So, M. (1988) DNA transformation leads to pilin antigenic variation in *Neisseria gonorrhoeae*. Nature (London) 336, 392-395.

Smith, H.W. and Lovell, M.A. (1985) Transduction complicates the detection of conjugative ability in lysogenic *Salmonella* strains. J. Gen. Microbiol. 131, 2087-2089.

Smith, N.H., Beltran, P. and Selander, R.K. (1990) Recombination of *Salmonella* phase-1 flagellin genes generates new serovars. J. Bacteriol. 172, 2209-2216.

Spratt, B.G., Zhang, Q.Y., Jones, D.M., Hutchison, A., Brannigan, J.A. and Dowson, C.G. (1989) Recruitment of a penicillin-binding protein gene from *Neisseria flavescens* during the emergence of penicillin resistance in *Neisseria meningitidis*. Proc. Nat. Acad. Sci. USA 86, 8988-8992.

Steffan, R.J., Goksoyr, J., Bej, A.K. and Atlas, R.M. (1988) Recovery of DNA from soils and sediments. Appl. Environ. Microbiol. 54, 2185-2191.

Stewart, G.J. and Carlson, C.A. (1986) The biology of natural transformation. Ann. Rev. Microbiol. 40, 211-235.

Stewart, G.J. and Sinigalliano, C.D. (1990) Detection of horizontal gene transfer by natural transformation in native and introduced species of bacteria in marine and synthetic sediments. Appl. Environ. Microbiol. 56, 1818-1824.

Threlfall, E.J. and Rowe, B. (1984) Antimicrobial drug resistance in salmonellae in Britain - a real threat to public health?, in "Antimicrobials and Agriculture". (Woodbine, M., Ed.), pp. 513-524. Butterworths, London.

van Elsas, J.D., Trevors, J.T., Starodub, M.E. and van Overbeek, L.S. (1990) Transfer of plasmid RP4 between pseudomonads after introduction into soil - influence of spatial and temporal aspects of inoculation. FEMS Microbiol. Ecol. 73, 1-11.

Whittam, T.S., Wolfe, M.L. and Wilson, R.A. (1989) Genetic relationships among *Escherichia coli* isolates causing urinary tract infections in humans and animals. Epidemiol. Infect., 102, 37-46.

Wilkins, B.M. (1988) Organization and plasticity of enterobacterial genomes. J. Appl. Bacteriol. Symp. Suppl. 65, 51S-69S.

Wray, C., McLaren, I., Parkinson, N.M. and Beedell, Y. (1987) Differentiation of *Salmonella typhimurium* 204c by plasmid profile and biotyping. Vet. Rec. 121, 514-516.

Young, J.P.W. and Wexler, M. (1988) Sym plasmid and chromosomal genotypes are correlated in field populations of *Rhizobium leguminosarum*. J. Gen. Microbiol. 134, 2731-2739.

Young, J.P.W. (1989) The population genetics of bacteria, in "Genetics of bacterial diversity" (Hopwood, D.A.and Chater, K. F. Eds.), pp. 417-438. Academic Press, London.

Zavarzin, G. A., Stackebrandt, E. and Murray (1991) A correlation of phylogenetic diversity in the Proteobacteria with the influences of ecological forces. Can. J. Microbiol. 37, 1-6.

Zeph, L.R., Onaga, M.A. and Stotzky, G. (1988) Transduction of *Escherichia coli* by bacteriophage P1 in soil. Appl. Environ. Microbiol. 54, 1731-1737.

APPLICATION OF GENETICALLY-MODIFIED MICROORGANISMS IN AGRICULTURE

G. Van Den Eede[1] and M. Van Montagu[2]

Commission of the European Communities[1]
Joint Research Centre
Isei-mtr, TP 634
I-21020 Ispra (VA) Italia

Laboratorium Genetika[2]
State University Ghent
K.L. Ledeganckstraat 35
B-9000 Ghent Belgium

INTRODUCTION

More than a decade ago, promising results stimulated researchers to cherish great ambitions on the applications of genetically-modified micro-organisms because the molecular tools were available to understand and modify the functioning of the genome, however, few examples can be found. A possible explanation for this delay, apart from a hesitant attitude of public and regulators, is the lack of relevant scientific data.

The entire process of constructing a genetically-modified organism that is to be released can be divided in two parts. The first is the construction in the laboratory of the genetic modification, the second is to release the organism into the environment in which there is an endogenous bacterial population, often containing the same or related species. Ideally, the organism should persist in this competing environment long enough to carry out its intended function, and short enough not to survive over an extended period. An intended release therefore should not be designed and based on pure molecular biological data but information about survival, persistence and environmental safety should also be considered. However, our knowledge of soil microbiology and ecology is too limited to encompass all the events that might occur in the environment and more is needed to develop a working strategy which would ensure success.

The aim of this chapter is to review the ways in which micro-organisms have been used traditionally in agriculture, how genetically-modified micro-organisms (GEMMOs) can contribute to improved agricultural practice and what gaps in our understanding still need to be filled. We will not discuss the use of viruses nor the role of biotechnology in bioremediation or food processing, since these are being dealt with by other contributors.

AGRONOMISTS ARE FAMILIAR WITH MICRO-ORGANISMS

For decades inoculations of soils with micro-organisms have been common agricultural practice. These aimed at increasing the availability of nitrogen or phosphorus for crop nutrition, or at eliminating biological agents such as insects, nematodes, fungi and viruses. A specific phenotypic trait that occurred naturally in a bacterial population was exploited, but molecular

The Release of Genetically Modified Microorganisms
Edited by D.E.S. Stewart-Tull and M. Sussman, Plenum Press. New York, 1992

biology offered the potential either to improve this character or to express it in a completely different background.

Inoculum Production for Crop Nutrition

Over many decades, **non-symbiotic** as well as **associative nitrogen-fixers** have been applied in agriculture.

The free living nitrogen fixers, *Azotobacter, Azospirillum* and *Herbaspirillum* for instance have been used in different ecosystems. The contribution of these bacteria to promote growth does not depend only on their capacity to convert free nitrogen into ammonia, but also upon the production of growth promoting hormones. Although the genetics of some of these bacteria is already well-understood, we believe that their beneficial impact on agriculture is too modest to expect relevant applications by modifying their agronomic properties. Probably, there is no economic relevance for using non-symbiotic nitrogen-fixers in agricultural conditions.

An exception can be made for plants growing on marginal soils. There is, for instance, an unquestionable impact when Kallar grass is grown on soils in Pakistan that contain extremely high salt concentrations. It has been demonstrated in the author's laboratory that association of Kallar grass with certain bacteria is indispensable for full performance, although the mechanism of the interaction is not well understood.

Symbiotic nitrogen-fixers, such as *Rhizobium, Bradhyrhizobium* and *Azorhizobium* have a recognized world-wide reputation as efficient nitrogen-fixers, that is, there is a clear economical benefit from the use of some of these in agricultural ecosystems. This is reflected in the sales figures: in 1987 inoculum production of *Bradhyrhizobium japonicum* was worth 18 million US $. Furthermore, there has been a unique experience in different ecosystems for several decades. Field-trials were initiated in Africa to test the impact on the quality and yield of rice using *Sesbania rostrata* inoculated with *Azorhizobium caulinodans* as a green manure. In terms of international acceptance it is meaningful that only a few years after the first reports, it was possible to co-organize a meeting in Senegal with participation of researchers from all continents, already experienced in field-trials with this bacterium.

Since the genetics of some of these species are well-understood, we believe that genetically-modified *Rhizobiaceae* species will find their way into agricultural use It is worthwhile to note in this context that many patent applications have been requested for modified symbiotic nitrogen-fixers.

Actinomycetes. The actinomycetes have value as symbiotic nitrogen-fixers. In sandy soils, for instance, an absolute soil stabilization can be obtained when exploiting the *Casuarinia-Frankia* symbiosis. Lack of knowledge about the genetics of the interaction, along with the difficulties in cultivating *Frankia* make it unlikely that a modified strain with a significant agronomical value can be expected in the near future.

Although commercialization of blue-green algae, such as the cyanobacterial symbiosis *Azolla-Anabaena,* has a rice growth-promoting effect the potentiality of exploiting genetically-modified organisms is also low.

Bacillus megaterium has been used in agricultural systems in Eastern Europe with the intention to solubilize phosphorus from the soil but the commercial product, phosphobacterin, is not available on the market. Probably the only potent microorganisms that make phosphorus available in soils are fungi, but their action is too little understood to expect commercialization or to envisage short term applications of modified species.

Micro-organisms have Controlled Biological Agents for Many Decades

For many years, our main weapon against pests were chemicals, but these had the disadvantage of being a potential threat to the environment. In addition, chemicals are often expensive and the target organisms may develop resistance. Biological pest control, apart from generally lacking these characteristics, has another advantage. An ecosystem has a finely-tuned balance between different types of organisms. Normally, a chemical kills indiscriminately, often wiping out indispensable members of the ecosystem. An example, for instance, is the widespread use of DDT on mosquito larvae which killed herons or reduced their reproduction capacity because they were at the end of the food-chain. Biological pest control attempts to deal with the pest without destroying the balance: bacteria could be used as a biopesticide to kill specifically mosquito larva. Even if such bacteria accumulate in the food-chain, this has no consequence on the growth of small and larger fish or herons. These considerations resulted in many countries setting up integrated pest management systems that often consider spraying as a last stopgap.

Bacillus thuringiensis, the old workhorse in biocontrol, has been marketed since the early fifties and shares about 1.5% of world insecticide sales worth 22 million US $ in 1987. It is now commercialized by several companies such as Abbott, Solvay, Sandoz, Novo Nordisk and in the Soviet Union by Glavmikrobioprum.

Bacillus thuringiensis (Bt), active against certain lepidopteran and dipteran pests, produces during sporulation a large proteinaceous crystal. This contains the insecticidal d-endotoxin that is non-toxic until solubilized and proteolytic cleavage in the alkaline insect gut. The toxin binds to specific receptors on epithelial cells and induces the formation of small pores. Through these pores ions and water enter the cells and cause cell swelling and lysis.

The standard commercial strain for controlling lepidopterous pests, HD1, was isolated from the pink bollworm. Different strains and varieties (for a review see Höfte and Whiteley, 1989) are being used to control different insects such as mosquitoes, the blackfly (which is the causative agent of river blindness), cabbage moth and a variety of beetles.

Bt is produced in large volumes by refined fermentation techniques but has still two drawbacks that give it only a small market niche. First, the host range is narrow and secondly, the preparation must be sprayed on the ecological site. A vigorous rainfall suffices to wash it off leaves or a climatological change might induce spore development.

Several plasmids carrying the endotoxin genes can be transferred by conjugation into one recipient *Bacillus* and thus expand the host range. Conjugation into another bacterium, such as a root-colonizing *Pseudomonas fluorescens* or into *Clavibacterium xyli*, which grows in the vascular tissue of corn, can trigger toxin production in a specific environment. Other approaches have used genetic modification techniques. Bt toxic proteins have two distinct structural domains: a highly variable one involved in insect specificity and a very conserved one involved in larvicidal activity. Therefore, it became possible to extend the host-range by constructing proteins that contain different insect-specific domains besides the larvicidal domain, that proved toxic to a wider range of insects.

The most dramatic genetic achievement came from the direct introduction of Bt toxin genes in the genomes of such plants as cotton, tomato or potato (Vaeck et al., 1987) to obtain *Manduca sexta* resistance. In field-trials, transgenic plants fare as well as plants treated with conventional insecticides. In cotton, for instance, both groups have about 4.0% damaged bolls, compared to 30% in the untreated controls.

No cases of resistance to Bt in the field have been observed. It is thought that this fortunate situation is due to the low persistence of *Bacillus thuringiensis* in the environment but this might change drastically when transgenic plants are widely used. Several generations of insects per year will then be continuously exposed to toxic crystal proteins, providing an ideal environment for the development of resistance (Höfte and Whiteley, 1989).

Several systems to study host-pathogen interactions have been developed. One assumes that the transmissibility and life-span of the pathogen do not differ between different pathogens. Another model divides the pathogen into two distinct subpopulations, one transmissible and short-lived, the other non-transmissible and long-lived. In *Manduca sexta*, for instance, the transmissible stages are those in which Bt resides as a spore on vegetative material that feeds the insect. All other forms of Bt can be considered as non-transmissible. However, when Bt is present in these reservoirs as a spore, it can cause infections if spores are translocated by biotic or abiotic agents. At equilibrium between host and pathogen communities there are two distinct types of population densities: stable oscillations or constant numbers. The prevailing stage is determined by the storage and release of the pathogenic forms in the reservoir. If this population is low, then the population density of pathogen and host will oscillate; higher levels result in constant host and pathogen populations (Hochberg, 1989).

This important interaction between the population size of pathogen and host does, of course, not apply to chemical insecticides and increases the power of biopesticides. Biocontrol strives at managing, rather than eradicating harmful pests, meaning that the user must be a more skillful observer and a more cautious practitioner.

Newer Routes in Protection Against Environmental Stresses and Biological Agents

Frost damage. *Pseudomonas syringae* has attracted the attention of researchers because of its role in ice nucleation. Ice nucleation is the event that precedes crystallization of a super-cooled aqueous solution: it consists of the formation of an ice crystal large enough to become an initiation point. Some organisms produce very accurate templates for the organization of the seed crystal and cause ice nucleation very close to the ice-water equilibrium. In *Pseudomonas*, the gene product of *ina*, is thought to be a structural component from the ice nucleus.

Ice nucleation active *Pseudomonas* (INA) as well as in^{a-} derivatives were avirulent and non-pathogenic after testing on 30 plant species and cultivars. When sufficient numbers of INA bacteria colonize plant leaves freeze damage occurs since nucleation starts at relatively high temperatures. These bacteria have been released to stimulate snow formation.

Interaction between INA and antagonistic bacteria can reduce the INA population up to 100-fold under field conditions, with this decreasing frost damage 30-95% relative to untreated plants. Also, the application of *ice-* bacteria resulted in the reduction of the population size of the nearly isogenic parental INA strain. The ice phenotype might provide a selective advantage where frost is likely to occur, but we are still far from understanding the nature of the selective pressures or the extent to which competition for limiting resources affects inter- and intra-specific fluctuations. Similarly, we know very little of the interactions between phytopathogenic bacteria and other microbial leaf inhabitants.

Control of tumor formation. *Agrobacterium tumefaciens* is a well-known bacterium in laboratory practice because of its use as the prefered vehicle to import foreign DNA into dicotyledons. The application of this bacterium in the construction of transgenic plants is the consequence of exploiting its pathogenic trait: tumor formation.

Two types of agrobacteria have been studied in detailed: *A.tumefaciens* and *A. rhizogenes*. The first induces tumorous growth at a wound in the stem of a variety of dicotyledonous plants (the crown gall) whereas the latter induces hairy root formation on wounded plant tissue (for a recent review, see Gelvin, 1990). This tumor formation results from the transfer of a piece of bacterial DNA, the T-DNA, to the plant cell, the random integration into the plant chromosome and the expression of the inserted DNA by the plant machinery. Consequently, normal growth is disturbed by the endogenous de novo production of plant hormones and cells will massively proliferate. Additionally, the plant synthesizes and excretes certain molecules, the so-called opines, that can be catabolyzed specifically by the *Agrobacterium* that was responsible for the DNA transfer. Thus, *Agrobacterium* forces a plant cell to propagate uninhibitedly and to synthesize a compound that *Agrobacterium* can utilize as a nitrogen and carbon source, with the creation of the perfect ecological niche (Gheysen et al., 1989).

Most of the virulence genes are clustered in two regions, outside the chromosome on the tumor-inducing or Ti plasmid. One set of genes, the *vir* genes, is responsible for transfering the second set of genes, located in the T-region, to the plant cell. This transfer region is well-defined and contains the genes necessary for the synthesis of plant hormones and opines and is bordered by a 25 bp short DNA sequence (Wang et al., 1984; Horsch and Klee, 1986). This border sequence delimits the portion of DNA that will be transfered, that is, all the sequences between the borders and only these will be transfered to the plant cell (Zambryski, 1988; Stachel and Zambryski, 1989).

Since transfer of DNA only occurs when *Agrobacterium* colonizes wounded plant tissue, an accurate interaction system with the plant host is developed. Acetoseringone is a rare chemical compound found in injured plant tissue, that is recognized in nanomolar concentrations by *Agrobacterium* where it activates the expression of the *vir* genes. These *vir* genes will mediate the transfer of the genes between the T-DNA borders to the plant cell where they become integrated.

Molecular biologists have shown that all the tumor-inducing genes within the T-DNA borders can be removed and substituted by any other sequence. These new inserted genes will now be accurately transfered to and integrated into the plant cell without inducing tumorous growth. Among many examples, this methodology has allowed the construction of transgenic plants that are resistant to insect attack by expressing the *Bacillus* endotoxin gene (Vaeck et al., 1987), that are resistant to herbicide by expressing a *Streptomyces* detoxifying enzyme (Thompson et al., 1987; De Greef et al., 1989), that are resistant to viruses by expressing the viral coat protein gene (Abel et al., 1986), that produce chemical compounds such as the neuropeptide Leu-enkephalin (Vandekerckhove et al., 1989) or the antibacterial drug magainin. Here, a plant pathogen has been disabled in its phytopathogenic properties and converted into a useful tool in plant molecular biology. Since *Agrobacterium* is eliminated during the procedure, its use does not entail any environmental risk.

In nature, where crown gall formation can have devastating effects, attempts were made to control the disease by using a non-tumorigenic *Agrobacterium radiobacter* strain, K84.This organism produces a bacteriocin, agrocin 84, which is taken up by susceptible agrobacteria by a permease carried in their Ti plasmid (Engler et al., 1975). *A.radiobacter* itself is resistant to agrocin and unable to induce crown gall. The mechanism of crown gall inhibition involves agrocin action and competition for wound colonization. The strain has been used on a very large-scale in efficient control of the crown gall disease in the field.

However, several scenarios are possible because of the widespread use of *A. radiobacter* :

1. In an environment in which large amounts of agrocin are being used, *A.tumefaciens* can mutate and form an altered permease. This mutant will resist the agrocin action, form crown gall and, following the opine concept, will have a selective advantage over the other agrobacteria. There will be a possibility that agrocin resistance spreads in the environment.

2. *A.tumefaciens* might occasionally transfer its Ti plasmid into *A. radiobacter*. Apart from the performance in terms of pathogenicity of this transconjugant, this will have no or few environmental consequences unless the bacterium is able to take over the niches previously occupied by *A. tumefaciens*.

3. *Agrobacterium radiobacter* might transfer the agrocin production and resistance genes to *A. tumefaciens*. The transconjugant will have a clear selective advantage and the effect of *A.radiobacter* will be limited to the competition for infection sites. This problem has been overcome by deleting the transfer functions of *Agrobacterium radiobacter*. This genetically-modified bacterium is now commercially available and presents one of the few examples of GEMMOs used in agriculture.

There are several projects now intending to use *Agrobacterium tumefaciens* as a biological weapon to control plant pests. Here, the chemotactic response of agrobacteria is being used to target biological control to wounds on plants. The strains used retain the *vir* genes, but have large deletions between the T-DNA borders.

MONITORING SURVIVAL, PERSISTENCE AND GENE SPREAD IN NATURE

Information on survival and persistence is critical in the assessment of the possible impact of released GEMMOs on natural ecosystems. Effective and safe release programmes require the development of sensitive detection methods to trace the parental organism, the GEMMO and the gene(s) that has/have been released. This is only possible by combining conventional techniques used by soil microbiologists and ecologists with molecular biological methods in a new discipline, molecular ecology.

Monitoring Survival and Persistence

There is an impressive battery of techniques available for the detection of the parental organism, and to discriminate between the parental and the modified organism. The Commission of the European Communities, in collaboration with the US Environmental Protection Agency and the US Department of Agriculture is providing an inventory of the existing techniques with an evaluation of their specificity, sensitivity, ease of use and time necessary, cost and effect of environmental media.
Among the newer techniques, the following may be listed:

1. Unusual phenotype tracking: the GEMMO contains a specific gene that distinguishes it easily from other micro-organisms. Examples are the resistance genes derived from transposons, genes for utilization of rare substrates or

chromogenic markers such as the β–galactosidase, the β-glucuronidase or the firefly luciferase which allow the visualization of the GEMMOs.

2. The use of nucleic acid probes and the great promise expected of the polymerase chain reaction and related techniques.
3. Immunological detection methods using mono- and poly-clonal antibodies, enzyme-linked immunosorbent assay or fluorescent antibodies along with many other techniques.

We want to focus on some particular issues that might come about when applying some of these methods:

• In tracking an unusual phenotype, proper gene expression in the soil is indispensable. Although this normally cannot pose technical difficulties, this aspect must be taken into consideration, especially when gene transfer in soil is assayed since there is no reason to assume *per se* that the gene will be expressed in the recipient host (see later).

• It is often argued that introduction of new genes causes a metabolic burden and decreases the viability of the bacterium in the soil. A solution is to introduce a single copy of the gene into the chromosome rather than to insert it on a plasmid that is present as multiple copies. A second solution would be to locate the gene downstream of specific regulatory sequences, such as the bacteriophage *p*L promoter under control of the temperature-sensitive lambda repressor, cI857.

• It is recognized that a positive or negative signal is not always a correct reflection of what occurs in nature. With β–glucuronidase for instance, a blue crystal is produced when the substrate, X-Gluc, is hydrolyzed and further oxidized. Therefore, to avoid false negatives, the substrate must sufficiently penetrate the cells and an oxidizing agent must be present. In contrast, false positives can be scored if a particular cell is producing β-glucuronidase and than spreading the crystals to other areas.

• In specific environmental samples, no methods are available to estimate accurately and reliably the total amount of living bacteria. Roszak and Colwell (1987) suggest that the currently employed standard methods for aquatic microbiology are inadequate to protect human health. Differences in culturable direct counts often have a magnitude of 10^6 to 10^7 and thus, Xu *et al.* (1982) proposed that the concept of "die-off" of bacteria in the environment was not valid. They introduced the concept of dormancy since the cells, remained viable although it was not possible to determine their numbers by culture methods.

Roszak and Colwell (1987) describe "rare" survival methods, such as pseudosenescence and reductive division, pointing out, however, that these might be more general than believed. Pseudosenescence, as shown for *Klebsiella aerogenes*, is the state in which the bacteria lose the ability to multiply as the result of certain stresses but remain completely functional as individuals. Reductive division refers to an increase in cell number without a significant increase in biomass, resulting in ultramicrobacteria with sizes less than 0.3µm. These are undoubtedly a portion of the species not recovered by standard culture methods and they may represent an early strategy in a linear progression of survival mechanisms. From their review, it can be concluded that in certain ecosystems, detection methods have only allowed the isolation"overfed" bacteria but that bacteria, when present below threshold levels, probably reside in the soil only able to grow on low C-levels. Consequently, bacteria are able to survive in large quantities in soil in such a form that is not (easily) detectable.

Ultimately, one is not solely interested in knowing the survival and persistence in the environment of a released micro-organism, but is merely interested in the spread of the introduced gene. There are several mechanisms described by which gene transfer occurs in soil and one can question what evolutionary barriers there are since "trans-kingdom sex" has been described for Gram-negative bacteria and plants as well as for Gram-positive bacteria and *Saccharomyces* or *Schizosaccharomyces*. In addition, we must not only take two species into consideration, but must realize that transfer occurs via different intermediate bacteria.

Although reliable methods exist that can suggest if gene transfer has occurred - Polymerase Chain Reaction for instance allows the detection of a specific nucleotide sequence in $1x10^{10}$ bacteria - a major difficulty remains in the precise identification of the potential recipient microbes. As it seems, there are barriers to the establishment and expression of transfered sequences, rather than to transmission itself: there is in nature no hindrance for DNA transfer, but selection is made against expression. In other words, tumor formation is restricted to the agrobacteria not because the Ti plasmid is not spread among soil bacteria but rather because the Ti plasmid genes are only correctly recognized by *Agrobacterium tumefaciens* itself. The extent and impact of horizontal gene transfer are difficult to determine and cannot be defined by just analyzing the gene products and nucleotide sequences.

CONCLUDING REMARKS

If we split the development of GEMMOs into two phases, that is, first the construction in the laboratory and secondly,the release in nature , then clearly, achievements in laboratories are enormous while the knowledge of comportment in nature is extremely limited.

Many attractive applications are being generated in laboratories. However, a major breakthrough will only come when, probably in a few years from now, a better understanding of the complex environmental processes will be obtained.

REFERENCES

Abel, P.P., Nelson, R.S., De, B., Hoffman, N., Rogers, S.G., Fraley, R.T., and Beachy, R.N. (1986). Delay of disease development in transgenic plants that express the tobacco mosaic virus coat protein gene. Science 232, 738-743.

Engler, G., Holsters, M., Van Montagu, M., Schell, J., Hernalsteens, J.-P., and Schilperoort, R. (1975). Agrocin 84 sensitivity: a plasmid determined property of *Agrobacterium tumefaciens*. Mol. Gen Genet. 137, 345-349.

De Greef, W., Delon, R., De Block, M., Leemans, J., and Botterman, J. (1989). Evaluation of herbicide resistance in transgenic crops under field conditions. Bio/technology 7, 61-64.

Gelvin, S.B. (1990). Crown gall disease and hairy root disease. A sledgehammer and a tackhammer. Plant Physiol. 92, 281285.

Gheysen, G., Herman, L., Breyne, P., Van Montagu, M., and Depicker, A. (1989). *Agrobacterium tumefaciens* as a tool for the genetic transformation of plants. In Genetic transformation and expression, L.O. Butler (Ed.). London, Intercept 6, 151-174.

Hochberg, M.E. (1989). The potential role of pathogens in biological control. Nature (London) 337, 262-265.

Höfte, H., and Whiteley, H.R. (1989). Insecticidal crystal proteins of *Bacillus thuringiensis*. Microbiol. Rev. 53, 242-255.

Horsch, R.B., and Klee, H.J. (1986). Rapid assay of foreign gene expression in leaf discs transformed by *Agrobacterium tumefaciens:* role of T-DNA borders in the transfer process. Proc. Natl. Acad. Sci. USA 83, 4428-4432.

National Research Council (1989). Field testing genetically modified organisms; framework for decisions. National Academy Press, Wahington D.C.

Roszak, D.B. and Colwell, R.R.(1987). Survival strategies of bacteria in the natural environment . Microbiol. Rev. 51, 365-379.

Stachel, S.E., and Zambryski, P.C. (1989). Generic trans-kingdom sex? Nature (London) 340, 190-191.

Thompson, C.J., Rao Movva, N., Tizard, R., Crameri, R., Davies, J.E., Lauwereys, M., and Botterman, J. (1987). Characterization of the herbicide-resistance gene bar from *Streptomyces hygroscopicus*. EMBO J. 6, 2519-2523.

Vaeck, M., Reynaerts, A., Hˆ fte, H., Jansens, S., De Beuckeleer, M., Dean, C., Zabeau, M., Van Montagu, M., and Leemans, J. (1987). Insect resistance in transgenic plants expressing modified *Bacillus thuringiensis* toxin genes. Nature (London) 328, 33-37.

Vandekerckhove, J., Van Damme, J., Van Lijsebettens, J., Botterman, J., De Block, M., Vandewiele, M., De Clercq, A., Leemans, J., Van Montagu, M., and Krebbers, E. (1989). Enkephalins produced in transgenic plants using modified 2S seed storage proteins. Biotechnology 7, 929-932.

Wang, K., Herrera-Estrella, L., Van Montagu, M., and Zambryski, P. (1984). Right 25-bp terminus sequences of the nopaline T-DNA is essential for and determines direction of DNA transfer from *Agrobacterium* to the plant genome. Cell 38, 455-462.

Xu, H.-S., Roberts, N., Singleton, F.L., Attwell, R.W., Grimes, D.J. and Colwell, R.R. (1982). Survival and viability of non-culturable *E. coli* and *Vibrio cholerae* in the estuarine and marine environment. Microb. Ecol. 8, 313-323.

Zambryski, P. (1988). Basic processes underlying *Agrobacterium*mediated DNA transfer to plant cells. Ann. Rev. Genet. 22, 1-30.

BIOREMEDIATION AND WASTE MANAGEMENT

Richard J.F. Bewley

Dames & Moore International
Blackfriars House
St Mary's Parsonage
MANCHESTER, M3 2JA,UK

INTRODUCTION

The treatment of waste materials, which arise from man's activities, by use of microorganisms has a long and well-documented history. Interest in the application of the technologies involved in such processes to problems of hazardous waste has increased in recent years with the growing legislative and economic pressures to develop destructive and cost-effective solutions for environmental clean-up.

There is a wide body of literature concerning both the transformation of xenobiotic organic molecules by microorganisms and various modified processes including industrial effluent treatment systems. The application of biological techniques to clean-up elements of the natural environment, such as contaminated land and groundwater is now a rapidly expanding field.

In contrast, however, published data concerning the full scale application of genetically modified microorganisms (GEMMOs) to the treatment of hazardous waste or contaminated environments is virtually non-existent, even for field-trial studies. This is a reflection of the current strict control over release of GEMMOs for such purposes, at least in the USA, and increasingly in Europe.

In this chapter the precedents set for the application of GEMMOs to the remediation of contaminated environments by the use of bioaugmentation techniques, and the relative merits of environmental engineering (versus genetic modification) to create the conditions necessary for achieving clean-up will be considered. It is intended to highlight the key issues involved in the use of both naturally occurring and genetically modified microorganisms in the field, in full scale application. In view of the accent of REGEM 2 to the release of GEMMOs, the emphasis will be orientated towards remediation of the natural environment, such as contaminated land and groundwater, rather than enclosed systems for waste treatment.

THE THEORETICAL BASIS FOR BIOREMEDIATION AND THE CASE FOR BIOAUGMENTATION

A cursory examination of a contaminated site raises two apparently conflicting issues: the ubiquity and metabolic diversity of microorganisms and the persistence of potentially degradable organic molecules. An understanding of the factors responsible for such persistence provides the key to the development and implementation of a successful remediation strategy. It is convenient to categorize such factors as those relating to the physico-chemical properties of the contaminant and those pertaining to the recipient environment which encompasses biological as well as physical and chemical components. The former includes all the properties relating to the chemical structure of the contaminant: generalizations concerning the effects of

The Release of Genetically Modified Microorganisms
Edited by D.E.S. Stewart-Tull and M. Sussman, Plenum Press, New York, 1992

33

chemical structure on biodegradation rates often tend to focus on the presence of specific functional groups or molecular fragments. For example, hydroxyl and carboxyl functional groups or benzene rings, usually tend to increase the rate of biodegradation, while halogen, nitro and sulphate functional groups usually decrease biodegradation (Dragun, 1988). In the case of petroleum hydrocarbons, alkanes tend to degrade more rapidly than aromatic hydrocarbons and within the alkanes, straight-chains are more susceptible than branched-chains to degradation (Atlas, 1981; Chakrabarti *et al*, 1989). The most recalcitrant xenobiotic molecules, such as halogenated pesticides, tend to be those containing chemical substituents not commonly encountered in unpolluted natural environments (Hardman, 1991). However, the occurrence of other relatively recalcitrant pollutants in natural systems should not be overlooked, for example some of the higher molecular weight polyaromatic hydrocarbons (Sims and Overcash, 1983).

The other properties of the contaminant which influence its persistence may be physical or temporal: one particularly important physical property is bioavailability. This is often directly attributable to solubility, although individually more soluble compounds may occur in a complex mixture of relatively insoluble higher molecular-weight compounds, for example in coal tar (Dragun, 1988). The contaminant concentration may be inhibitory to biodegradation at both ends of the scale. Not only are high concentrations of pollutants often directly toxic, but bioavailability may be reduced simply through the mass of waste providing a barrier difficult to penetrate even by a microorganism: once again, coal tar is a good example. Conversely, the occurrence of contaminants at very low concentrations may reduce the potential for biodegradation and there are several reports of thresholds for biodegradation in natural water (Alexander, 1985; Goldstein *et al*, 1985). Such thresholds may be influenced by competition with other organic substrates (Swindoll *et al*, 1988). Temporal aspects of the contamination could also influence its persistence in the natural environment. It is possible that in the case of a fresh spill, insufficient time may have elapsed for the development of an appropriate microflora capable of its degradation.

The physical and chemical properties of the environment which influence the degradation of pollutants include availability of nutrients, pH, soil structure and texture, moisture content, redox potential, presence of a suitable electron acceptor and presence of toxic compounds, such as heavy metals. These were discussed in detail in several reviews (Balba and Bewley, 1991, Bewley 1986, Dupont *et al*, 1989) and need not be elaborated further.

In the development of a bioremediation strategy for a specific contaminated site, one of the key questions that should be addressed is to what extent will clean-up be achieved by optimising all of the above factors pertaining to the contaminant and the physico-chemical properties of the environment in which it is distributed. This issue can be addressed through careful assessment of these site specific factors and properly designed bench scale treatability studies.

Based upon the observations discussed above, it is possible to postulate that the case for bioaugmentation, which is used here to refer to the addition of specific strains of microorganisms for the purpose of enhancing bioremediation, will be greatest in the following circumstances:

•the treatment of highly recalcitrant compounds which are degraded very slowly even under optimal conditions by the natural flora;

•the treatment of compounds present in relatively high or very low concentrations;

•the treatment of sites contaminated with recent spillages of organic compounds;

•the treatment of environments with a physico-chemical characteristic inhibiting the activity of the natural microflora;

•faster and more effective clean-up by respectively accelerating the rate of degradation and/or enhancing the extent of degradation (i.e. achieving lower residual levels);

•where it is essential to maximise reliability of the treatment process.

In certain instances, a case can be made for the addition of a GEMMO as opposed to an organism which has been selected as a result of natural processes; these will be discussed later. It is necessary first, however, to examine case studies in which the inclusion of specific strains of microorganisms in particular remediation strategies has been tested against the process of natural stimulation alone.

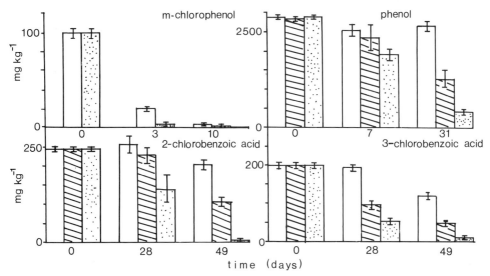

Fig. 1 Effects of abiotic versus microbial additions on the degradation of contaminants in soil. Various soils were subjected to a range of treatments and monitored for pollutant concentration. Treatments were in 1.0kg capacity soil pots and consisted of: open histograms, control (water and mixing only); hatched histograms, abiotic (inorganic nutrient addition, water and mixing only); dotted histograms, nutrients plus microorganisms. Incubation was at 15°C, 15% moisture with regular mixing. Error bars indicate 95% confidence limits. Reproduced from Ellis and Bewley (1990) by permission of E & F. N. Spon.

BIOAUGMENTATION: EXAMPLES OF THE USE OF INOCULATED ORGANISMS FOR POLLUTANT TREATMENT

Unfortunately, for many of the projects which have used bioaugmentation techniques in the field, it has not been possible to test the effects of additions of say nutrients alone or to establish untreated controls and compare the results of these against the full treatment including inoculants. This is because such projects are often full-scale treatments designed to produce a cleaned-up site, in the shortest possible time within very tight economic constraints, rather than academic research projects. However valuable for future bioremediation research and development, unfortunately, it is not possible in most cases to expect clients to allow "controls" of contaminated soil and groundwater to be established alongside full scale treatment projects, let alone finance the associated costs of such work! Rather than a comprehensive state of the art review, the following is intended to focus on a few specific case studies, both bench-scale and in the field, which have assessed the requirement for the addition of specific strains of microorganisms in the formulation of a treatment strategy.

Case Study 1: Bench Scale Treatability Testing

The reduction of a range of pollutant compounds in various soil systems using either proprietary nutrients alone or proprietary nutrients with specific microbial inocula relative to a control is shown in Figure 1 (Ellis and Bewley, 1990). Substantial reductions were obtained by the addition of specific nutrients, but in most cases both the rate and extent of degradation were enhanced by bioaugmentation.

Reduction of 2-chlorobenzoic acid from 245 to 105 mg/ kg (P<0.05) was obtained after seven weeks of incubation by the addition of nutrients alone. However, the addition of *Pseudomonas putida* at a concentration of 4 x 10⁶ cells/g soil resulted in a statistically significant reduction after only four weeks. Moreover, a lower final concentration of 2.0 mg/kg, was achieved with this organism after seven weeks compared with the treatment with nutrients alone. Given the occurrence of phenols as natural products in soil systems (Dragun, 1988), the enhancement achieved through bioaugmentation in this study is perhaps more surprising; chlorophenols tended to be readily degraded. The question arises of course, as to the implications of such findings for field application. Laboratory studies of this nature may in fact under-estimate or over-estimate the potential benefits of including specific microbial strains in the treatment process, because they are generally conducted under optimal conditions for biodegradation. In the field, however, such differences may be less, if the applied organisms are unable to function effectively against the rigors of the prevailing physico-chemical conditions. On the other hand, the application of particularly hardy strains may increase the overall enhancement of degradation.

Case Study 2: Field Application

The case of bioaugmentation techniques versus natural stimulation was exemplified in the first full scale microbial land reclamation contract undertaken in the U.K., at the 10 ha Greenbank gas works site (Bewley *et al*, 1989). This project involved the biological treatment of approximately 30,000 m^3 of soil, contaminated with coal tar and phenols, together with the encapsulation of approximately 12,000 m^3 of material contaminated with metals and complex cyanides. Prior to full-scale implementation of the treatment process a prototype treatment bed was established to test the most effective combinations of microbial inocula, nutrients and surfactants, as determined from laboratory experiments. The results of this treatment are given in Table 1. The conclusions from this trial are that application of a specific polyaromatic hydrocarbon (PAH) degrading isolate ERB002 (*Pseudomonas putida*) was necessary to achieve statistically significant reductions both in 'total' PAH content as measured by the sum of the 16 PAH compounds designated priority pollutants by the U.S. EPA and in concentrations of some specific individual PAH's. There were no reductions in PAH content obtained by addition of nutrients or surfactant alone, or in the control.

In fact, PAH cconcentration actually increased slightly over the treatment period in these two schemes; this was probably a reflection of the tilling method which resulted in the break-up and redistribution of lumps of coal tar.

Unfortunately, it was not possible to continue the trial after the eight week period. The objective of the trial had been to establish whether the microbial treatment could reduce total PAH concentrations below the 10,000mg/kg target required for fulfilment of the clean-up criteria and this was demonstrated by the success of treatment 2A. Evidence for the achievement of reductions in total PAH content was also supported by treatment 2B, where lower starting concentrations of total PAH were also reduced significantly over the eight week period; the limits of the treatment process in this example are unknown. In the full-scale application of this treatment, independent validation of re-instated soil indicated that all concentrations of PAH's were well below the target concentration of 10,000kg/kg and in fact the mean total of PAH concentrations across the site was 148 mg/kg on completion of the project (Bewley *et al*, 1989). Total concentrations of phenols were also reduced across the site to below 5 mg/kg from starting levels in excess of 500 mg/kg in places. It is not claimed that such reductions were achieved wholly as a result of bioremediation, as the process involved physical processing and homogenization of the contaminated material. The field trial results in Table 1 do, however, indicate the potential element of PAH reduction that may be achieved through microbial processes.

One rather interesting fact which emerges from the above field trial is that isolate ERB002 (*Pseudomonas putida*) appears more effective than isolate ERB001 (*Pseudomonas fluorescens*) in reducing concentrations of PAH. Both had given encouraging results in microcosm experiments (Bewley *et al*, 1989). The *Ps. fluorescens* strain was an indigenous isolate from the Greenbank gas works site isolated in a mineral salt medium supplemented with naphthalene as sole carbon source, whereas the *Pseudomonas putida* strain was a naturally occurring isolate taken from a culture collection (NCIB 9816). Both had demonstrated PAH-degrading capacity in soil microcosm experiments (Bewley *et al*, 1989). At this site at least, the non-indigenous isolate was clearly able to function better than the re-inoculated indigenous organism.

As discussed above, it is not known to what extent such differences would be maintained over an extended treatment period. The usefulness of adding such organisms as strain ERB002 however is their apparent acceleration of the treatment process, which may have profound economic implications for the acceptability of bioremediation as a viable treatment technology. These organisms have been referred to as "vanguard" microorganisms on account of the hypothesis that they are responsible for catalysing initial rate-limiting biochemical reactions (Bewley *et al*, 1991). If this is the case, it is not only necessary but also desirable that they have a temporary competitive advantage when introduced into the soil system. This

TABLE 1

CONCENTRATIONS OF POLYAROMATIC HYDROCARBONS AFTER 0 AND 8 WEEKS IN CONTAMINATED SOIL PLACED IN A TREATMENT BED WITH AND WITHOUT APPLICATION OF MICROORGANISMS, NUTRIENTS AND SURFACTANT

TREATMENT	1		2A		2B		3		4	
Microorganism added	ERB001		ERB002		ERB002		None		None	
Nutrients added	Yes		Yes		Yes		Yes		No	
Surfactant added	Yes		Yes		Yes		Yes		No	
PAH coal tar constituent[b]	0 wks	8 wks	0 wks	8 wks	0 wks	8 wks	0 wks	8 wks	0 wks	8 wks
Acenaphthene	11	0	0	0	0	0	0	0	1	0
Fluorene	233	279	504	212	148	166	127	98	113	57
Phenanthrene	519	532	716	490	358	361	151	284	150	168
Anthracene	212	206	295	184	153	105	57	106	62	57
Pyrene	702	388*	833	366*	524	209*	224	227	170	161
Fluoranthene	3,271	1,104*	3,664	1,019*	2,271	647*	1,049	637	614	479
Benz (a) anthracene	397	365	558	353*	306	177	167	257	155	175
Chrysene	469	596	597	515	290	238	183	439	186	296
Benz (a) pyrene	98	58	159	70*	100	45*	45	55	59	43
Benz (b) fluoranthene	229	264	552	244*	244	127	108	285	167	174
Benz (k) fluoranthene	190	153	446	260	225	129	177	223	152	116
Dibenz (a,h) anthracene[a]	3,200	3,155	3,836	3,751	3,156	2,749	1,111	2,539	950	2,438
Indeno (1,2,3-cd) pyrene	232	217	316	177*	221	136	121	134	146	110
Coal tar as measured by total PAH compounds	9,763	7,317	12,476	7,641*	7,996	5,089*	3,520	5,284	2,925	4,274

a Includes also benz (ghi) perylene
b Figures represent mean of four samples
* Indicates a statistically significant reduction (p<0.05) as measured by a 't' test

From Bewley et al, 1989. Reproduced by permission of the Society for Chemical Industry

was probably the situation at the Greenbank site where nutrients were added to the system simultaneously with the introduced organisms. The addition of relatively large quantities of readily available organic and inorganic nutrients to the liquid broth containing the pseudomonads may have allowed at least a temporary predominance within the soil treatment bed, whereas an active microflora is likely to have completed the degradation process (Heitkamp *et al*, 1988).

THE CASE FOR RELEASE OF GEMMOS AS PART OF A BIOAUGMENTATION STRATEGY

Having examined some specific examples where bioaugmentation can significantly enhance bioremediation, it is now possible to focus on how GEMMOs may provide further advantages. Sayler and Day (1991) gave five major instances where the contribution of GEMMOs can make a significant impact on bioremediation and waste management:

•through improvement of biochemical performance and versatility of microbial strains involved;
•through improvements in predictability of bioremediation, e.g. by use of gene probes;
•through improvements in the hardiness of pollutant degrading organisms in their application to contaminated environments, e.g. as better scavengers, or through chemotactic action;
•through cellular control to control residual organisms, e.g. self destruct processes;
•through the recovery of strains from the environment with improved or unrealized performance in bioremediation, e.g. use of *in-situ* recipient strains to accumulate degrading genes and serve as an *in-situ* reservoir for other indigenous microbes.

From an ecological perspective, it is clear that there are particular examples where improvement of physiological performance and versatility, together with improvements in environmental hardiness could have a particularly significant impact on bioremediation. The use of modified strains to attack particularly high concentrations of the contaminant is case in point, but oligotrophic strains capable of growth using low concentrations of organic substrates are probably of greater importance as more stringent regulations require ever decreasing target concentrations (Piotrowski, 1991). Indirect advantages could also result in terms of the use of organisms which require lower concentrations of supplementary nutrients, for example in nitrate-sensitive areas.

It may be argued, therefore, that genetic modification has a potential role either in the inclusion of enhanced degradative capacity in natural strains normally adapted to function in environments which are suboptimal for most microorganisms, or by enhancing the ability of natural strains of efficient pollutant degraders to function under conditions of environmental stress.

There are many examples of the potential use of GEMMOs in the degradation of particularly xenobiotic compounds such as highly halogenated polymeric organic molecules. One such example concerns the possibility of achieving degradation of polychlorinated biphenyls (PCBs), (Dean-Ross, 1987). Under natural conditions, there are two major pathways for degradation of aromatics. In the case of the *ortho* pathway, where cleavage occurs between two hydroxylated ring carbons, enzymes are active on both chlorinated and non-chlorinated aromatic compounds, whereas in the case of the *meta* pathway, where cleavage occurs between hydroxylated and non-hydroxylated ring carbons, enzymes only attack non-chlorinated aromatic compounds. Co-metabolism has been demonstrated as a mechanism for PCB degradation by soil microorganisms supplied with biphenyl (Focht and Brunner, 1985), but these organisms contain the enzymes which mediate the meta-fission pathway of ring cleavage. A possible role of genetic modification is, therefore, the combination of these two pathways to obtain a single organism capable of growth on PCBs.

Another area of control within a biochemical pathway where genetic modification could play a significant role is in the development of strains to target intermediate compounds resulting from initial degradation by the natural microflora. More specific examples of intermediates will be discussed later. However, even in the field of oil hydrocarbon degradation, about which there is a vast body of literature (e.g. Atlas, 1977, 1981; Bossert &

Bartha, 1984; Bartha, 1986; Walker *et al*, 1976; Colwell and Walker, 1977), there is much scope for improving the overall efficiency of the process in terms of achieving mineralization rather than simply biodegradation (Bewley *et al*, 1990). Fatty acids derived from the biodegradation of alkanes, for example, may be inhibitory to microbial activity and their accumulation could limit the degradation process (Atlas &, Bartha, 1973a). An important area of potential research, therefore, is to develop strains of organisms capable of effectively targeting such compounds and achieving overall more effective mineralization.

There are of course many examples of such possibilities, but against these have to be weighed the potential problems and pitfalls involved which are discussed later. First, however, a number of bench-scale studies have been published which provide some useful insight into the possibilities and limitations of GEMMOs and examples of these are discussed below.

Monitoring of GEMMOS and Natural Transconjugants in Experimental Systems: Further Case Studies

While recognizing that there is limited evidence to draw from concerning the success or otherwise of GEMMOs in full-scale waste treatment systems, there are some interesting examples from bench scale systems. It is obviously important that such systems should be representative microcosms of proposed full-scale treatment, but with relatively simple apparatus it is possible to investigate potential effects of introduction of GEMMOs on well-established waste treatment processes, such as activated sludge. *Pseudomonas putida* UWC1 harboring plasmid pD10 which enables the organism to utilize 3-chlorobenzoate as sole carbon source, was introduced into such a laboratory-scale activated sludge unit for the purpose of monitoring survival and activity (McClure *et al*, 1989).

The organism survived for more than eight weeks following introduction but did not enhance biodegradation. However, bacteria present in the activated sludge populations acted as recipients for plasmid pD10 and degraded 3-chlorobenzoate faster than the original strain UWC1 (pD10) in batch culture. In a further study (McClure *et al*, 1991), the survival of this strain appeared to be adversely affected by the development of a natural 3-chlorobenzoate-degrading microflora. However, two 3-chlorobenzoate-degrading organisms derived from the activated sludge, one a naturally-occurring isolate AS2, the other *Pseudomonas putida* ASR2.8, a transconjugant harboring plasmid pD10, not only survived for long periods, but enhanced degradation of 3-chlorobenzoate when re-introduced into the model system. These studies, therefore, indicate that addition of a GEMMO to a waste treatment system may not accelerate degradation *per se*, even though inoculation of a naturally-occurring isolate with similar degradative capacity may indeed enhance the process. However, they also demonstrate the possible mechanism of achieving potentially enhanced degradation indirectly through transfer of the degradative gene to members of the natural flora. Much more study is required to examine to what extent this can indeed be achieved in particular waste treatment systems, where environmental conditions can at least be reasonable managed and more importantly, in the natural environment.

A microcosm representing an aquifer system was used to monitor the maintenance and stability of introduced genotypes carrying the plasmids TOL (pWWO) in a groundwater microbial community (Jain *et al*, 1987). By colony hybridization with gene probes of various specificity, successful maintenance of the introduced isolates and the plasmid-borne, catabolic genes were demonstrated. This was an encouraging result, although the removal of toluene and chlorobenzene was as great in similar microcosms which had not been inoculated, but had received nutrient in the form of a heat-killed bacterial suspension. These microcosm systems were maintained at 22°C, whereas large tracts of natural water, both surface or ground-water which may become contaminated, have relatively low ambient temperatures. The use of psychrotrophic bacteria indigenous to these environments as recipient strains for biodegradative genes from mesophilic organisms therefore holds open the possibility of achieving enhanced bioremediation both for relatively cold environments and industrial waste-water effluents.

Successful transfer of the TOL plasmid pWWO by conjugation from the mesophile *Pseudomonas putida* PaW1 to the naturally-occuring psychrotroph *Pseudomonas putida* Q5 allowed the latter to utilize and degrade toluate (1,000mg/l) as sole carbon source, at temperatures as low as 0°C (Kolenc *et al*, 1988). While such studies hold open many exciting possibilities, the potential problems of introducing GEMMOs into the natural environment including the possible risks associated with such release should not be overlooked. Some of these potential problems are common to any bioremediation project involving bioaugmentation, others relate specifically to release of GEMMOs. These are explored below.

BIOAUGMENTATION - POTENTIAL PROBLEMS AND PITFALLS

The potential hazards associated with the release of GEMMOs are discussed elsewhere in this volume and will not be elaborated here. In this section, some potential problems of release will be considered in relation to the specific fields of waste treatment and bioremediation.

Contaminated natural environments often contain a diversity of organic materials in addition to the target contaminants, which will support the activity of a complex range of microbial communities. Establishment of the vanguard organism, whether a naturally-occurring isolate or a GEMMO in such an environment will require a degree of competitive advantage, albeit only temporary. As discussed in the examples above, this is often difficult to achieve. In the field of remediation of contaminated environments, this problem is compounded by variability. Variability in both the biological and physico-chemical aspects of the environment and in the nature and concentration of the contaminant presents the following difficulties:

•targeting a released organism to contaminated microsites with differences in physical and chemical characteristics which will affect the performance of the organism (a) directly in terms of its physiology and (b) indirectly through shifts in the competitive advantage of the natural established flora;

•monitoring and validating the remediation process.

Variability should never be underestimated in biological treatment. It exists on a micro-scale where, for example, the influence of pH or metal concentration on the physiology of a bacterial cell may be dependent upon its proximity to a clay particle, or on a macro-scale where for example the ground of a contaminated industrial site may contain rubble, slag, soil, concrete, cobbles, paper and timber.

The situations where a released organism such as a GEMMO is most likely to be successful are where such variability can be reduced by modifying physico-chemical factors to suit the specific physiological and ecological requirements. It is easier to clean-up sub-surface contamination where the contaminant can be mobilized from the sub-surface, for example by use of a dispersant or surfactant (Elis *et al*, 1990), and then treated in a closed system, such as a waste-water treatment plant, than in a truly *in-situ* situation where adjustment of physico-chemical conditions within the ground-water can be far less easily modified.

Within soil, the reduction of variability is also maximized in a closed system such as a high solids slurry reactor. Here the greater degree of homogeneity achieved allows better potential control over the degradation process mediated by an introduced organism than can be effected than say in a treatment bed. Although such systems are clearly more expensive, this has to be balanced against the alternative costs of treatment, particularly for very recalcitrant, halogenated xenobiotics, precisely the types of compounds where there is a greater requirement for bioaugmentation and potentially, for the use of GEMMOs.

For many situations, however, such closed systems are not applicable, on economic grounds alone. Even under standard *ex-situ* treatment though, the use of advanced rotovation techniques such as spaders operational to 60cm or greater depth, (Ellis *et al*, 1990) can substantially reduce variability, at least on a macro-scale. As the treatment proceeds, this reduction in variability, which occurs within a bed, allows the taking of fewer samples for monitoring contaminant concentration in order to achieve an acceptable sampling error. This reduction of variability is therefore of major significance in terms of the monitoring and validation of the process because of facilitating a proper statistical assessment of bioremediation on a before and after basis. This allows for associated cost reductions in sampling procedures.

Project managers of bioremediation contracts are acutely aware that analytical costs associated with monitoring and validating the process constitute a substantial proportion of the total budget, so that the cost reductions which ensue from homogenization of the treated material may have a significant impact on the economic viability of this technology.

One of the concerns regarding the application of microbial processes for the degradation of organic pollutants is the possibility of production of intermediates, particularly if these are more toxic than the parent compound. It is important to assess this on a case by case basis.

Certainly for some organic pesticides there is well-documented evidence for such mechanisms (Alexander, 1977), but attention has also been focused on other groups of contaminants amenable to biodegradation such as the PAHs. Laboratory studies have identified intermediate production from PAHs containing both lower (e.g. Evans et al, 1965) and higher numbers of benzene-rings (Gibson et al, 1975). These intermediates include dihydrodiols, diols, diones, or carboxylic acid PAH derivatives. Certain fungi may also transform PAHs to more toxic or carcinogenic metabolites such as trans diol epoxides (Thakker et al, 1985).

However, in evaluating the implications of these findings, it is important to distinguish between results of pure culture studies and those involving inoculation of organisms into natural environments where bioremediation may be achieved by a combination of the introduced species and the natural community, the case of "vanguard" organisms.

If we now examine microcosm studies, where transformations by combinations of specific strains of organisms and consortia from the natural community, a rather different picture emerges. During studies of PAH transformation in sediments with ^{14}C-labelled PAH, concentrations of intermediates never rose to a significant fraction of the total ^{14}C present (Herbes and Schwall, 1978). These findings were supported in a later study by Heitkamp et al, (1988), concerning the degradation of PAHs containing three or more aromatic rings in sediment microcosms.

Both authors concluded that the rate-limiting step for the biodegradation of PAHs was the initial ring oxidation after which subsequent degradation, by indigenous microorganisms, proceeds rapidly with little to no intermediate accumulation. This may also include co-metabolism (co-oxidation), (Keck et al, 1989). This lends support, therefore, to the vanguard microorganism strategy discussed above.

As has been discussed, PAHs do occur naturally in soil, whereas other groups of truly xenobiotic molecules such as chlorinated hydrocarbons, may represent more of a case for the use of a GEMMO. It is under these circumstances that more consideration needs to be given to the formation of potentially toxic intermediates, which may have significant ecological implications.

There is evidence, at least in microcosm systems, that the introduction of a GEMMO with a specific biodegradative capacity could have a detrimental effect on the activities of the natural soil microflora. Introduction of a genetically modified Pseudomonas putida PP0301 (pRO103) to soil supplemented with 500mg/kg of 2,4-dichloro-phenoxyacetate (2,4-D) (500mg/g) resulted in the accumulation of the metabolite 2,4-dichlorophenol (2,4-DCP; Short et al, 1991). At the same time, there was a significant decrease in the numbers of fungal propagules and a reduction in the rate of carbon dioxide evolution, whereas bacterial numbers did not appear to be affected. Evidence that the production of 2,4-DCP was responsible for such changes was supported by demonstrations of the toxicity of this intermediate to fungi, both in in-situ and in in-vitro studies. Such results indicate the potential for release of pollutant-degrading GEMMO, to influence the activities, ecology and population dynamics of microorganisms responsible for biogeochemical cycling.

ENVIRONMENTAL ENGINEERING OR GENETIC MODIFICATION?

Given these considerations, is the case for use of GEMMOs overstated? There are certainly instances where modifying physico-chemical environmental properties of the environment has as substantial effect on contaminant removal, as the introduction of specific organisms. A case in point is the use of oleophilic fertilizers (Atlas & Bartha, 1973b; Olivieri et al, 1976), which have been applied recently in the Arctic in the case of the Exxon Valdez spill (Sveum & Ladousse, 1989). The use of such a formulation which will dissolve in the oil and thus prevent excessive loss of nutrients to the water represents a dramatic example of an improvement in the efficiency of bioremediation without inoculation. Successful chemical manipulation of the environment is also illustrated by the use of co-metabolism as a mechanism for bioremediation, for example in the provision of methane as substrate for methanotrophs capable of degrading a range of halogenated aliphatic compounds (Leahy et al, 1989; Wilson and Wilson, 1987).

An example of how environmental engineering may be complementary to the application of a specific isolate, is in the use of the white rot fungus Phanerochaete chrysosporium. The capacity of this organism for degradation of a variety of structurally diverse contaminants particularly polychlorinated aromatics, has been well documented (e.g. Bumpus et al, 1985;

Eaton, 1985; Bumpus and Aust, 1987; Mileski *et al*, 1988). However, the organism is normally associated with decaying wood, rather than with soil. Field-trials presently being conducted by J. Waid and Dames & Moore International to test the applicability of white rot fungi for treatment of PCB contaminated soil have incorporated a suitable organic substrate such as plant residues inoculated with selected strains of white rot fungi into the contaminated soil. This represents an example of engineering the environment to favour the proliferation of a particular species of a pollutant degrading organism.

The necessity for a careful evaluation of physico-chemical interactions and the use of inoculated organisms for bioremediation however, is no better illustrated than in the application of surfactants.

Low bioavailability may be a key factor limiting the degradation of hydrocarbons, and the use of surfactants to enhance mobilization of tar and oil residues has been employed (e.g. Bewley *et al*, 1989, 1990; Ellis *et al*, 1990). However, the use of surfactants may be detrimental to bioremediation because of inherent toxicity, substrate competition or prevention of the adherence of microorganisms to the contaminant, as in the case of oil degradation.

Mineralization of hexadecane in heptamethylnonane by *Arthrobacter* sp was inhibited by Triton X-100, by the prevention of bacterial adherence to the heptamethylnonane-water interface whereas this surfactant apparently increased the rate and extent of naphthalene degradation (Efroymson and Alexander, 1991). This was probably attributable to the relatively high partitioning of naphthalene into the aqueous phase (compared to hexadecane). The enhancement of degradation may then have arisen from the lack of spatial and nutrient limitations that occur in association with the solvent water interface.

The surfactants Triton X-100 (BDH, U.K.) and Tensoxid S50 (Tensia, Liege, Belgium) were particularly effective at desorbing Aroclor 1242 from sand. However, Tensoxid S50 exerted varying degrees of inhibition on the growth of several bacteria capable of degrading various PCB congeners. Triton X-100, while generally less inhibitory to growth of these isolates apparently reduced their activity as an alternate carbon source (Viney & Bewley, 1990). Under these circumstances such treatments are best implemented as separate strategies (McDermott *et al*, 1989).

Conversely, an example of the integration of physico-chemical techniques for enhancing biological mechanisms of waste treatment, is seen in the use of degradable plastics. In this example, light or temperature will generate free radicals when photo- or pro-oxidants are incorporated into polyethylene films. These initiate oxidation and cause associated physico-chemical changes in the material which may become more amenable to biological degradation. Lee *et al* (1991) have demonstrated significant cleavage of water-soluble, high-molecular-weight, chemically oxidized polyethylene residues of starch-containing plastics by lignin degrading Streptomyces.

CONCLUSIONS

Genetic modification could potentially make a major contribution to the remediation of contaminated environments and treatment of toxic waste. It is important to emphasize that it is still largely a potential treatment technology and only one of several in the field of bioremediation. There are for example, at least three possible potential bioremediation strategies for PCB's:

(i) aerobically by bacteria for the lower chlorinated congeners   (Bedard *et al*, 1986), and with biphenyl as co-substrate (Focht and Brunner, 1985);

(ii) aerobically with the ligninase enzyme system of the white rot fungus *Phanerochaete chrysosporium* (Eaton, 1985);

(iii) anaerobically, which is particularly effective for the higher chlorinated PCB congeners (Quensen *et al*, 1990).

One, or a combination, of these mechanisms may be used to develop a bioremediation strategy for a PCB contaminated site which may be as effective and more acceptable than through the release of a GEMMO. The least controversial use of GEMMOs in the field of bioremediation and waste treatment is in the production of cell free enzyme preparations such as for organophosphate wastes, although such strategies may have limited applications (Payne *et al*, 1989). The next stage involves the use of GEMMOs in a closed system, for example a bioreactor or high solids soil slurry system which do not actually involve release into the natural environment. The third scenario, that of actual release must be carefully evaluated on a case by case basis with realistic models and microcosm systems, and a thorough evaluation of the potential impact on both physical and biological receptors. In the U.K., the requirement for risk assessment from such release has been embodied in the Environmental Protection Act (1990). There is, however, a need for more clearly-defined protocols and guidelines to be established. The success of bioremediation can only be fully realized if the following factors are understood:

• the heterogeneity of the natural environment;
• the site specificity of the technology;
• the requirement for integration with physical and chemical techniques.

Not only will the case for using a GEMMO need to be convincingly demonstrated in terms of negligible risk but its potential for success will require a thorough understanding of the subtleties, complexity and variability of the components of our natural world.

Disclaimer: The views expressed in this chapter are those of the author and do not necessarily represent those of Dames & Moore International.

REFERENCES

Alexander, M. (1977). "Introduction to Soil Microbiology". J. Wiley & Sons, New York.
Alexander, M. (1985). Biodegradation of organic chemicals. Environ. Sci. Technol. 18, 106-111.
Atlas, R.M. (1977). Stimulated petroleum biodegradation. Crit. Rev. Microbiol. 5, 371-386.
Atlas, R.M. (1981). Microbial degradation of petroleum hydrocarbons: on environmental perspective. Microbiol. Rev. 45, 180-209.
Atlas, R.M. and Bartha, R. (1973a). Inhibition by fatty acids of the biodegradation of petroleum. Antonie van Leeuwenhoek J. Microbiol. Serol. 39, 257-271.
Atlas, R.M. and Bartha, R. (1973b). Stimulated biodegradation of oil slicks using oleophilic fertilisers. Environ. Sci. Technol. 7, 538-541.
Balba, M.T. and Bewley, R.J.F. (1991). Organic Contaminants and Microorganisms, in "Organic Contaminants in the Environment". (Jones, K. C, Ed.) pp 237-274. Elsevier, Barking, U.K.
Bartha, R. (1986). Biotechnology of petroleum pollutant biodegradation. Microb. Ecol. 12, 155-172.
Bewley, RJ.F. (1986). A microbiological strategy for the decontamination of polluted soil, in "Contaminated Soil" (Assink, J.W. and van den Brink, W.J., Ed.) pp 759-768. Martinus Nijhoff, Dordrecht, The Netherlands.
Bewley, R., Ellis, B., Theile, P., Viney, I., Rees, J. (1989). Microbial Clean-Up of Contaminated Soil. Chem. Ind. (London),778-783.
Bewley, R.J.F., Ellis, B., Rees, J.F. (1990). Development of a microbiological treatment for restoration of oil contaminated soil. Land Deg. & Rehabil. 2, 1-11.
Bewley, RJ.F., Sleat, R., Rees, J.F. (1991). Waste treatment and pollution clean-up, in "Biotechnology/The Science and the Business". (Moses, V. and Cape R.E., Ed.) pp 507-519. Harwood Academic Publishers, London.
Bossert, I. and Bartha, R. (1984). The fate of petroleum in soil ecosystems, in "Petroleum Microbiology" (Atlas, R.M., Ed.) pp 435-473. Macmillan, New York.
Bumpus, J.A. and Aust, S.D. (1987). Biodegradation of DDT [1,1,1-trichloro-2,2-Bis (4-chlorophenyl) Ethene] by the White Rot Fungus *Phanerochaete chrysosporium* Appl. Environ. Microbiol. 53, 2001-2008.
Bumpus, J.A., Tien, M., Wright, D. and Aust S.D. (1985). Oxidation of persistent environmental pollutants by a white rot fungus. Science 228, 1434-1436.
Colwell, R.R. and Walker, J.D. (1977). Ecological aspects of microbial degradation of petroleum in the marine environment. Crit. Rev. Microbiol. 5, 423-445.
Chakrabarti, T., Subrahmanyam, P V.R. and Sundaresan, B.B. (1989). Biodegradation of recalcitrant industrial wastes, in "Biotreatment systems Vol II" (Wise, D.L., Ed.) pp 171-234. CRC Press, Inc., Boca Raton, Fla.
Dean-Ross D. (1987). Biodegradation of Toxic Wastes in Soil. ASM News 53, 490-492.

Department of the Environment (1990). Environmental Protection Act 1990 Chapter 43. HMSO, London.

Dragun, J. (1988). "The Soil Chemistry of Hazardous Materials", pp 325-445. HMCRI, Silver Spring MD, USA.

Dupont, R.R., Sims, R.C., Sims J.L. and Sorensen, D.L. (1989). In-Situ biological treatment of hazardous waste-contaminated soils, in "Biotreatment Systems Volume II" (Wise D.L., Ed.) pp 23-94. CRC Press Inc., Boca Raton, Fla.

Eaton, D.C. (1985). Mineralisation of polychlorinated biphenyls by *Phanerochaete chrysosporium* a ligninolytic fungus. Enzyme Microb. Technol. 7, 194-196.

Efroymson, R.A. and Alexander, M. (1991). Biodegradation by an *Arthrobacter* species of hydrocarbons partitioned into an organic solvent. Appl. Environ. Microbiol. 57, 1441-1447.

Ellis B., Balba, M.T. and Theile, P. (1990). Bioremediation of oil contaminated land. Environ. Technol. 11, 443-455.

Ellis, B. and Bewley R.J.F. (1990). Biotreatment of contaminated land, in "Microbiology in Civil Engineering "(Howsam, P., Ed.) pp 231-240. E. & F.N. Spon, London.

Evans, W.C., Fernley, H.N. and Griffiths, E. (1965). Oxidative metabolism of phenanthrene and anthracene by soil pseudomonads. Biochem J. 95, 819-831

Focht, D.D. and Brunner, W. (1985). Kinetics of biphenyl and polychlorinated biphenyl metabolism in soil. Appl. Environ. Mcrobiol. 50, 1058-1063.

Gibson, D.T., Jerina, D.M., Yagi H. and Yeh, H.J.C. (1975). Oxidation of the carcinogens benzo(a)pyrene and benzo(a)anthracene to dihydrodiols by a bacterium. Science 189, 295-297.

Goldstein, R.M. Mallory,L.M. and Alexander,M. (1985). Reasons for possible failure of inoculation to enhance biodegradation. Appl. Environ.Microbiol. 50,977-983

Hardman, D. (1991). Microbial pollution control: a technology in its infancy. Chem. Ind. (London), 244-246.

Heitkamp, M., Freeman, J.P., Miller, D.W. and Cerniglia, C.E. (1988). Pyrene degradation by a Mycobacterium sp: Identification of ring oxidation and ring fission products. Appl. Environ. Microbiol. 54, 2556-2565.

Herbes, S.E. and Schwall, L.R. (1978). Microbial transformation of polycyclic aromatic hydrocarbons in pristine and petroleum-contaminated sediments. Appl. Environ. Microbiol. 35, 306-316.

Jain, R.K., Sayler, G.S., Wilson, J.T., Houston, L. and Pacia, D. (1987). Maintenance and stability of introduced genotypes in groundwater aquifer material. Appl. Environ. Microbiol. 53, 996-1002

Keck, J., Sims, R.C., Coover, M., Park, K. and Symons, B. (1989). Evidence for co-oxidation of polynuclear aromatic hydrocarbons in soil. Wat. Res. 23, 2467-1476.

Kolenc, R.J., Inniss, W.E., Glick, B.R., Robinson, CW. and Mayfield, C.I. (1988). Transfer and expression of mesophilic plasmid-remediated degradative capacity in a psychotrophic bacterium. Appl. Environ. Microbiol. 54, 638-641.

Leahy M.C., Findlay, M. and Fogel, S. (1989). Biodegradation of chlorinated aliphatics by a methanotrophic consortium in a biological reactor, in "Biotreatment - The Use of Microorganisms in the Treatment of Hazardous Materials and Hazardous Wastes", Proceedings of the 2nd National Conference, pp 3-9 HMCRI, Washington, DC.

Lee, B., Pometto, A.L., Fratzke, A., Bailey, T.B. Jr. (1991). Biodegradation of degradable plastic polyethylene by *Phanerochaete* and *Streptomyces* species.

McClure, N.C., Weightman, A.J. and Fry, J.C. (1989). Survival of *Pseudomonas putida* UWCI containing cloned catabolic genes in a model activated sludge unit. Appl. Environ. Microbiol. 55, 2627-2634.

McClure, N.C., Fry, J.C. and Weightman, A.J. (1991). Survival and catabolic activity of natural and genetically engineered bacteria in a laboratory-scale activated sludge unit. Appl. Environ. Microbiol. 57, 366-373.

McDermott, J.B., Unterman, R., Brennan, M.J., Brocks, R.E., Mobley, D.P., Schwartz, C.C. and Dietrich, D.K. (1989). Two strategies for PCB soil remediation:biodegradation and surfactant extraction. Environ. Prog. 8, 46-51.

Mileski, G.J., Bumpus, J.A., Jurek, M.A. and Aust, S.D. (1988). Biodegradation of Pentachlorophenols by the White Rot fungus *Phanerochoraete chrysosporium*. Appl. Environ. Microbiol. 54, 2885-2889.

Olivieri, R., Bacchin, P., Robertiello, A., Oddo, N., Degen, L. and Tonolo,A. (1976). Microbial degradation of oil spills enhanced by a slow-release fertiliser. Appl. Environ. Microbiol. 31, 629-634.

Payne, G F., Coppella, S.J. and DelaCruz, N. (1989). Genetic engineering approach to treating organophosphate wastes, in "Biotreatment - the use of Microorganisms in the Treatment of Hazardous Materials and Hazardous Wastes", proceedings, of the 2nd National Conference, pp 129-133. HMCRI, Washington, DC.

Piotrowski, M.R. (1991). Bioremediaton. Hazmat World, 4, 47-49.

Quensen, J.F., Boyd, S.A. and Tiedje, J.M. (1990). Dechlorination of four commercial polychlorinated biphenyl mixtures (aroclors) by anaerobic microorganisms from sediments. Appl. Environ. Microbiol. 56, 2360-2369.

Sayler, G. and Day, S.M. (1991). Bioremediation. Hazmat World 4, 51-53.

Short, K.A., Doyle, J.D., King, R.J., Seidler, R.J., Stotzky, G. and Olsen, R.H. (1991). Effects of 2,4-dichlorophenol, a metabolite of a genetically engineered bacterium, and 2,4-dichlorophenoxyacetate on some microorganism-medicated ecological processes in soil. Appl. Environ. Microbiol. 57,412-418..

Sims, R.C. and Overcash, M.R. (1983). Fate of polynuclear aromatic compounds (PNA's) in soil-plant systems. Res. Rev. 88, 1-67.

Sveum, P. and Ladousse, A. (1989). Biodegradation of oil in the Arctic:Enhancement by Oil Soluble Fertilizer Applications in "Proceedings of 1989 Oil Spill Conference", pp 439-446, American Petroleum Institute, Washington DC.

Swindoll, C.M., Aelion, C.M. and Pfaender, F.K. (1988). Influence of inorganic and organic nutrients on aerobic biodegradation and on the adaptation response of subsurface microbial communities. Appl. Environ. Microbiol. 54, 212-217.

Thakker, D.R., Yagi, H., Levin, W., Wood, A.W., Cooney, A.H. and Jerina, D.M. (1985). Polycyclic aromatic hydrocarbons:metabolic activation to ultimate carcinogens, in "Bioactivation of Foreign Compounds" (Anders, M.W., Ed.), pp 177-192. Academic Press, Inc., New York.

Viney, I. and Bewley, R.J.F. (1990). Preliminary studies on the development of a microbiological treatment for polychlorinated biphenyls. Arch. Environ. Contam. Toxicol. 19, 789-796.

Walker, J.D., Colwell, R.R. and Petrakis, L. (1976). Biodegradation rates of components of petroleum. Can. J. Microbial. 22, 1209-1213.

Wilson, J.T. and Wilson, B.H. (1987). Biodegradation of halogenated aliphatic hydrocarbons, United States Patent Number 4713343.

PROGRESS IN THE GENETIC MODIFICATION AND

FIELD-RELEASE OF BACULOVIRUS INSECTICIDES

Robert D. Possee, Linda A. King[1], Matthew D. Weitzman[1]
Susan G. Mann[1], David S. Hughes[1], Iain R. Cameron, Mark L. Hirst
and David H.L. Bishop

NERC Institute of Virology and Environmental Microbiology
Mansfield Road
Oxford OX1 3SR, U.K.

School of Biological and Molecular Sciences[1]
Oxford Polytechnic
Gipsy Lane
Headington, Oxford OX3 0BP, U.K.

INTRODUCTION

Baculoviruses are invertebrate-specific pathogens which have been used as alternatives to chemical agents for the control of some insect pests. They have a large, double-stranded, circular DNA genome (ca. 80-200 kilobase pairs). This is packaged within a rod-shaped nucleocapsid structure, further enclosed within a lipoprotein envelope and ultimately occluded by a crystalline matrix largely comprising a single protein, polyhedrin (28 Kdal), which serves to protect the virus in the environment (Figure 1). In general, baculoviruses infect the larval stage in insect development and cause significant mortality and morbidity.

The objective of this chapter is to review the history of field-release experiments with genetically modified *Autographa californica* nuclear polyhedrosis virus (AcNPV) in the UK. and also to summarize the progress whch has been made to improve the effectiveness of these agents as biological insecticides. Bishop *et al.* (1988) described field-release experiments with genetically-modified AcNPV that had been performed in the period 1986-1987 and also gave an extensive summary of the rationale for the use of baculoviruses as biological insecticides, and these points will not be reiterated here. The reader is refered to the previous publication (Bishop *et al.*, 1988) and others (e.g. Podgewaite, 1985; Entwistle and Evans, 1985; Huber, 1986; Evans and Entwistle, 1987) for information relevant to the use of baculoviruses as insecticides. Further details are also provided in these papers of the structure and replication strategies of baculoviruses. Another excellent review, by Blissard and Rohrmann (1990), summarizes the recent literature about baculovirus gene structure and function and the molecular aspects of the mode of replication of these viruses.

The Release of Genetically Modified Microorganisms
Edited by D.E.S. Stewart-Tull and M. Sussman, Plenum Press, New York, 1992

47

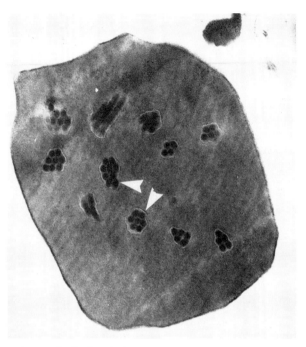

Fig 1. Electron micrograph of a baculovirus polyhedron. Note the virus particles (white arrows) containing multiple nucleocapsids.

PAST FIELD RELEASE EXPERIMENTS WITH BACULOVIRUS INSECTICIDES

Genetically Marked, Polyhedrin-Positive AcNPV

The programme to investigate the use of genetically-modified baculoviruses in the field was initiated in 1986. In the first study, a synthetic oligonucleotide marker was inserted within a non-essential region of the AcNPV genome to produce a polyhedrin-positive, genetically marked virus. This modification remained stable after fifty rounds of virus replication in insect cells. Further risk assessment analyses included an examination of the effect of the change on host range and physical stability of the virus. It was concluded that the genetic modification caused no changes in the phenotype of the baculovirus. The results from these tests were submitted to the appropriate regulatory authorities (see Bishop *et al.*, 1988 for full details) and permission was given for the field-trial to be conducted. The marked virus was released to the environment by feeding polyhedra to susceptible *Spodoptera exigua* (small mottled willow) larvae and placing these insects on sugar beet plants within an insect-proof enclosure. Uninfected insects were placed in a similar plot to serve as a control. The virus-infected larvae died within a seven-day period. Thereafter, marked virus could be recovered from the soil in the enclosure until the site was disinfected with formalin six months later. After this treatment infectious virus could no longer be recovered from the site. This trial demonstrated that it was possible to release a genetically altered virus into the environment, monitor its survival over a period of months and ultimately inactivate residual virus at the release site.

Genetically Marked, Polyhedrin-Negative AcNPV

In the second phase of the field-trials, conducted in 1987, a polyhedrin-negative (crippled or self-destruct) virus was used with a second, unique synthetic marker in place of the polyhedrin gene-coding region and promoter elements. The object of this study was to determined whether the persistence of a baculovirus insecticide in the environment could be limited by removal of the protective polyhedrin protein from the virus particles. Appropriate risk assessment analyses were again performed and the results submitted to the regulatory authorities to request permission for the field-trial. The virus was released after infecting *S. exigua* larvae in the laboratory with non-occluded virus and then placing them on the sugar beet plants in the enclosed facility used for the 1986 study. After seven days all of the larvae had died. Two weeks post-release, infectious virus could not be identified on the plant surfaces, or in soil samples. This showed that the baculovirus insecticide could be modified to reduce its persistence in the environment. The experiment was conducted in the early autumn when fortuitously, fine weather prevailed.

Polyhedrin-Negative Virus Containing the Beta-Galactosidase Gene

In 1988 and 1989 two further field-release experiments were conducted in Oxford, with a recombinant AcNPV containing a copy of the *Escherichia coli* beta-galactosidase (*lacZ*) coding region under the control of the polyhedrin promoter. The objective of these studies was to demonstrate that a polyhedrin-negative virus could be used to produce a foreign gene product in insect larvae in the field. The foreign gene chosen was not expected to have any effect on the outcome of the virus infection in insects. Essentially similar results were obtained in the 1988 and 1989 release experiments and a summary of both trials is presented below.

The recombinant virus was constructed by standard techniques to insert a copy of the *E. coli lacZ* coding region *in lieu* of the polyhedrin gene coding sequences. Effectively, the virus was identical to the recombinant, non-occluded baculovirus used in the 1987 field study, with the addition of the bacterial sequences. It was expected to behave in the same way in the field-release experiment, that is, to kill the insect larvae and fail to persist on leaf surfaces or in the soil. The AcNPV polyhedrin promoter was preserved to enable high-level synthesis of the beta-galactosidase protein to be achieved in insect cells. Figure 2 shows that the amount of beta-galactosidase protein produced by the recombinant virus matched that of the normal polyhedrin protein synthesized by unmodified virus. After repeated passage of the virus in cell culture and insects, it was found to remain stable and to continue to produce the beta-galactosidase.

The virus was subjected to the routine risk assessment programme established in the previous studies (host-range testing etc.) before permission was sought and granted, for the field-release experiment. This was conducted in a similar way to the previous two release experiments. About 400 *S. exigua* larvae were fed with the virus in the laboratory, before they were placed onto sugar beet plants in the contained experimental area. A similar number of uninfected larvae were placed in another area within the contained facility. The fate of the laboratory virus-infected larvae was monitored by testing larvae, foliage and soil samples recovered from the site for the presence of the recombinant virus. Materials returned to the laboratory were fed to highly susceptible *Trichoplusia ni* (cabbage looper) larvae to detect the recombinant virus.

In summary, both the 1988 and 1989 field trials were unsatisfactory. First, few of the virus-infected larvae placed in the facility died from virus infection. The majority of the larvae pupated by the end of the experimental period. Secondly, there was equal damage to the plants that were infested with virus-infected larvae as those that received the uninfected larvae. These results may have been a consequence of the cold weather during the post-release period, in contrast with the warm weather during the 1987 trial with the genetically marked, non-occluded baculovirus. Reduced temperatures slow down the rate of larval development and almost certainly have an effect on virus replication. Although bioassays before release indicated that a single dose of the same virus inoculum killed *S. exigua* larvae, subsequently reared on diet in the laboratory at 18°C. Any change in the course of infection such as low temperatures, can be expected to reduce the progress of infection, unless an overdose of the virus is used. In fact double infections were employed to provide as much virus as possible; two successive doses of undiluted virus fed to *S. exigua* in the laboratory before release. However, this proved inadequate in the subsequent field-study. It was concluded that polyhedrin-negative virus, although suitable for laboratory analyses is not suitable for field-work unless the species is highly-susceptible.

Future field-release experiments should make use of the polyhedrin-positive baculovirus insecticides. The 1986 study used a marked, occluded AcNPV. It demonstrated that larval deaths occurred as the result of virus infection and that virus persisted *in situ* in cadavers, in the soil and on the plants. Virus was not detected outside the area of the barriered plot within

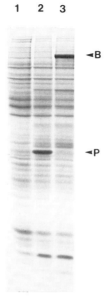

Fig. 2. Synthesis of beta-galactosidase by a polyhedrin-negative AcNPV containing the *lac*Z coding region under the control of the polyhedrin promoter. *Spodoptera frugiperda* (fall army worm) cells (lane 1) were inoculated with AcNPV (lane 2) or the polyhedrin-negative virus containing the *lac*Z coding sequences (lane 3). The cells were harvested at 30 hours post-infection, fractionated in a denaturing polyacrylamide gel and stained with Coomassie blue.

thefacility. Also at the end of the study the plot was successfully decontaminated by formaldehyde treatment. This confirms that it is safe to use the occluded form of the virus, since despite virus persistence, the site may be cleared of infectious virus at the end of the study, or if the study has to be prematurely terminated.

IMPROVING BACULOVIRUS INSECTICIDES

The field-release experiments, described above, used genetically-modified baculoviruses with relatively minor alterations to the virus genome, which produced no enhancement of their insecticidal properties. The addition of synthetic marker elements, or the removal of the polyhedrin gene, permitted investigation of the persistence of the viruses after release to the environment. The next major step in this programme will be to test genetically-modified baculoviruses improved with respect to their ability to control the insect pests. The approach favored by most workers in this area has been to insert genes encoding insect-specific hormones, enzymes or toxins into the baculovirus genome with the expectation that synthesis of the foreign protein will serve to reduce the time taken to kill the insect larvae or at least reduce the amount of feeding damage. The use of hormones and enzymes, particularly when taken from the target pest insect species, offers the advantage that the virus will only be producing, albeit at a higher level, the same proteins that the insect larvae normally synthesizes during the course of its development. The chances of the insect becoming resistant to its own gene products are considered remote, although as outlined below, it may not be quite so easy to overcome the tightly-regulated developmental processes. Furthermore, the use of insect genes present less of a problem in assessing the safety of these recombinant viruses, since the genes are not "foreign" to the target insect.

The use of toxin genes, however, is attractive since many of these proteins are very fast-acting in their native form and may be expected to have a similar effect when produced by the baculovirus. Conversely, there is probably a greater risk of the insect developing resistance to the foreign toxin. We consider that, this risk is small, given that the insect would also have to become resistant to the effects of the virus infection.

It is premature to conclude which approach offers the most satisfactory way of improving baculovirus insecticides. Both are deserving of future study and the examples described below atest to the interest of both methods.

Baculoviruses Containing Insect Hormones and Enzymes

Diuretic hormone. The control of water balance in insects is influenced by the action of diuretic and anti-diuretic hormones (Maddrell, 1986). It is proposed that this balance could be disrupted if a recombinant baculovirus produced elevated levels of either hormone in infected larvae. A diuretic hormone (DH) has been isolated from *Manduca sexta* (tobacco hornworm) and the protein sequence determined (Kataoka *et al.*, 1989). The hormone consists of a 41-amino acid protein, which is amidated at the C-terminus. This protein sequence was used to deduce the appropriate DNA coding region and design and construct a synthetic oligonucleotide. A signal peptide coding sequence from the *Drosophila melanogaster* (fruit fly) cuticle protein was added to the 5' end of the DH coding region to facilitate the secretion of the hormone from virus-infected cells. This synthetic gene coding region was introduced into the *Bombyx mori* (silk worm) NPV genome, in place of the polyhedrin gene coding region, under the control of the polyhedrin gene promoter. This served to produce a polyhedrin-negative virus (Maeda, 1989). The virus could be propagated in *B. mori* cells in culture and behaved in a similar way to the wild-type, unmodified virus. When insect larvae were injected with the recombinant, polyhedrin-negative virus a decrease in haemolymph volume was noted during collection of samples. The reduction in haemolymph volume was estimated to be about 30% compared with mock-infected larvae or larvae infected with wild-type virus. Furthermore, insect larvae infected with the recombinant virus died one day earlier than larvae infected with the unmodified virus. The production of the DH in insect larvae was infered from analysis of mRNA from virus-infected insects. Assessment of the levels of protein production in infected insects was complicated by the small size of the expected hormone. A reverse-phase column was used to isolate the DH, which was assayed by using newly emerged adults of *Pieris rapae* (small white butterfly) (Kataoka *et al.*, 1989). However, direct detection of the hormone was not accomplished by standard techniques, such as protein gel electrophoresis.

Juvenile hormone esterase. The processes that regulate the metamorphosis of insect larvae into pupae present another suitable target for recombinant baculoviruses. In the last larval instar there is a reduction in the titre of juvenile hormone (JH) which has been shown to initiate metamorphosis and terminate feeding (deKort and Granger, 1981). This decrease is accompanied by a substantial increase in the levels of juvenile hormone esterase (JHE). The JHE hydrolyzes the chemically-stable, conjugated methyl ester of JH to the JH acid (Hammock, 1985). Inhibition of the JHE results in the production of overgrown insect larvae, because the JH level remains high enough to keep the insect in the feeding stage (Sparks and Hammock, 1980). Conversely, it was proposed that if sufficient JHE was produced early in the insect developmental stages, then JH would be inhibited, resulting in the abolition of feeding and possible premature pupation (Hammock *et al.*, 1990). The JHE from *Heliothis virescens* (tobacco budworm) was isolated by affinity purification and partially sequenced (Abdel-Aal and Hammock, 1986; Hanzlik *et al.*, 1989). This information provided the basis for designing oligonucleotide probes which were used to isolate clones from a complementary DNA library made from fat body mRNA (Hanzlik and Hammock, 1987). Thereafter, a full-length clone, encoding the JHE, was inserted into a baculovirus transfer vector, utilizing the AcNPV polyhedrin promoter, and this was used to derive a polyhedrin-negative baculovirus (Hammock *et al.*, 1990). The recombinant virus, AcRP23.JHE, produced JHE in insect cell culture and in *T. ni* larvae. When the virus was fed to first instar *T. ni* larvae a reduction in feeding activity was observed, in comparison with untreated control larvae or larvae treated with a virus lacking the JHE coding sequences. The levels of JHE measured in the haemolymph of the AcRP23.JHE-infected insects matched those normally recorded at ecdysis, but were only 10% of those normally seen in the last larval instar. The reduction in feeding was only observed when the recombinant virus was fed to first instar larvae. Later larval instars did not show a reduction in the rate of weight gain. Several explanations for this observation have been proposed. First, in later larval stages the levels of JHE produced may be unable to overcome hormone biosynthesis. Secondly, the production of a virus gene-encoded ecdysteroid UDP-glucosyl transferase (O'Reilly and Miller, 1989) may reduce the effects of

JHE. Finally, the JHE is extremely unstable *in vivo* when produced by the recombinant virus, or in its natural form.

The lack of effect on the later larval instars would appear to reduce the prospects for the successful use of this enzyme in a recombinant baculovirus insecticide. It is not likely that insect pests could consistently be targeted at the neonate stage. However, it is encouraging that an insect gene can be used to modify the efficacy of a baculovirus insecticide. Undoubtedly, progress will be made to increase the levels of JHE attained in virus-infected insects and to improve the stability of the enzyme *in vivo*.

Baculoviruses Containing Insect-Specific Bacterial Toxin Genes

***Bacillus thuringiensis* delta endotoxin.** A gene which is particularly attractive for expression in a baculovirus insecticide is the *Bacillus thuringiensis* delta endotoxin. This is an insect-specific toxin normally produced by the Gram-positive bacterium to form crystalline inclusions. *B. thuringiensis* strains have been reported with activities against a wide range of insect species (Burges, 1981). The delta endotoxin is produced by the bacterium as a 130 Kdal protoxin, which is inactive. This is cleaved in the midgut after ingestion by the insect, to form a 62 Kdal active toxin. The presence of the active toxin in the midgut serves to cause an immediate reduction in feeding (Heimpel and Angus, 1959) and is thought to generate pores in cell membranes, leading to disruption of the osmotic balance and cell lysis (Knowles and Ellar, 1987; reviewed by Hofte and Whiteley, 1989). The natural bacterial product has been used as an insecticide (Luthy *et al.*, 1982; Cunningham, 1988) and thus is already accepted as a safe product for use in the environment.

Two studies have reported the insertion of *B. thuringiensis* delta endotoxin genes into the AcNPV genome (Merryweather *et al.*, 1990; Martens *et al.*, 1990). Essentially the same results were recorded in both reports. A full-length copy of the endotoxin coding sequences was inserted into the baculovirus genome in place of the polyhedrin gene-coding region, to derive polyhedrin-negative viruses which produced the endotoxin. Martens *et al.* (1990) demonstrated that the protein produced in insect cells formed large crystals in the cytoplasm. These crystals had an ultrastructure similar to that of the native toxin synthesized by the bacterium. When insect larvae were fed recombinant virus-infected cell extracts there was an immediate reduction in feeding activity and the larvae eventually died. This mimicked the natural effect of the endotoxin.

Merryweather *et al.* (1990) also produced a polyhedrin-positive virus by inserting the endotoxin coding sequences upstream of the polyhedrin gene, under the control of a copy of the AcNPV p10 promoter. This facilitated the removal of endotoxin from virus and thus permitted standard bioassays to be performed to assess the efficacy of the virus. However, when insects were given purified polyhedra in a bioassay there was no improvement in the effectiveness of the virus insecticide. This was attributed to the fact that only the full-length protoxin was produced in insect cells. This must be cleaved to derive biologically-active toxin. Since this process normally occurs in the gut of insect larvae, it must be concluded that the protoxin produced by the virus does not enter the gut in the course of the infection. Furthermore, the baculovirus-derived toxin was only detected as an intracellular protein. To have a biological effect it must be exported from the cell. Future studies should address the question as to whether the addition of a suitable signal peptide sequence to the toxin would facilitate secretion of the recombinant product from the cell. Despite the disappointing results achieved so far, the two studies (Merryweather *et al.*, 1990; Martens *et al.*, 1990) have demonstrated that it is possible to synthesize a bacterial toxin in an insect cell in an active form.

Baculoviruses Containing Insect-Specific Neurotoxin Genes

The use of an insect-specific toxin, such as the *B. thuringiensis* delta endotoxin, which acts in the gut of the target insect appears to offer poor prospects for improving the efficacy of baculovirus insecticides at the present time. A more attractive proposition is the use of insect-specific neurotoxins. Two reports have recently been published which describe the production of a mite neurotoxin gene (Tomalski and Miller, 1991) and a scorpion neurotoxin gene (Stewart *et al.*, 1991) by recombinant baculoviruses in infected insects. Both examples clearly demonstrated biological activity of the neurotoxins and an improvement in the effectiveness of the baculovirus insecticide.

Baculoviruses containing an insect-specific scorpion neurotoxin. The North African (Algerian) scorpion, *Androctonus australis*, produces a venom containing an *insect-specific* neurotoxin (Zlotkin *et al.*, 1971). The neurotoxin affects the sodium conductance of neurons, producing a presynaptic excitatory effect which leads to paralysis and death (Walther *et al.*, 1976; Teitelbaum *et al.*, 1979). In the study reported by Stewart *et al.* (1991) a synthetic oligonucleotide was constructed, based on the protein sequence of the scorpion neurotoxin determined by Darbon *et al.* (1982). A copy of the AcNPV glycoprotein (gp) 67 signal peptide coding sequence (Whitford *et al.*, 1989) was also added to the 5' end of the coding region, to facilitate secretion of the neurotoxin by insect cells. This assembly was inserted into AcNPV, upstream of the polyhedrin gene, to derive a polyhedrin-positive virus. After infection of cells in culture with recombinant virus, a secreted form of the toxin was detected by injection of *Musca domestica* adults with toxin purified from cell culture medium by HPLC. The recombinant virus was assessed in bioassays with *T.ni* to determine the LD_{50} and ST_{50} values. There was a small, but significant reduction in the LD_{50}, and a 25% reduction in the ST_{50}. When further tests were performed to assess the effect of the virus on the feeding behaviour of insect larvae, it was calculated that a 50% reduction in damage to cabbage leaves was achieved. If this result could be repeated in the field the recombinant baculovirus containing the scorpion toxin gene would offer a real advantage over the unmodified AcNPV. This remains to be tested.

Baculoviruses containing a mite insect-specific neurotoxin. In a similar study to that described by Stewart *et al.* (1991), Tomalski and Miller (1991) inserted a mite (*Pyemotes tritici*) neurotoxin gene (TxP-1) into the AcNPV genome. The toxin gene was isolated from a cDNA expression library produced with mite mRNA and identified with a polyclonal anti-TxP-1 serum. The full-length cDNA clone was sequenced to determine that the toxin is probably 291 amino acids with a molecular mass of 33 Kdal. The toxin gene was expressed initially with the baculovirus system, in place of the polyhedrin coding region, but under the control of the polyhedrin gene promoter. Proteins specific for the toxin were identified in cell lysates and the culture medium of cells infected with the recombinant virus. When 5th instar *T. ni* larvae were injected with the non-occluded virus the insects became paralysed by the second day. Larvae injected with the unmodified AcNPV remained active. An occluded form of the virus was also produced, with the polyhedrin gene function retained to facilitate polyhedra production. This virus, or AcNPV was fed to neonate *T.ni* and the insects monitored for signs of paralysis. At three days post-infection 60% of the larvae fed with the recombinant virus were paralysed or dead. In comparison, about 5% of the insects fed AcNPV showed similar symptoms. By day four the relative proportions were about 82% and 10%. All insects, in both groups, were dead by day six. These results are very encouraging, but it has yet to be demonstrated that the virus can reduce the rate of larval feeding, although this may be infered since the insects are paralysed by the action of the toxin.

RISK ASSESSMENT OF BACULOVIRUSES CONTAINING FOREIGN GENE CODING SEQUENCES

The general principles governing the risk assessment of baculovirus insecticides have been previously outlined (Bishop *et al.*, 1988) and will not be repeated here. One of the most important aspects is whether the addition of foreign DNA affects the host-range of the recombinant virus insecticide. The procedures whereby this information may be gathered are now well-established and involve the collection of wild insect species for challenge with the virus in the laboratory. Given that viruses are now being produced with improved characteristics, these host-range tests remain important. For instance, a species which was previously only marginally susceptible to the unmodified virus, may be killed more efficiently by the recombinant baculovirus. However, in preliminary studies with the virus containing the scorpion toxin this has not been confirmed. The host range of the virus remained unchanged.

The persistance of the recombinant baculovirus in the environment is likely to be similar to that of the wild-type virus, given that the additional foreign DNA sequences do not affect the formation of mature polyhedra. Therefore, the results from the earlier studies with genetically marked viruses may be used to predict the consequences of releasing a baculovirus containing a foreign gene. Recombination between baculoviruses may occur in nature, but probably only between related viruses after co-infection of the same host. The concern is whether the foreign gene in the recombinant baculovirus insecticide may be able to move to a virus with a different host range and thus affect a wider selection of species. Tests conducted in our laboratory showed that it was not possible to "rescue" a polyhedrin-negative AcNPV with a U.K. baculovirus, *Panolis flammea* (pine beauty moth) NPV polyhedrin gene. In these tests the chances for recombination were maximized by performing a co-transfection with AcNPV DNA and a plasmid containing the PfNPV polyhedrin gene. This ensured that cells were likely to receive both virus and plasmid DNA. In parallel experiments it was shown that if the PfNPV polyhedrin promoter is deliberately introduced into the AcNPV genome it would function efficiently (Cameron and Possee, 1989). This demonstrated that the failure to derive polyhedrin-positive virus was not a consequence of the inability of the heterologous promoter to direct transcription.

Probably the most serious concern of those undertaking the field release experiments and of parties concerned with assessing the safety of these trials, is whether the recombinant protein produced in the virus-infected insect will present a hazard to other species, particularly mammals. This is particularly the case when a toxin gene has been inserted into the virus genome. It is important, therefore, to be sure that the toxin in question is insect specific. This can be readily ascertained in laboratory studies, by injection, ingestion and inhalation studies with small mammals. The LD_{50} values determined from these studies may be used to estimate the dose of toxin which would have an effect on humans. Given a knowledge of the yield of the toxin in virus-infected insects it will also be possible to estimate the toxin load in the environment, after release of the recombinant virus. This estimate, together with the LD_{50} information will permit the proposers and regulators alike to determine whether man or other mammals would be likely to receive a harmful dose of toxin in the environment.

A further question to be addressed is whether the toxin gene inserted into the baculovirus could mutate to yield a product which might be active against mammals. Analysis of protein sequence databases to identify closely-related proteins is the best way to investigate this possibility. Calculations may then be made to determine the number of nucleotide changes which would be required to derive a mammalian-active protein. Given that the rate of DNA mutation at a given nucleotide is constant, a reasonably accurate estimate may be made of the likelihood that these changes could occur.

GENETIC MODIFICATION OF OTHER BACULOVIRUSES

For most of the studies described above, AcNPV has been used as the system of choice for performing field release experiments, or investigating the feasibility of modifying the efficacy of baculovirus insecticides. The availability of a permissive cell-culture system *in vitro* has made it a relatively simple matter to derive recombinant viruses. Unfortunately, AcNPV cannot be used to control every insect species, for example, the larvae of *P. flammea* in Scotland. The homologous virus *P. flammea* (Pf) NPV has been used to control this pest (Entwistle and Evans, 1987) but to date there is no permissive cell culture system available to facilitate genetic manipulation of the virus. Fortuitously, the related baculovirus isolated from *Mamestra brassicae* (cabbage moth) can also be used to control *P. flammea* insects (Doyle *et al.*, 1990). This virus shares 70% homology with PfNPV, as measured by DNA:DNA hybridization (Possee and Kelly, 1988). There is a semi-permissive cell system which permits propagation of MbNPV *in vitro*. While this system has been available for several years (R.D. Possee, unpublished data), it has not been suitable for constructing recombinant viruses because the MbNPV DNA cannot be used to transfect *M. brassicae* cells by calcium phosphate precipitation. This problem has only been solved recently with the use of lipofectin (Felgner *et al.*, 1987), which is a much more efficient reagent for introducing DNA into insect cells.

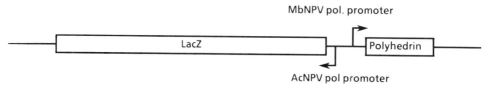

Fig 3. Genetic organization of MbNPV containing the *lac* Z coding sequences. The *lacZ* coding region was inserted upstream of the intact MbNPV polyhedrin gene, under the control of the AcNPV polyhedrin gene promoter.

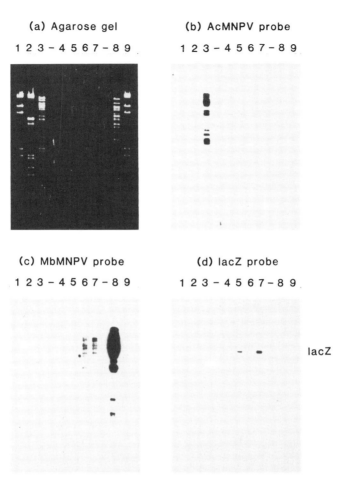

(a) Agarose gel   (b) AcMNPV probe

1 2 3 - 4 5 6 7 - 8 9   1 2 3 - 4 5 6 7 - 8 9

(c) MbMNPV probe   (d) lacZ probe

1 2 3 - 4 5 6 7 - 8 9   1 2 3 - 4 5 6 7 - 8 9

lacZ

Fig. 4. Southern blot hybridization analysis of DNA extracted from *M. brassicae* cells infected with MbNPV or MbNPV.*lacZ* (as described in Figure 3). (a) Ethidium bromide-stained agarose gel (0.8%) of DNA extracted from *M. brassicae* cells infected with MbNPV (lanes 4 and 6) or MbNPV.*lacZ* (lanes 5 amd 7), DNA of AcNPV (lane 3) and MbNPV (lane 8) digested with *Hind*III. Lambda DNA digested with *Hind*III (lanes 1 and 9) and *Hind*III and *Eco*RI (lane 2) was used as size markers. (b) Southern blot of gel shown in (a) hybridized with nick translated AcNPV DNA. (c) Hybridization of the filter with nick translated MbNPV DNA. (d) Hybridization with a *lacZ*-specific nick translated probe.

Preliminary studies have shown that lipofection can be used to transfect *M. brassicae* cells in culture with MbNPV DNA (S.G. Mann, L.A. King and R.D. Possee, unpublished data). The resulting infection is identical with that obtained with virus particles obtained from MbNPV-infected insect haemolymph.

Plasmid transfer vectors for MbNPV were constructed in a manner similar to those derived for AcNPV. These contained a copy of the *E.coli lacZ* coding region inserted upstream of the MbNPV polyhedrin gene. The *lacZ* coding region was placed under the control of the well-characterized AcNPV polyhedrin promoter (Figure 3). This was predicted to function in MbNPV because it had already been shown that the MbNPV polyhedrin promoter works efficiently in AcNPV (Cameron and Possee, 1990). Duplication of the MbNPV polyhedrin promoter was avoided to minimize the likelihood of recombination between homologous sequences. The plasmid transfer vector was co-transfected with infectious MbNPV DNA into *M. brassicae* cells and incubated for seven days before staining the culture with X-gal. Further incubation for three to four days resulted in the development of a blue colour, indicating the presence of beta-galactosidase. Subsequent attempts to plaque purify the recombinant virus were hampered by the inability to passage sequentially the virus in cell culture. It has proved necessary to feed the virus-infected cells to *M. brassicae* insect larvae and subsequently derive haemolymph for reinfection of cells *in vitro* (Weitzman, 1991). This procedure, although cumbersome, facilitates isolation of the recombinant virus. In Figure 4 a DNA hybridization analysis of the recombinant virus is presented to demonstrate the presence of the *lac*Z sequences within the virus genome. This virus could be used in a field-release experiment in a forest ecosystem to monitor the spread and persistance of a recombinant baculovirus insecticide. Alternatively, it provides a starting point for the derivation of other recombinant viruses, containing genetic markers, other foreign genes etc., since the replacement of the *lac*Z sequences will result in the formation of virus lacking the facility to produce beta-galactosidase.

SUMMARY

The prospects for the improvement of baculovirus insecticides by genetic modification are excellent. It is possible to insert genes encoding insect-specific neurotoxins into the baculovirus genome and derive a virus which can control insect pests more effectively in laboratory tests. The results with insect-specific hormones and enzymes have been less successful but have shown some cause for optimism. While most studies have concentrated on the use of AcNPV, more recent studies have demonstrated that other baculoviruses may be modified. These viruses will be of use in field-trials in other ecosystems, such as forests. There is a need for continued risk assessment studies with these reagents. Caution should be exercised in determining whether the modified baculoviruses should be released to the environment. The results to date suggest that recombinant baculoviruses may be safely used in field-release experiments.

ACKNOWLEDGEMENTS

We thank Mr C. F. Rivers, Dr P. H. Sterling, Mrs M. E. K. Tinson and Mr T. Carty for their assistance in collecting and/or rearing the insects used in virus host-range studies and the field-release experiments. We thank Dr J. I. Cooper and his staff for providing the sugar beet plants. The advice of Mr P. F. Entwistle and Dr J. S. Cory is gratefully acknowledged. M. D. Weitzman and D. S. Hughes were supported by NERC studentships and M. L. Hirst was supported by the Department of the Environment.

REFERENCES

Abdel-Aal, Y.A.I. and Hammock, B.D. (1986) Transition state analogs as ligands for affinity purification of juvenile hormone esterase. Science 233, 1073-1076.

Bishop, D.H.L., Entwistle, P.F., Cameron, I.R., Allen, C.J. and Possee, R.D. (1988) Field trials of genetically-engineered baculovirus insecticides, in "The Release of Genetically Engineered Micro-organisms" (Sussman, M., Collins, C.H., Skinner, F.A. and Stewart-Tull, D.E., Eds.), pp. 143-179. Academic Press, New York and London.

Blissard, G.W. and Rohrmann, G.F. (1990) Baculovirus diversity and molecular biology. Ann Rev. Entomol. 35, 127-155.

Burges, H.D. (Ed.) (1981) "Microbial Control of Pests and Plant Diseases 1970-1980". Academic Press, New York and London.

Cameron, I.R. and Possee, R.D. (1989) Conservation of polyhedrin gene promoter function between *Autographa californica* and *Mamestra brassicae* nuclear polyhedrosis viruses. Virus Res. 12, 183-200.

Cunningham, J.C. (1988). Baculoviruses: their status compared to *Bacillus thuringiensis* as microbial insecticides. Outlook on Agriculture 17, 10-17.

Carbon, H., Zlotkin, E., Kopeyen, E., van Rietschoten, J., and Rochat, H. (1982) Covalent structure of the insect toxin of the North African scorpion *Androctonus australis* Hector. Int. J. Pept. Protein Res. 20, 230-330.

Doyle, C.J., Hirst, M.L., Cory, J.S. and Entwistle, P.F. (1990)   Risk assessment studies: Detailed host range testing of wild-type cabbage moth, *Mamestra brassicae* (Lepidoptera: Noctuidae), nuclear polyhedrosis virus. Appl. Env. Microbiol. 56, 2704-2710.

Entwistle, P.F. and Evans, H.F. (1985) Viral Control, in "Comprehensive Insect Physiology, Biochemistry and Pharmacology", (Gilbert, L.I. and Kerkut, G.A., Eds.), vol 12, pp. 347-412. Pergamon Press, Oxford.

Evans, H.F. and Entwistle, P.F. (1987) Viral Diseases, in "Epizootiology of insect diseases", (Fuxa, J.R. and Tanada, Y., Eds.), pp. 257-322. John Wiley, New York.

Felgner, P.L., Gadek, T.R., Holm, M., Roman, R., Chan, H.W., Northrop, J.P., Ringold, G.M. and Danielsen, M. (1987) Lipofectin: A highly efficient lipid-mediated DNA-transfection procedure. Proc. Natl. Acad. Sci. USA. 84, 7413-7416.

Hammock, B.D. (1985) Regulation of juvenile hormone titer: degradation, in "Comparative Insect Physiology Biochemistry and Pharmacology" (G.A. Kerkut and L.I. Gilbert, Eds.), pp. 431-472. Pergamon Press, New York.

Hammock, B.D., Bonning, B.C., Possee, R.D., Hanzlik, T.N. and Maeda, S. (1990) Expression and effects of the juvenile hormone esterase in a baculovirus vector. Nature 344, 458-461.

Hanzlik, T.N. and Hammock, B.D. (1987) Characterization of affinity-purified juvenile hormone esterase from *Trichoplusia ni*. J. Biol. Chem. 262, 13584-13591.

Hanzlik, T.N., Abdel-Aal, Y.A.I., Harshman, L.G. and Hammock, B.D. (1989) Isolation and sequencing of cDNA clones coding for junvenile hormone esterase from *Heliothis virescens*. J. Biol. Chem. 264, 12419-12425.

Heimpel, A.M. and Angus, T.A. (1959) The site of action of crystalliferous bacteria in Lepidoptera larvae. J. Insect Pathol. 1, 14-31.

Hofte, H. and Whiteley, H.R. (1989) Insecticidal proteins of *Bacillus thuringiensis*. Microbiol. Revs. 53, 242-255.

Huber, J. (1986) Use of baculoviruses in pest management programs, in "The biology of baculoviruses. II. Practical applications for insect control", (Granados, R.R. and Federici, B.A., Eds.), pp. 181-202, CRC Press, Boca raton, Florida.

Katoaka, H., Troetschler, R.G., Li, J.P., Kramer, S.J., Carney, R.L. and Schooley, D.A. (1989) Isolation and identification of a diuretic hormone from the tobacco hornworm, *Manduca sexta*. Proc. Natl. Acad. Sci. USA 86, 2976-2980.

Knowles, B.H. and Ellar, D.J. (1987) Colloid-osmotic lysis is a general feature of the mechanism of *Bacillus thuringiensis* delta-endotoxins with different insect specificities. Biochem. Biophys. Acta 924, 509-518.

de Kort, C.A.D. and Granger, N.A. (1981) Regulation of juvenile hormone titre. Ann. Rev. Entomol. 26, 1-28.

Luthy, P., Cordier, J.L. and Fischer, H.M. (1982) *Bacillus thuringiensis* as a bacterial insecticide: basic consideration and applications, in "Microbial and Viral Pesticides" (Kurstak, E., Ed.), pp. 35-74. Marcel Dekker, New York.

Maddrell, S. (1986) in "Insect Neurochemistry and Neurophysiology" (A.B. Borkovec and D.B. Gelman, Eds), pp. 79-90. Humana, Clifton. NJ.

Maeda, S. (1989) Increased insecticidal effect by a recombinant baculovirus carrying a synthetic diuretic hormone. Biochem. Biophys. Res. Commun. 165, 1177-1183.

Martens, J.W.M., Honee, G., Zuidema, D., van Lent, J.W.M., Visser, B. and Vlak, J.M. (1990) Insecticidal activity of a bacterial crystal protein expressed by a recombinant baculovirus in insect cells. Appl. Environ. Microbiol. 56, 2764-2770.

Merryweather, A.T., Weyer, U., Harris, M.P.G., Hirst, M., Booth, T. and Possee, R.D. (1990) Construction of genetically engineered baculovirus insecticides containing the *Bacillus thuringiensis* subsp. *kurstaki* HD-73 delta endotoxin. J. Gen. Virol. 71, 1535-1544.

O'Reilly, D.R. and Miller, L.K. (1989) A baculovirus blocks insect molting by producing ecdysteroid UDP-glucosyl transferase. Science 245, 1110-1112.

Podgewaite, J.D. (1985) Strategies for field use of baculoviruses, in "Viral Insecticides for Biological Control (K. Maramorosch and K.E. Sherman, Eds), pp. 775-797. Academic Press, New York.

Possee, R.D. and Kelly, D.C. (1988) Physical maps and comparative DNA hybridization of *Mamestra brassicae* and *Panolis flammea* nuclear polyhedrosis virus genomes. J. Gen. Virol, 69, 1285-1298.

Sparks, T.C. and Hammock, B.D. (1980) Comparative inhibition of the juvenile hormone esterases from *Trichoplusia ni, Tenebrio monitor,* and *Musca domestica.* Pestic. Biochem. Physiol. 14, 290-302.

Stewart, L.M.D., Hirst, M., Lopez-Ferber, M., Merryweather, A.T., Cayley, P.J. and Possee, R.D. (1991) Construction of an improved baculovirus insecticide containing an insect-specific toxin gene. Nature 352, 85-88.

Teitelbaum, Z., Lazarovici, P. and Zlotkin, E. (1979) Selective binding of the scorpion venom insect toxin to insect nervous tissue. Insect Biochem. 9, 343-346.

Tomalski, M.D. and Miller, L.K. (1991) Insect paralysis by baculovirus-mediated expression of a mite neurotoxin gene. Nature 352, 82-85.

Walther, C., Zlotkin, E. and Rathmayer, W. (1976) Action of different toxins from the scorpion *Androctonus australis* on a locust nerve-muscle preparation. J. Insect Physiol. 22, 1187-1194.

Weitzman, M.D. (1991) Characterization of two strains of *Panolis flammea* nuclear polyhedrosis virus. Ph.D Thesis, Oxford Polytechnic.

Whitford, M., Stewart, S., Kuzio, J. and Faulkner, P. (1989). Identification and sequence analysis of a gene encoding gp67, an abundant envelope glycoprotein of the baculovirus *Autographa californica* nuclear polyhedrosis virus. J. Virol. 63, 1393-1399.

EXPLOITATION OF GENETICALLY-MODIFIED MICROORGANISMS IN THE FOOD INDUSTRY

Michael Teuber

Institute of Food Science
Laboratory of Food Microbiology
Swiss Federal Institute of Technology
CH-8092 Zurich

INTRODUCTION

Microorganisms traditionally used for the production of fermented food have been the subject of genetic modification for a series of purposes for three main reasons:

i. they are generally recognized as safe (GRAS),
ii. there is ample experience to handle them on a large industrial scale, and
iii. there is a great market potential.

At the 4th European Congress on Biotechnology 1987 in Amsterdam, I reported on "The use of genetically-manipulated microorganisms in food: opportunities and limitations" (Teuber, 1987). At that time, it was obvious that all genera of microorganisms employed for food fermentations were investigated regarding methods for their genetic manipulation. Now, vectors and transformation and expression systems for homologous and heterologous genes are available for almost all important genera: *Acetobacter* (Murooka *et al.*, 1981), *Gluconobacter* (Condon *et al.*, 1991), *Zymomonas* (Drainas *et al.*, 1990), *Brevibacterium* (Martin *et al.*, 1987), *Lactobacillus* (Zink *et al.*, 1991), *Lactococcus* (van de Guchte *et al.*, 1991), *Leuconostoc* and *Pediococcus* (Luchansky *et al.*, 1988), *Propionibacterium* (Rehberger and Glatz, 1990), *Staphylococcus* (Götz, 1990), *Streptococcus* (Somkuti and Steinberg, 1988), *Candida* (Sugisaki *et al.*, 1985), *Kluyveromyces* and *Saccharomyces* (Hollenberg and Strasser, 1990), *Aspergillus, Monascus, Mucor* and *Penicillium* (Esser and Mohr, 1990).

In contrast to the situation 4 years ago, some genetically modified microorganisms and their products are now actually exploited by the food industry. They have been legalized by certain countries and as a consequence are successfully being introduced into the market. This chapter reviews the products on the market and those developments which may be close to an industrial application. Safety issues and consumer concerns will be briefly discussed.

FOOD ENZYMES PRODUCED WITH GENETICALLY-MODIFIED MICROORGANISMS: CHYMOSIN

A food processing enzyme (chymosin E.C.3.4.23.4) made with the aid of genetically-modified microorganisms has rather quietly entered the market and consequently the human food chain. This is used to coagulate the casein of milk during cheese-making. Enzyme extracts (calf rennet) are normally prepared from the fourth stomach (*abomasum*) of suckling calves. However, since calves are being slaughtered at an ever-increasing age the content of

The Release of Genetically Modified Microorganisms
Edited by D.E.S. Stewart-Tull and M. Sussman, Plenum Press, New York, 1992

59

Table 1. Steps in the production of chymosin with genetically-modified microorganisms

1. isolation of prochymosin m-RNA

2. transcription into cDNA

3. cloning into expression (& secretion) vector

4. transformation of suitable microorganisms
   Escherichia coli K12
   Kluyveromyces lactis
   Aspergillus niger subsp. awamori

5a. expression of prochymosin and formation of inclusion
   bodies *(E. coli)*

5b. expression and secretion of prochymosin (fusion protein)
   (K. lactis & A.niger)

6a. extraction and purification of prochymosin inclusion bodies,
   chemical conversion into autocatalytically activatable form
   (E. coli)

6b. purification of prochymosin (*K. lactis, A. niger*)

7. autocatalytical conversion of prochymosin to chymosin at pH 2

8. formulation of commercial preparations (dilution, addition of
   NaCl and preservatives, standardization)

9. microbiological & enzymatic quality control eg. absence of
   genetically engineered microorganisms

available chymosin in the stomach is decreasing. On the other hand, cheese production is still growing on a world-wide basis with an annual output of more than 13.6 million metric tons per annum (FAO Production Yearbook 1987). Since microbial rennets (eg. from *Mucor* spp.) do not have the same specificity, they lead to cheeses with a different taste, especially in long-ripened products. The availability of appropriately genetically-modifed microorganisms was a timely event to provide some of the 50.000 metric tons of chymosin needed each year for cheesemaking. An in depth review has been published (Teuber, 1990a).

The route of synthesis is given in Table 1. The first important step was to prove that the nucleotide sequence of the produced cDNA (Nishimori *et al.*, 1982) was coincident with the amino acid sequence obtained by classical protein sequencing (Foltmann *et al*., 1979), which was the case. For all three commercially available preparations, prochymosin cDNA was chosen for cloning and expression since prochymosin is enzymatically inactive but can easily be transformed autocatalytically into enzymatically active chymosin by incubation at pH 2.0. During this activation, an aminoterminal peptide of 42 amino acids is liberated. Fusion-proteins of prochymosin with 10 to 11 N-terminal amino acids of e.g. *E.coli* beta-galactosidase or *Aspergillus* glucoamylase are still autocatalytically activatable giving rise to chymosin which is biochemically indistinguishable from bovine chymosin.

Another advantage of the treatment at pH 2.0 is the depurination of any remaining recombinant vector DNA present in the product. In transformation experiments, the resulting DNA is biologically inactive. In addition, most of the contaminating proteins in the chymosin extracts obtained from the genetically-modified microorganisms are denatured at pH 2.0. On the basis of the proven genetic identity, it is not surprising that the chemical, immunological

Table 2. Recombinant chymosin preparations available on the market

Chymosin isomer	Producing microorganisms	Brand name/ producing company
B	*Kluyveromyces lactis*	MAXIREN[R]/ Gist Brocades (Delft)
B	*Aspergillus niger*	CHYMOGEN[R]/ Genencor - Chr. Hansen (San Franscisco/Copenhagen)
A(C)	*Escherichia coli* K12	CHY-MAX[R]/ Pfizer (New York)

and technological properties investigated were quite comparable to calf rennet. The main difference was the much higher purity of recombinant chymosin as compared with calf rennet: at least 60% chymosin as against ca. 2%, respectively, by protein estimation. Twenty-eight and ninety days feeding studies with rats at a 100,000-fold the dose expected for average human consumption did not reveal any toxic effects. Genetically-modified microorganisms used for the production of recombinant chymosin were not detected in the tested preparations.

The preparations described in Table 2 have been applied in cheese making for the following traditional products: Appenzell, Asiago, Caciotta, Camembert, Cheddar, Colby, Creszenza, Edam, Emmental, Feta, Gouda, Grana Padano, Italico, Manchego, Montasio, Mozzarella, Parmigiano Reggiano, St. Paulin and Tilsit cheeses. These experiments were performed in recognized dairy research laboratories all over the world including the following countries: Australia, Canada, Switzerland, Germany, Denmark, Spain, France, Great Britain, Greece, Ireland, Italy, Holland and the United States.

The history of FDA approval for the production of chymosin from genetically-modified microorganisms has recently been published (Flamm, 1991), he arrived at the following conclusions in summarizing the safety and environmental aspects:

"After a comprehensive review of the information in the published literature, FDA concluded that the calf chymosin gene cloned into *E . coli* K-12 is expressed, that the gene product is chemically and biologically indistinguishable from calf chymosin, and that impurities in the microbial chymosin preparation do not make the preparation unsafe. As for any food ingredient that is new or made by a new method, FDA reviewed the manufacturing method to determine product purity and identity. From its review of this chymosin preparation and other new biotechnology-derived food ingredients, the Food and Drug Administration is finding that the review of the safety, purity, and identity of these products is fundamentally no different from that of analogous products derived from unmodified or traditionally modified organisms. It is also finding that biotechnology provides powerful tools for resolving safety questions."

Insiders from the US estimate that the *E. coli* K-12 chymosin has conquered about 20% of the 40 to 50 million dollar milk coagulant market since the approval by FDA on March 23, 1990 (DHHS, 1990).

The food laws of countries other than the USA as for example the UK, Portugal or Germany may not require such a review. However, the new German gene technology law requires that release of genetically-modified microorganisms requires legal approval. In that case it is important to demonstrate that a fermentation-derived chymosin does not contain any living or active ingredients. Chymosin produced with *Kluyveromyces lactis* was approved in Switzerland in 1988. Australia has also recently legalized the use of chymosin. In view of the clear safety situation, none of the countries mentioned requires specific labelling of cheese to indicate that it has been produced with fermentation-produced chymosin.

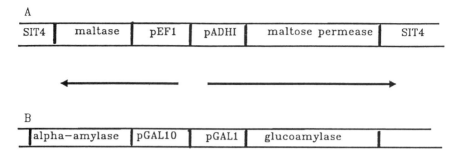

Figure 1. Expression cassettes for *S. cerevisiae*
A. Maltose system (Osinga *et al.*, 1988)
B. Amylase/glucoamylase system (Hollenberg & Strasser, 1990)
p: promoter; ⟶ : direction of translation

GENETICALLY-MODIFIED MICROORGANISMS FOR FOOD FERMENTATIONS: YEASTS AND LACTIC ACID BACTERIA

Microorganisms when used traditionally for food fermentations are generally recognized as safe (GRAS), a concept of the Food Law of the USA. Many traditional products, however, are made with mixed cultures, components which provide for different functions such as acid production, proteolysis, aroma formation etc. The tasks of interest in food biotechnology can be summarized as:

i.optimization of the function(s) of a single component culture, and
ii.combination of properties from different biological systems in one microorganism.

The following developments with genetically-modified *Saccharomyces cerevisiae* strains are ready for the market or close to completion:

S.cerevisiae with a Constitutive Maltose Utilization

This genetically-modified yeast has been constructed in the laboratory of Gist Brocades (Delft, Holland) which is one of the leading baker's yeast companies in the world. The purpose was to obtain a strain with a derepressed maltose utilization system which ferments maltose equally well in sweet and lean doughs. Normally, glucose, fructose and mannose repress the maltose system composed of a maltose permease and a maltase. The strain has been approved for the use by the responsible authorities in 1990 in Great Britain as baker's yeast.

The invention was described in detail in a European Patent Application (Osinga *et al.*, 1988). Since this yeast is inevitably released during use and is even consumed by man under certain conditions, the following self-imposed restrictions in the genetic modification work have been applied:

- origin of the employed genetic material from within the species *S. cerevisiae*,

- the complete removal of foreign DNA (eg. antibiotic resistance genes) necessary for the construction work,

- the limitation of synthetic DNA linkers to an absolute minimum,

- insertion of termination signal cassettes (in frame-shift) at the end of the inserted maltose permease and maltase genes to avoid any possibility of expression of fusion proteins,

- the incorporation of the construct into the yeast chromosome to obtain genetic stability, and to avoid uncontrolled dissemination of the contruct by gene transfer.

The construction protocol is schematically shown in Table 3, the structure of the construct in Figure 1A. The strain ferments sucrose and glucose very well, and maltose fermentation measured as carbon dioxide production increases by 18 to 28%, depending on time; it is not repressed by glucose. The company claims that this yeast in addition to its use in bread-making can also be used for the alcoholic fermentation of mashes containing the relevant sugars. The company has also provided data for the British authorities to show that the genetically-modified yeast behaves in the environment (dough, bread, sewage, river- and sea-water) like the parent yeast.

S. cerevisiae which Grows on Starch as Sole Carbon Source

An old dream of fermentation biotechnologists has materialized through genetic modification techniques: the direct conversion of starch into ethanol by *S.cerevisiae*. Hollenberg's laboratory which was already involved in the early work on the cloning and expression of chymosin in *Kluyveromyces* has achieved expression and excretion of a *Schwanniomyces occidentalis* amylase and glucoamylase in and from *S.cerevisiae* (Dohmen *et al.*, 1990; Hollenberg and Strasser, 1990). Whereas the *Schwanniomyces* amylase gene (AMY1) was expressed and excreted from its own promoter, the *Schwanniomyces* glucoamylase gene (GAM1) had to be coupled to a *S.cerevisiae* promoter in order for it to be expressed and excreted in *S.cerevisiae*.

Table 3. Steps in the construction of a genetically-modified strain of *Saccharomyces cerevisiae* with a constitutive maltose utilization system. (Osinga *et al.*, 1988)

1.	Selection and preparation of the chromosomal site for the later incorporation of the constitutive maltose utilization system: sporulation induced transcribed gene (SIT4)
1.1	Cloning of a suitable SIT4 DNA fragment
1.2	Insertion of a gentamycin resistance gene under the direction of a yeast promoter (G418r) into the cloned SIT4 fragment
1.3	Transformation of wild-type strain with linear SIT4-G418r DNA to yield pairing with homologous SIT4 allele and gene-replacement in one allele
1.4	Selection of gentamycin resistant transformants and repetition of transformation procedure for the second allele
2.	Preparation of the constitutive maltose utilization cassette
2.1	Cloning of the structural genes of the maltose and maltose permease from wild-type *S. cerevisiae*
2.2	Cloning of two strong *S. cerevisiae* promoters: alcohol dehydrogenase (pADHI) and translation elongation factor (pEFalphaA)
2.3	Assembly of the maltose utilization system as a cassette as shown in figure 1 with synthetic safety valve stop codons downstream of the maltose and maltose permease genes, respectively
2.4	Incorporation of the maltose utilization cassette into the proper SIT4 DNA fragment
3.	Incorporation of the constitutive maltose cassette into the yeast genome
3.1	Cotransformation of gentamycin resistant yeasts (see 1. above) with linear DNA of the constitutive maltose-SIT4 cassette and a phleomycin resistance plasmid pUT332
3.2	Gene replacement of SIT4-G418r by the maltose-SIT4 cassette
3.3	Repetition of this step for the second allele
3.4	Curing of the phleomycin resistance plasmid

Optimal expression and excretion was achieved when both genes were assembled in a cassette each with its own *S.cerevisiae* promoter (see Figure 1B). Under identical conditions, the genetically-modified *S.cerevisiae* cells produced 5% more extracellular amylase and 75% more glucoamylase activity than the wild-type *Schwanniomyces*. The expression cassette could be stably integrated into a centromere plasmid which was also used for the transformation procedure. The glucoamylase excreted by *S.cerevisiae*, however, seemed to be modified differently from the original *Schwanniomyces* enzyme. Since both enzymes are completely inactivated by heating at 60°C for 10 min, they could be useful in the brewing industry. In addition, a combination of both enzymes would produce a low carbohydrate beer. Safety considerations regarding this yeast have not yet been published, however, both *Saccharomyces cerevisiae* and *Schwanniomyces occidentalis* are regarded as safe organisms.

S. cerevisiae Secreting Legume Lipoxygenase

The aim of the project is to construct strains of baker's yeast that express and secrete lipoxygenase (E.C.1.13.11.2) with genes from *Pisum sativum* (Casey *et al.*, 1990). Lipoxygenase is thought to bleach wheat-flour carotenoids and xanthophylls and to oxidize the protein SH-groups in wheat gluten to enhance dough rheology and stability. Pea lipoxygenase cDNA was linked to the yeast invertase signal peptide. This cassette was put under the control of different yeast promoters (GAL1 and PGK). Transformation, expression and excretion were achieved, however, the amount of lipoxygenase excreted has still to be increased (Knust, 1991). Again, it is too early for detailed safety discussions which probably would follow similar argumentation lines as for *S.cerevisiae* with the constitutive maltose system (see above).

S. cerevisiae Secreting Barley Beta-1,3-1,4-Glucanase

Polymer beta-1,3-1,4-glucanes are released during the malting process from the barley endosperm cell walls. Under mechanical stress, highly polymer beta-glucane gels (up to 300 mg per litre of wort) may be formed which limit the filtration capacities of filtration units in breweries. If not removed, beta-glucanes give rise to undesirable haze in beer at low storage temperatures. Addition of purified beta-glucanases (e.g. from *Bacillus, Aspergillus* or *Trichoderma* species) is costly and not allowed in countries like Germany which have a strict purity law for beer.

To achieve degradation of beta-glucanes in wort, the gene of a barley beta-1,3-1,4-glucanase was cloned into a yeast vector coupled to a signal peptide sequence, the ADH1-promoter and the tryptophane 1-terminator of *S. cerevisiae* (Jackson *et al.*, 1986). The 2 micro metre plasmid vector also contained a resistance against the aminoglycoside antibiotic G418. Under selective conditions, transformed yeast cells excreted barley beta-glucanase in amounts

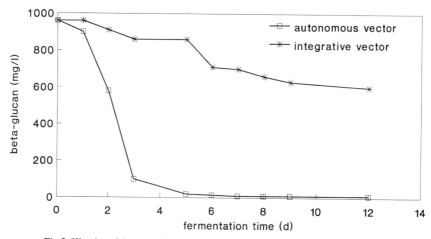

Fig.2. Kinetics of the degradation of beta-glucan (from Berghof and Stahl, 1991)

that were able to depolymerize up to 1000 mg beta-glucanes per litre of wort within one week of fermentation. If incorporated stably into the yeast chromosome (without the antibiotic resistance gene), expression of the barley beta-glucanase was diminished, however, still high enough to degrade the natural beta-glucane content of wort within 2 weeks of fermentation thereby increasing filter standing time by 50% (Berghof and Stahl, 1991). The kinetics of this process are shown in Figure 2.

Another tasks for yeasts in brewing which has been achieved by genetic modification of the brewer's yeast is a more rapid reduction of the bad-tasting diacetyl in beer (Gjermansen *et al.*, 1988; Goelling and Stahl, 1988). For this purpose, the alpha-acetolactate decarboxylase gene from *Acetobacter pasteurianus* was cloned into *S.cerevisiae* (Stahl and Goelling, 1990).

Lactic Acid Bacteria.

Species of the genera *Lactobacillus, Lactococcus, Leuconostoc, Pediococcus* and *Streptococcus* are traditionally used for lactic fermentations of milk, meat, vegetables, wine, beer and silage. The genetics of these bacteria with *Lactococcus* in the first place has made such rapid progress during the last decade that genetic modification is getting close to practical application (Teuber, 1990; deVos, 1990b). This development was initiated by the observation that most of the technologically important functions of lactococci are coded for on plasmids (see Table 4). Such plasmids have been used to construct suitable vectors for genetic reorganization studies. In addition, two routes of natural gene transfer are available: conjugation and transduction. *In vitro* transformation is possible with protoplasting followed by regeneration, and by electroporation. Shuttle vectors for *Lactococcus, Bacillus subtilis* and *E.coli* have been developed. Heterologous genes cloned and expressed in *Lactococcus* include chloramphenicol acetyltransferase (*Staphylococcus aureus*), ribosome methylase (*S.aureus*), aminoglycoside phosphotransferase (*S.aureus*), tetracycline resistance (*Enterococcus faecalis*), beta-galactosidase (*E.coli*), lysozyme (*Gallus gallus gallus*), chymosin (*Bos taurus*), and thaumatin (*Thaumatococcus daniellii*) (review see Teuber *et al.*, 1991).

Gene expression in *Lactococcus* is now so well understood that genetic modification can be performed. Expression is not the problem, but excretion may be a limiting factor (van de Guchte *et al.*, 1991). The other genera of the lactic acid bacteria are not as thoroughly investigated. However, the European Community has launched within the BRIDGE programme a large T-project with some 25 laboratories from academic, governmental and industrial research institutions participating. The next three years will therefore yield a wealth of new information on the genetics, genetic modification and biotechnological application of these bacteria. Unfortunately, the EC was unable due to lack of funds to support a project on the safety aspects of genetically-modified lactic bacteria, a project which had been running for two years within the BAP programme (Beringer, 1990). However, a similar project on a national basis with five laboratories involved is still continuing in Germany into 1992. It is being financed by the Federal Ministry for Research and Technology, Bonn (Teuber, 1990; Wagner *et al.*, 1990; Vogel *et al.*, 1990; Klein *et al.*, 1990; Ludwig *et al.*, 1990).

Table 4. Genes and functions coded for on plasmids in lactococci

1.	Lactose metabolism: enzyme II and factor III of lactose phosphotransferase system, phophobeta-galactosidase, tagatose-6-P isomerase, tagatose-6-P kinase, tagatose-1,6-P aldolase
2.	Citrate metabolism: citrate permease
3.	Proteolytic system: beta-casein specific, cell wall associated protease, maturation protein for the protease
4.	Sucrose-6-P hydrolase
5.	Slime production in ropy strains
6.	Bacteriocin production and immunity
7.	Several bacteriophage insensitivity phenomena

The main topics of genetic modification work on lactic acid bacteria in all different food areas are:

- stabilization of technological functions
- the construction of bacteriophage-resistant starter cultures
- the establishment of starter cultures with high proteolytic activities to accelerate cheese ripening and aroma production
- the construction of starter cultures excreting bacteriocins and peptide antibiotics (e.g. nisin, diplococcin, lactacin, pediococcin, acidophilin, helveticin) to inhibit *Clostridium, Listeria* and other harmful pathogens in fermented food and fodder
- the construction of amylase-containing lactic bacteria for silage fermentations.

SAFETY CONSIDERATIONS AND CONSUMER CONCERNS REGARDING THE APPLICATION OF GENETICALLY-MODIFIED MICROORGANISMS AND THEIR PRODUCTS IN FOOD

The German gene technology law of 1990 states in the 1st paragraph that its purpose is to protect life and the health of man, animals and plants, as well as to protect the environment in its entirety including material goods against potential dangers of genetic modification processes and products, and to avoid the generation of potential dangers (FRG 1990). In as much, it is clear that genetically-modified microorganisms and their products to be used in or as food, should be and have actually been developed with caution under carefully controlled conditions. (see section on chymosin and baker's yeast above).

The specific safety issues for products made with genetically-modified microorganisms may be easier to judge than genetically-modified microorganisms themselves which are added later to food or fodder. For the products, it suffices in my opinion to show that they are free of genetically-modified microorganisms and biologically-active recombinant DNA. Otherwise, like any food or food additive it must not contain toxic or otherwise harmful substances. If it is identical with a traditional compound, the new product should be evaluated regarding safety and nutritional value with the conventional, analogous food product as a standard. This concept was applied for the approval of fermentation-produced chymosin by the FDA in the USA (Flamm, 1991). This evaluation procedure may serve as a model for other compounds, other governments and other countries.

For genetically-modified microorganisms directly used as food, the situation is more complicated since a new organism has been created. On the basis of case to case evaluation, these following points might be followed:

1. exact knowledge of the function(s) incorporated or modified
2. prove that this function is correctly expressed
3. the nucleotide sequence of the incorporated DNA should be determined
4. the insertion site into the genome or a known vector should be determined
5. unwanted dissemination of the inserted DNA into the genome of other organisms by horizontal gene transfer should be checked and controlled
6. the behavior of the genetically-modified microorganisms in food and in the environment (biosphere, water, soil, air) should be anticipated and eventually evaluated
7. the absence of toxicity and pathogenicity must be shown, however, by analogous methods to those used for the genetically-unmodified parent strain.

For public acceptance, it will be very important that governments, the European Community, and world-wide agencies like FAO and WHO (1991) speak one language. Concerns of consumers and consumer organizations or parties are mainly centered on the following issues and must be taken serious by the scientific community:

- potential toxicity and pathogenicity of genetically-modified microorganisms
- potential hazards to the environment

- displacement of natural by artificial food
- increased advantage for big, multinational food companies
- increased technology gap between industrialized and developing countries.

Whereas the arguments relating to health and environment may be handled by a proper information and education policy at least for the majority of the public, the political issues remain issues to be addressed by most of the scientific community.

REFERENCES

Berghof, K., and Stahl, U. (1991) Verbesserung der Filtrierbarkeit von Bier durch Verwendung beta-glucanaseaktiver Brauhefe. Bioengineering vol.7,No. 2, 27-32.

Beringer, J.E. (1990) Risk assessment. In: Biotechnology R&D in the EC (A. Vassarotti and E. Magnien eds.) Volume I Catalogue of BAP Achievements, Elsevier, Amsterdam, pp.75-80.

Casey, R., von Wettstein, D., and Petersen, A. (1990) Construction of a baker's yeast excreting legume lipoxygenase during production of bread dough. In: Biotechnology R&D in the EC (A. Vassarotti and E. Magnien eds.) Volume II, Detailed Final Report of BAP Contractors, Elsevier, Amsterdam, pp. 333-337.

Condon, C., FitzGerald, R.J., and O'Gara, F. (1991) Conjugation and heterologous gene expression in *Gluconobacter oxydans* ssp. *suboxydans*. FEMS Microbiology Letters 80, 173-178.

Department of Health and Human Services, Food and Drug Administration (1990) Direct food substances affirmed as generally recognized as safe; chymosin enzyme preparation derived from *Escherichia coli* K-12. Final Rule. Federal Register vol. 55, no. 57, Friday 23 March 1990, pp. 10932-10935.

Dohmen, R.J., Strasser, A.W.M., Dahlems, U.M., and Hollenberg, C.P. (1990) Cloning of the *Schwanniomyces occidentalis* glucoamylase gene (GAM1) and its expression in *Saccharomyces cerevisiae*. Gene 95, 111-121.

Drainas, C., Sahm, H., and Typas, M.A. (1990) Genetic manipulation of the anaerobe *Zymomonas mobilis* for fermentation of fruit juices. In: Biotechnology R&D in the EC (A. Vasarotti and E. Magnien eds.) Volume II, Detailed Final Report of BAP contractors. Elsevier, Amsterdam, pp.365-370.

Esser, K., and Mohr, G. (1986) Integrative transformation of filamentous fungi with respect to biotechnological application. Process Biochemistry 21, 153-159.

FAO/WHO (1991) Strategies for assessing the safety of foods produced by biotechnology. World Health Organization, Geneva 1991, Report of a joint FAO/WHO Consultation, 59 pages. ISBN 92 4 15 61 45 9.

Federal Republic of Germany (1990) Gesetz zur Regelung von Fragen der Gentechnik vom 20. Juni 1990. Bundesgesetzbl. Jahrg. 1990, Teil I, pp. 1080-1095.

Flamm, E.L. (1991) How FDA approved chymosin: a case history. Biotechnology 9, 349-351.

Foltmann, B., Pedersen, V.B., Kauffman, D., and Wybrandt, G. (1979) The primary structure of calf chymosin. J. Biol. Chemistry 254, 8447-8456.

Gjermansen, C., Kiellandbrandt, M.C., Holmberg, S., Petersen, J.G.L., Nilssontillgren, T., and Sigsgaard, P. (1988) Towards diacetyl-less brewer's-yeast - Influence of Ilv2 and Ilv5 mutations. J.Basic Microbiology 28, 175-183.

Goelling, D., and Stahl, U. (1988) Cloning and expression of an alpha-acetolactate decarboxylase gene from *Streptococcus lactis* subsp. *diacetylactis* in *Escherichia coli*. Appl. Environm. Microbiol. 54, 1889-1891.

Götz, F. (1990) Applied genetics in the Gram positive bacterium *Staphylococcus carnosus*. Food Biotechnology 4, 505-514.

Hollenberg, C.P., and Strasser, A.W.M. (1990) Improvement of baker's and brewer's yeast by gene technology. Food Biotechnology 4, 527-534.

Jackson, E.A., Ballance, G.M., Thomsen, K.K. (1986) Construction of a yeast vector directing the synthesis and release of barley (1,3-1,4)-beta-glucanase. Carlsb. Res. Commun. 51, 445-458.

Klein, J., Henrich, B., Ulrich, C., and Plapp, R. (1990) Entwicklung eines Modellsystems zur Untersuchung der biologischen Sicherheit gentechnisch veränderter Lactobacillen in der Fermentation von Lebensmitteln. In: Biologische Sicherheit (PBEO-Jülich ed.) Vol. 2, Jülich, pp. 107-114.

Knust, B. (1991) Construction of baker's yeast secreting plant seed lipoxygenases. Abstr. Meeting on Yeast, genetics and molecular biology, San Francisco, May 23-28, 1991.

Luchansky, J.B., Muriana, P.M., and Klaenhammer, T.R. (1988) Application of electroporation for transfer of plasmid DNA to *Lactobacillus, Lactococcus, Leuconostoc, Listeria, Pediococcus, Bacillus, Enterococcus* and *Propionibacterium*. Molecular Microbiology 2, 637-646.

Ludwig, W., Betzl, D., and Schleifer, K.H. (1990) Biologische Sicherheit gentechnisch veränderter Mikroorganismen in der Lebens- und Futtermittelbiotechnologie. Entwicklung von Nachweismethoden. In: Biologische Sicherheit (PBEO-Jülich ed.) Vol. 2, Jülich, pp. 115-126.

Martin, J.F., Santamaria, R., Sandoval, H., Del Real, G., Mateos, L.M., Gil, J.A., and Aguilar, A. (1987) Cloning systems in amino acid-producing corynebacteria. Biotechnology 5, 137-146..

Murooka, Y., Takizawa, N., and Hatada, T. (1981) Introduction of bacteriophage Mu into bacteria of various genera and intergeneric gene transfer by RP4::Mu. J. Bacteriol. 145, 358-368.

Nishimori, K., Kawaguchi, Y., Hidaka, M., Uozumi, T., and Beppu, T. (1982) Nucleotide sequence of calf prorennin cDNA in *Escherichia coli*. J.Biochem. 91, 1085-1088.

Osinga, K.A., Beudeker, R.F., Van der Plat, J.B., and De Hollander, J.A. (1988) New yeast strains providing for enhanced rate of fermentation of sugars, a process to obtain such yeasts and the use of these yeasts. EP 0 306 107 A2. Filed: 01.09.88, published: 08.03.89 Bulletin 89/10.

Rehberger, T.G., and Glatz, B.A. (1990) Characterization of *Propionibacterium* plasmids. Appl.Environm.Microbiol. 56, 86 4-871.

Somkuti, G.A., and Steinberg, D.H. (1988) Genetic transformation of *Streptococcus thermophilus* by electroporation. Biochimie 70, 579-585.

Stahl, U., and Goelling, D. (1990) Isolation and expression of the alpha-acetolactate decarboxylase gene of *Acetobacter pasteurianus*. In: Bulletin of the European Brewery Convention (EBC), J.R.M. Hammont (ed.), pp. 83-94.

Sugisaki, Y., Sakaguchi, K., Yamasaki, M., Tamura, G. (1985) Transfer of DNA killer plasmids from *Kluyveromyces lactis* to *Kluyveromyces fragilis* and *Candida pseudotropicalis*. J. Bacteriol. 164, 1373-1375.

Teuber, M. (1987) The use of genetically-manipulated microorganisms in food: opportunities and limitations. Proc. 4th European Congress on Biotechnology Amsterdam, June 14-19, 1987 (O.M. Neijssel, R.R. van der Meer and K.Ch.A.M. Luyben eds.) Vol. 4, Elsevier, Amsterdam 1987, pp. 383-391.

Teuber, M. (1990a) Production of chymosin (EC 3.4.23.4) by microorganisms and its use for cheesemaking. Bulletin of the International Dairy Federation (Brussels) 251, 3-15.

Teuber, M. (1990b) Strategies for genetic modification of lactococci. Food Biotechnology 4, 537-546.

Teuber, M. (1990c) Biologische Sicherheit gentechnisch veränderter Mikroorganismen in der Lebensmittel- und Futtermittelbiotechnologie. In: Biologische Sicherheit (PBEO-Jülich ed.) Vol. 2, Jülich, pp. 89-94.

Teuber, M., Geis, A., and Neve, H. (1991) The genus Lactococcus. In: The Prokaryotes, A Handbook on the Biology of Bacteria, Ecophysiology, Isolation, Identification, Applications 2nd Edition (A. Balows, H.G. Trüper, M. Dworkin, W. Harder and K.H. Schleifer eds.), Springer-Verlag New York 1991, Vol 2, pp. 1482-1501.

van de Guchte, M., Kok, J., and Venema, G. (1991) Gene expression in *Lactococcus lactis*. FEMS Microbiology Reviews, in press.

Vogel, R.F., Gaier, W., Knauf, H.J., and Hammes, W.P. (1990) Verhalten gentechnisch veränderter Bakterien bei der Herstellung von Rohwurst aus Fleisch - Ein Risiko-Modell auf der Grundlage von horizontalem Gentransfer. in: Biologische Sicherheit (PBEO-Jülich ed.) Vol. 2, Jülich, pp. 101-106.

Vos, W.M. de (1990) Application of genetic manipulation to lactic cultures. Proceedings 23rd International Dairy Congress, Oct. 8-12, 1990,Montreal, Dairying in a Changing World, Vol. 2, pp. 1596-1603.

Wagner, E., Brückner, R., and Götz, F. (1990) Entwicklung von Sicherheitsvektoren bei *Staphylococcus carnosus* und Unters uchung von Stabilität und Mobilität dieser Vektoren in der Fermentation unde während der Rohwurstreifung. In: Biologische Sicherheit PBEO-Jülich ed.) Vol. 2, Jülich, pp. 95-100.

Zink, A., Klein, J.R., and Plapp, R. (1991) Transformation of *Lactobacillus delbrückii* ssp. lactis by electroporation and cloning of origins of replication by use of a positive selection vector. FEMS Microbiology Letters 78, 207-212.

LIVE ATTENUATED SALMONELLAE: ORAL VACCINES FOR SALMONELLOSIS AND COMBINED VACCINES CARRYING HETEROLOGOUS ANTIGENS

Carlos E. Hormaeche[1], Gordon Dougan[2] and Steve N. Chatfield[3]

Division of Microbiology and Parasitology[1]
Department of Pathology
Tennis Court Road
Cambridge CB2 1QP, U.K.

Department of Biochemistry[2],
Imperial College of Science, Technology and Medicine
Wolfson Laboratory
London SW7 2AY

Vaccine Research Group[3]
Medeva Group Research
Imperial College of Science, Technology and Medicine
Wolfson Laboratory
London SW7 2AY, U.K.

SUMMARY

Live attenuated salmonellae are protective, and are candidate vaccines against invasive salmonella infections in man and animals. Different attenuating mutations have been described, and more than one can be incorporated in a vaccine for added safety. Combined salmonella vaccines express target carbohydrate and protein antigens or epitopes from viruses, bacteria and eukaryotic parasites, either within or on the surface of the cell, as capsules, fimbriae, or in the flagellin. Humoral, secretory and cellular responses to the recombinant antigens have been demonstrated. Experimental protection against diseases including streptococcal infection, tetanus, influenza and malaria has been obtained, and such hybrid strains may offer important advances in immunization against infectious diseases.

INTRODUCTION

Vaccination continues to be a very cost-effective means of prevention of disease (Robbins 1990). The successes obtained with conventional vaccines offer great hope that new improved preparations may offer a means of controlling other forms of infective disease. In particular, the great advance in recent years has been in the development of novel live attenuated vaccines which can, in some cases, confer much better protection than conventional killed vaccines. Whereas this superior efficacy of live vaccines has been known for many years, modern recombinant DNA technology now offers the possibility of introducing defined mutations into the genome, thus attenuating pathogenic organisms in a

The Release of Genetically Modified Microorganisms
Edited by D.E.S. Stewart-Tull and M. Sussman, Plenum Press, New York, 1992

rational way, creating safe, non-reverting vaccines which can be produced cheaply. Many of these novel live vaccines can be administered orally, eliminating the risk of needle-stick contamination. A further major advance has been the use of live attenuated vaccines as combined vaccines carrying recombinant antigens from a variety of pathogens (Dougan, Smith and Heffron, 1988; Levine et al. 1983; Hone and Hackett, 1989; Hackett, 1990; Chatfield, Strugnell and Dougan, 1989), opening the possibility of developing not only improved delivery systems for existing vaccines, but also a means of mass immunization against diseases for which vaccine development has so far proved impracticable. A major case in point are diseases caused by protozoan and metazoan parasites (Blackwell and Miller, 1990; Murray, 1989). Recombinant DNA technology is providing an increasing number of candidate target immunogens for a variety of parasites (Playfair et al.. 1990), for which salmonellae could provide an effective delivery system.

IMMUNITY TO SALMONELLA

The mechanisms of immunity to invasive salmonellosis (e.g. typhoid fever and its laboratory analogue, mouse typhoid) involve both humoral and cell - mediated immunity (CMI). Salmonellae are generally believed to be facultative intracellular pathogens (although this has recently been questioned, (Hsu 1989) which can survive inside macrophages of the reticuloendothelial system (RES). In lethal infections salmonellae grow extensively in the RES, with death probably due to endotoxin poisoning. In sublethal infections, salmonellae only reach moderate levels in the tissues, further growth being halted by a first T cell independent host response in which TNFα is an important mediator (Mastroeni et al. 1991; Tite, Dougan and Chatfield 1991). Salmonellae are eventually cleared from the tissues by a host response requiring CD4+ T cells (Nauciel, 1990).

Post-infection immunity in experimental animals is solid (Collins 1974) and second attacks of typhoid fever are uncommon (although they have been documented; Marmion, Naylor and Stewart, 1953). Attempts to recreate this immunity with killed vaccines have met with limited success; in experimental animals, recovery from infection or immunisation with attenuated salmonellae provides very much stronger protection than does vaccination with killed organisms. This difference is believed to be due to the fact that whereas killed vaccines can only elicit a humoral immune response, live salmonella vaccines can in addition induce potent CMI. Passive transfer of immune serum confers much less protection than does vaccination with live organisms, which is of high level and long-lasting (Collins 1974; Eisenstein, Killar and Sultzer 1984; Hormaeche 1991).

The ability to elicit a CMI response is believed to be central to the protection conferred by live salmonella vaccines, but the nature of this response is still very unclear. The identity of the protective immunogen(s) is in doubt, and it is not known why only live vaccines will elicit CMI. Delayed hypersensitivity to crude salmonella extracts does not correlate well with protection (Hormaeche et al. 1981; Killar and Eisenstein, 1986). Other areas of uncertainty are the reason for the lack of effectiveness of some live vaccines (Smith et al. 1984), and why it is that whereas vaccine - induced protection can be adoptively transferred with macrophages from immune animals, it has proved so difficult to transfer protection to virulent challenge with T cells (Maskell et al., 1987a; Eisenstein and Sultzer, 1983).

SALMONELLA VACCINES

Vaccination with killed salmonellae has been practised for many years. Killed typhoid vaccines confer significant but limited protection to humans; this low efficacy, together with frequently encountered unpleasant side reactions, has prompted the search for improved preparations. Vaccination against typhoid fever with a purified preparation of the virulence - associated outer surface Vi antigen can confer significant protection, with reduced reactogenicity (Robbins et al., 1988). However, the proven efficacy of live vaccines in experimental systems has prompted a search for improved live attenuated preparations for human and veterinary use. As the agent of human typhoid fever, S.typhi, will not cause invasive disease in mice unless injected intraperitoneally in mucin, much work has been done on mouse typhoid with S.typhimurium and S.enteritidis, which cause invasive disease believed to have many points in common with human enteric fevers.

The first live salmonella vaccines were prepared by traditional methods such as chemical or U.V. mutagenesis. These strains had an impaired capacity for replication in normal culture conditions *in vitro* which would also show a reduction in virulence. However, the undefined nature of these vaccines made them difficult to subject to quality control and has also raised fears that they may revert to virulence. Salmonellae with an impaired ability to grow at 37°C are attenuated, and they can confer protection (Fahey and Cooper, 1970). However, in a study of the persistence of temperature-sensitive (TS) salmonellae in mice, we observed that TS mutants of virulent salmonellae would grow subcutaneously and cause local necrotic lesions, and could cause septic arthritis in mice injected intravenously (Hormaeche *et al.*, 1981). This suggests that at least some TS mutants of virulent strains can grow in the tissues and cause lesions in areas of reduced body temperature, and casts doubts as to their safety as vaccines for general use.

Streptomycin dependency was found to attenuate virulence, and a streptomycin dependent *S.typhi* mutant was developed for human immunization. The mutant protected human volunteers from challenge with virulent *S.typhi* (Reitman, 1967; Levine *et al.*, 1976). Unfortunately, the effectiveness of the strain was found to be seriously affected by lyophilization, and its development was discontinued.

Modern approaches to rational strain attenuation are aimed at the inactivation of defined loci rather than the multiple undefined lesions introduced by chemical mutagenesis. This approach has led to the identification of several loci which will attenuate to different levels, sufficient for strains to be used as vaccines. Some of these are not in loci coding for specific virulence genes which determine the difference in invasiveness of salmonellae vs. e.g. *Escherichia coli,* but in housekeeping genes whose inactivation causes impaired resistance to host defences and/or impaired ability to multiply in the tissues.

In a search for mutations which would yield effective vaccines in mice, Germanier and Fürer (1971) found that *galE* mutants were effective vaccines in mice. The gene encodes uridine 5'-diphosphate-(UDP)-glucose-4-epimerase, which is responsible for normal synthesis of UDP-galactose from UDP glucose. UDP galactose is normally required for the incorporation of galactose into the core LPS. Mutants show defective LPS and are phenotypically rough. Their reduced virulence is believed to be due to increased serum sensitivity and inability to resist cellular defences. Ty21a is an NTG induced *galE* mutant of *S.typhi* which proved to confer significant protection in human volunteers. It has been tested extensively in the field, and is innocuous; however, the protection it confers is not complete, and protection is best following more than one dose (Germanier and Furer, 1975; Gilman *et al.,* 1977; Levine *et al.*, 1983; Hone and Hackett, 1989; Hackett, 1990; Black *et al.*, 1990).

However, the value of developing further *galE* mutants for human use has been recently brought seriously into question. It is now clear that the attenuation of the Ty21a strain may not be related to its *galE* lesion at all, but is probably due to another unspecified mutation(s) caused by the two rounds of NTG treatment used in its preparation. Gal^+ revertants of Ty21a are not virulent (Silva-Salinas *et al.*, 1985). More to the point, a strain of *S.typhi* with a precise *galE* lesion and also lacking the Vi antigen was found to retain virulence for human volunteers (Hone *et al.*, 1988). This contrasts with the finding by the same group that similar precisely engineered mutants of *S.typhimurium* were avirulent and protective in mice (Hone *et al.*, 1987). Some *galE* mutants of *Salmonella choleraesuis* are also virulent for mice (Nnalue and Stocker, 1986). The existing Ty21a oral typhoid vaccine is nonetheless innocuous, continues to be tested in the field, and has been used as the basis for development of combined vaccines for human use (see below). It may also be possible to improve its efficacy by restoring the missing Vi antigen (Crysz, 1989).

There is a search for further improved live salmonella vaccines. Early investigations into the effect of auxotrophy on the virulence of *S.typhi* in mice (Bacon, Burrows and Yates, 1951) showed that mutants defective in para-aminobenzoic acid (PABA) or purine biosynthesis were avirulent in mice. This work was followed up by Hoiseth and Stocker (1981) who demonstrated that PABA requirement resulting from stable mutations in gene *aroA* were also attenuating, and protected against challenge with the virulent parent strain. The *aroA* gene is located at 19 min on the salmonella genetic map (Sanderson and Roth, 1988) and encodes 5-enolpyruvylshikimate-3-phosphate synthetase, the penultimate enzyme of the early common prechorismate pathway leading to the synthesis of chorismate and the aromatic amino acids, PABA, and 2,4-dihydroxybenzoate. Aro mutants are dependent on these compounds for growth *in vitro*. Mammals do not possess this pathway, and it is probable that it is the limited availability of PABA *in vivo* which is responsible for the attenuation of *aro* mutants, since mutations in *pabB* (encoding the p-aminobenzoate synthase large subunit) are as attenuated as *aroA* mutations (Stocker 1990). A possible mechanism for the low virulence of *aroA* mutants could be the lack in the host of fMet-tRNA$^{met}_f$, required as the precursor the formyl- methionine residue necessary for protein synthesis (Stocker 1990). *Aro* mutants grow very slowly in the tissues and are markedly attenuated; we have been unable to give a lethal dose to mice by the oral route. They have been found to be effective in mice (Hoiseth and Stocker, 1981; Stocker, 1988), sheep (Mukkur *et al.*, 1987), cattle (Smith *et al.*, 1984; Jones *et al.*, 1991) and chickens (Cooper *et al.*, 1990, Barrow *et al.*, 1990). Aro *S.typhi* mutants have been constructed as candidate human live oral typhoid vaccines, and they are under evaluation in humans (Chatfield *et al.* 1991*a*; Dougan *et al.*, 1991; Tacket *et al.*, 1992a, 1992b).

It is highly likely that it is the PABA requirement which is responsible for the attenuation of *aro* mutants. This unique mechanism of attenuation could confer on the strain a limited invasive potential even in the face of lowered host defences. This would be a desirable safety consideration if these strains are ever to be used in the field in areas where immunodeficiency is likely to be encountered, e.g. due to AIDS. We have recently reported that *aro* mutants do not display markedly increased invasiveness in conditions which lead to increased invasiveness of wild type salmonellae, i.e. in mice given a dose of sublethal irradiation, or in male (CBA/NXBALB/c)F1 mice displaying the sex-linked *xid* agammaglobulinemia (Izhar *et al.*, 1990). We have also shown that administration of anti-TNFα antiserum will enhance the virulence of wild type salmonellae, but not that of an *aro* mutant (Tite, Dougan and Chatfield, 1991). *AroA* salmonellae are also attenuated to the same level in salmonella susceptible, endotoxin - unresponsive C3H/HeJ mice carrying the *lps*^{d} defect (Killar and Eisentein 1986), in mice pretreated with silica (an antimacrophage agent) or cyclophosphamide, and in the immunodeficient *scid* mice (Stocker 1990).

Live vaccines would have a greatly reduced probability of reversion to virulence if they harboured more than one attenuating lesion, especially if they mapped at widely differing points of the chromosome. Mutations at *purA* (causing a requirement for adenine or adenosine) are markedly attenuating (Macfarland and Stocker 1989), and Stocker constructed strains of *S.typhi* harboring *aroA* and *purA* mutations; these were innocuous in human volunteers (Levine *et al.*, 1987, 1990). However, the attenuation caused by *purA* may be too great for optimum immunogenicity (Sigwart *et al.*, 1989; O'Callaghan *et al.*, 1988a, 1990*a*). We have found that *purA* mutants had reduced persistence in the tissues as compared to *aroA* mutants, and showed reduced priming for an anti-salmonella T-cell response and lower macrophage activation (O'Callaghan *et al.* 1990*a*). We therefore investigated the possibility of introducing multiple mutations in the aromatic pathway (Miller *et al.*, 1989; Dougan *et al.*, 1988; Fairweather *et al.*, 1990*a*).

Mutations at *aroD* (5-dehydroquinate dehydratase, (Miller *et al.* 1989) and *aroC* (chorismate synthetase; (Dougan *et al.* 1988) are also attenuating, and strains harboring different combinations of *aroA*, *aroC* and *aroD* mutations were constructed. Single and double mutants behaved in a similar manner in terms of attenuation and *in vivo* persistence and all were excellent oral vaccines. Strains of *S.typhi* harboring mutations in more than one gene of the aromatic pathway are currently under evaluation in humans, with encouraging results (Dougan *et al.* 1991)

We have recently shown that *S.typhimurium* strains harboring stable mutations in *ompR*, a positive regulator of porin expression, are attenuated both orally and parenterally in mice (Dorman *et al.* 1989; Chatfield *et al.* 1991*b*). Orally-immunised mice are also well protected against challenge with the virulent parent strain. *OmpR* forms a two gene operon with *envZ*. It has been proposed that *envZ*, an inner membrane protein (Forst *et al.* 1987), acts as an environmental sensor (in this case responding to osmolarity) and transmits signals to *ompR*. The OmpR protein then modulates transcription of various genes including those encoding the major porins ompC and ompF.

Lesions in genes encoding adenylate cyclase and cAMP receptor protein (*cya crp* mutants, (Curtiss *et al.*, 1988) affect carbohydrate and aminoacid metabolism; the mutants are effective vaccines. *PhoP* mutants have impaired resistance to phagocyte bactericidal mechanisms, and show reduced survival within host macrophages, they are effective vaccines (Groisman and Saier, 1990; Miller, Mekalanos and Pulkkinen, 1990). We have recently reported that salmonellae harboring lesions in the salmonella analog of *htrA*, an *E.coli* heat stress protein gene, are avirulent and immunogenic (Johnson *et al.*, 1990, 1991, Chatfield *et al.* 1991c).

The new vaccines described above are all derived from mutations in what may be considered housekeeping genes, many of which are complemented by their analogs from *E.coli*. It might perhaps be desirable to devise methods of strain attenuation based on the inactivation of specific virulence loci unique to salmonellae. The advances being made in research on the genetic determinants of virulence in salmonellae will make it possible to investigate the feasibility of this approach (Dougan 1989; Curtiss *et al.*, 1988; Barrow, 1990). Approximately 60 to 120 genes are required for virulence by the intraperitoneal route, and 200 to 500 by the oral route (F. Heffron; cited in Curtiss, 1990).

EXPRESSION OF FOREIGN ANTIGENS IN SALMONELLAE

To be effective, combined salmonella vaccines must express the recombinant antigen, presumably in undegraded form, at a level sufficient to induce an immune response. The great similarities in the cellular and molecular biology between *Salmonella* and *E.coli,* the commonly used host for expression of recombinant proteins, greatly facilitates this task as many of the genetic manipulations involved can be applied easily to both species. A manipulated DNA sequence which expresses in *E.coli* will normally also express in *Salmonella*, making it possible to clone and characterise a potentially useful antigen in *E.coli* before transferring it into salmonella vaccine strains for evaluation.

Foreign DNA can be introduced into salmonellae by transformation, conjugation, transduction or electroporation. Most live vaccines currently under investigation (except *galE* strains like Ty21a) are fully smooth, and a major concern is to use procedures that do not select for O-rough derivatives with defective LPS, which would be unsuitable as vaccines. *E.coli* derived DNA is also subject to restriction when transferred into *Salmonella*. These problems can be overcome by electroporating into the vaccine strain, either directly or via an intermediate salmonella host such as LB5010, a *galE S.typhimurium* which is r^-m^+ for all three host restriction-modification systems. These strains are easily transformable with *E.coli* DNA, and will also allow replication of the O-specific transducing phage P22 which will yield smooth transductants (Brown *et al.* 1987; Chatfield *et al.* 1991a).

EXAMPLES OF MULTIVALENT VACCINE STRAINS

Results from several groups have shown that salmonellae can be expected to induce humoral, local secretory and cell-mediated immune responses to recombinant foreign antigens from viruses, bacteria and parasites (Brown *et al.*, 1987; Hackett, 1990; Curtiss *et al.*, 1989; Dougan, Smith and Heffron, 1988; Stocker 1990). One of the early examples of construction of a hybrid strain was the introduction of K88, the fimbrial adhesin from porcine enteropathogenic *E.coli* strains, into a virulent strain of *Salmonella*, resulting in modulation of its virulence (Smith and Linggood, 1971). K88 has since been extensively used in carrier studies, facilitated by the fact that K88 is normally expressed on plasmids. A *galE* strain of *S.typhimurium* was used by Stevenson and Manning (1985) as a combined oral vaccine, eliciting circulating and intestinal antibody to K88 in mice. More recently Dougan *et al.* (1986) introduced K88 into an *aroA S.typhimurium* and demonstrated its presence on the bacterial surface. The strain elicited serum antibodies to K88 in mice, and the construct protected mice from virulent salmonellae - an important consideration when contemplating the use of these constructs in the field.

LT-B, the immunodominant, non-toxic subunit of heat-labile enterotoxin from *E.coli,* has also been used in combined vaccines. It has been expressed in *aroA* strains of both *S.typhimurium* and *S.enteritidis*. These strains elicited high levels of antibodies to LT-B in the serum and also as secretory IgA in the gut (Clements *et al.*, 1988; Maskell *et al.*, 1987); the antibodies neutralized the LT holotoxin, and the mice were protected from challenge with virulent salmonellae (Maskell *et al.*, 1987). LT-B has also been expressed in the *galE* Ty21a human typhoid vaccine strain; such strains could be the basis of potential bivalent combined vaccines for typhoid and *E.coli*-related diarrheas (Clements and El-Morshidy, 1984).

Other potential combined vaccines for diarrheal diseases have been constructed with Ty21a. The *Shigella sonnei* form 1 polysaccharide antigen was expressed by introduction of the 120 MDa shigella plasmid (Formal *et al.* 1981). The strain also expressed the salmonella O-9 and O-12 LPS antigens, and induced antibodies to both salmonella and shigella antigens in mice; variable results were obtained when tested in human volunteers (Black *et al.*, 1987, Herrington *et al.*, 1990). A Ty21a strain expressing *Vibrio cholerae* antigens also conferred partial protection from cholera (Tacket *et al.*, 1990). Type and group O antigens from *Shigella flexneri* have also been expressed in Ty21a (Baron *et al.*, 1987).

Antigens from less closely related pathogens have also been expressed in salmonellae. *S.typhimurium cya crp* mutants lacking the virulence plasmid expressing the *Streptococcus sobrinus* SpaA colonization protein have been constructed as a possible dental caries vaccine. Mice fed this strain developed good DTH to the streptococcal antigen but the secretory response was not high, although it was increased by boosting (Curtiss *et al.*, 1988). Oral vaccination with an *aroA S.typhimurium* expressing the *Streptococcus pyogenes* M protein protected mice from parenteral challenge with streptococci (Poirier, Kehoe and Beachey, 1988; Stocker, 1990). Other combined vaccines have induced systemic and local secretory responses to the *Bordetella pertussis* filamentous haemagglutinin (Molina and Parker, 1990), *Brucella abortus* outer membrane proteins (Stabel *et al.*, 1990), and meningococcal group B outer membrane proteins (Tarkka *et al.*, 1989). Antigens from *Mycobacterium leprae* have also been expressed in *cya crp asd* salmonella vaccine strains (Clark-Curtiss, 1990). The immunogenicity of salmonellae expressing the 69 kDa protein of *Bordetella pertussis*, a candidate vaccine antigen, is under investigation (Fairweather *et al.*, 1990*a*)

Hepatitis B virus hybrid nucleocapsid antigen/pre-S2 particles have been stably expressed in *aroA* salmonella vaccines; the construct is immunogenic in mice following a single dose (Schödel, Millich and Will, 1990; Schödel and Will, 1990). We have expressed the simian immunodeficiency virus P27 *gag* antigen (Strahan, Kitchin and Hormaeche, 1992) and *Schistosoma mansoni* (Taylor *et al.*, 1986) antigens in salmonellae. Herpes simplex type 1 virus glycoprotein D (HSV GD) antigens have been expressed in *aroA* salmonellae (Bowen *et al.*, 1990; M. Izhar and C.E. Hormaeche, unpublished data). Mice immunized with *aro* salmonellae expressing HSV GD antigens, either as the whole protein intracellularly or as an epitope in the salmonella flagellin, produced humoral antibody to the HSV GD antigen (M. Izhar and C.E. Hormaeche, to be published).

Stocker and his colleagues have devised an elegant system for expressing foreign epitopes as part of the salmonella flagellin; epitopes from cholera enterotoxin B subunit, hepatitis B virus surface protein, *Streptococcus pyogenes* type-5 M protein, HIV surface protein gp160, and murine cytomegalovirus immediate-early protein p89 have been expressed with this system; all except CMV (not tested) were immunogenic in mice; the construct expressing the streptococcal epitope protected mice from challenge with streptococci (Newton, Jacob and Stocker, 1989; Wu *et al.*, 1989; Stocker, 1990). More recently, an N-terminal 15 amino acid epitope of M protein has been inserted into salmonella flagellin (Newton *et al.* 1991) Mice were protected against a *Streptococcus pyogenes* type 5 challenge by prior immunization with a salmonella strain expressing the epitope. The strain has the advantage that it does not carry the cross reactive epitopes of the whole protein.

Mice immunized with a salmonella vaccine expressing a membrane protein of *Francisella tularensis* showed a reduction in the bacterial load in the tissues after challenge with virulent organisms (Sjösted, Sandström and Tärnvik, 1990). We have constructed a hybrid strain of *S.typhimurium aroA* expressing the C fragment of tetanus toxin in soluble form. C fragment is the 50 kDa carboxy terminal fragment of the toxin which has previously been shown to protect mice against challenge with tetanus toxin. This hybrid strain protected mice from challenge with tetanus holotoxin when given parenterally and importantly also when given orally (Fairweather *et al.* 1990*b*). We have recently constructed a *S.typhi* Ty2 *aroA aroC* expressing tetanus toxin fragment C (Chatfield *et al.* 1991*a*). We consider this strain to be a candidate bivalent typhoid/tetanus vaccine ready for preliminary tests of its safety and immunogenicity in human clinical trials.

Vaccination with salmonellae expressing recombinant influenza A virus nucleoprotein protected mice from challenge with the virus; the mice developed virus-specific CD4+ cytotoxic T cells; spleen cells challenged with virus *in vitro* proliferate, producing IFN-γ and IL-2 (Tite *et al.*, 1990).

Sadoff and his colleagues have protected mice from challenge with malarial parasites following immunization with attenuated salmonellae expressing the circumsporozoite antigen of *Plasmodium berghei;* the mice develop cytotoxic CD8 T cells, as do mice vaccinated with salmonellae expressing the *Plasmodium falciparum* CS antigen (Sadoff *et al.*, 1988, Aggarwal *et al.*, 1990).

PROBLEMS IN VACCINE DEVELOPMENT

Expression of the recombinant antigen stably in undegraded form in a salmonella vaccine strain can pose problems (reviewed by Charles and Dougan,1990). Many workers have expressed the foreign antigen from conventional low-copy number vectors commonly used in *E.coli,* resulting in expression in the cytoplasm. Mice immunized with salmonellae (and human typhoid patients) develop antibodies to both intracellular and outer membrane proteins, and the response to the latter is particularly strong (Brown and Hormaeche, 1989). It is not clear whether this strong antibody response is due to the surface location of these proteins, the amount on the cell, or their intrinsic immunogenicity. An attenuated *Salmonella* expressing *E.coli* beta-galactosidase will elicit both humoral and cellular immunity to the beta-galactosidase (Brown *et al.,* 1987; Brown and Hormaeche, 1989). It may be that expression on the cell surface will be required for some antigens, and systems for expression of foreign antigens in outer membrane proteins such as LamB (Leclerc *et al.,* 1989; O'Callaghan *et al.,* 1990*b*) and PhoE [Agteberg *et al.,* 1990] have been described. The *E.coli* K1 polysaccharide has been expressed on *aroA* salmonellae as a surface capsule (O'Callaghan *et al.,* 1988b)

High level unregulated expression of a recombinant protein from a high copy number vector may prove deleterious to the salmonella vaccine strain and cause loss of the plasmid vector in the absence of antibiotic selective pressure. The use of antibiotics in vaccine strains may be undesirable, and several alternative methods have been devised to overcome this problem. Expression vectors containing genes essential to the carrier organism can reduce plasmid instability (Nakayama, Kelly and Curtiss, 1988). Another strategy is to incorporate the foreign gene into the salmonella chromosome, eliminating the need for a plasmid vector and reducing the gene copy number to 1 which may in some cases reduce instability of the foreign DNA. Different systems have been described which can incorporate the foreign gene into the salmonella chromosome by homologous recombination, by means of cloning vectors with flanking regions of homology to host genes such as *his* (Hone *et al.,* 1988; Clements and Cárdenas, 1990) or *aroC;* (Strugnell *et al.,* 1990); with the latter system, integration into the chromosome in itself creates an attenuating lesion. Another recently described expression system uses an invertible DNA sequence causing spontaneous segregation of cells with very high levels of expression of the recombinant antigen; the system induced antibodies to cholera toxin CT-B (Yan, Reuss and Meyer, 1990).

Regulation of the level of expression is an important consideration. Ideally, the gene would be maximally expressed once the salmonella has reached its final destination inside macrophages. Heffron and his colleagues have recently shown that salmonella stress proteins are expressed to a high level during growth inside macrophages (Fields, Groisman and Heffron, 1989; Buchmeier and Heffron, 1990); the promoters of these or other *in vivo* inducible proteins could be useful for construction of *in vivo* expression systems.

REFERENCES

Aggarwal, A., Kumar, S., Jaffe, R., Hone, D., Gross, M., Sadoff, J (1990) Oral salmonella:malaria circumsporozoite recombinants induce specific CD8+ cytotoxic T cells. J. Exp. Med. 172:1083-1090.

Agteberg, M., Adriaanse, H., Lankhof, H., Meloen, R., and Tommassen, J. (1990) Outer membrane protein PhoE protein of *Escherichia coli* as a carrier for foreign antigenic determinants: immunogenicity of epitopes of foot - and - mouth disease virus. Vaccine 8:85-91.

Baron, L.S., Kopecko, D.J., Formal, S.J., Seid, R., Guerry, P., and Powell, C. (1987) Introduction of *Shigella flexneri* 2a type and group antigens genes into oral typhoid vaccine strain *Salmonella typhi* Ty21a. Infect. Immun. 55:2797-2801.

Barrow, P.A. (1990) Immunity to experimental fowl typhoid in chickens induced by a virulence plasmid cured derivative of *Salmonella gallinarum*. Infect. Immun. 58:2283-2288.

Barrow, P.A., Hassan, J.O., Lovell, M.A., and Berchieri, A. (1990) Vaccination of chickens with *aroA* and other mutants of *Salmonella typhimurium* and *S.enteritidis*. Res. Microbiol. 141:851-853.

Black, R.E., Levine, M.M., Clements, M.L., Losonsky, G., Herrington, D., Berman, S., and Formal, S.B. (1987) Prevention of shigellosis by a *Salmonella typhi - Shigella sonnei* bivalent vaccine. J. Infect. Dis. 155: 1260-1265.

Black, R.E., Levine, M.M., Ferreccio, C., Clements, M.L., Lanata, C., Rooney, J., Germanier, R., and the Chilean Typhoid Committee (1990) Effect of one or two doses of Ty21a *Salmonella typhi* vaccine in enteric-coated capsules in a controlled field trial. Vaccine 8:81-84.

Bowen, J.C., Alpar, O., Philpotts, O., Roberts, I.S., and Brown, M.R.W. (1990) Preliminary studies on infection by attenuated *Salmonella* in guinea pigs and on expression of herpes simplex virus. Res. Microbiol. 141: 873- 877.

Brown, A., and Hormaeche, C.E. (1989) The antibody response to salmonellae in mice and humans studied by immunoblots and ELISA. Microbial Pathogenesis, 6:445-54.

Brown, A., Hormaeche, C.E., Demarco de Hormaeche, R.A., Dougan, G., Winther, M., Maskell, D., and Stocker, B.A.D. (1987) An attenuated *aroA S.typhimurium* vaccine elicits humoral and cellular immunity to cloned beta galactosidase in mice. J Infect Dis, 155:86-92.

Buchmeier, N., and Heffron, F. (1990) Induction of salmonella stress proteins upon infection of macrophages. Science 248:730-732.

Charles, I., and Dougan, G. (1990) Gene expression and the development of live enteric vaccines. TIBTECH 8:117-121.

Chatfield, S.N., Strugnell, R.A., and Dougan, G. (1989) Live salmonellae as vaccines and carriers of foreign antigenic determinants. Vaccine 7:495-498

Chatfield, S.N., Fairweather, N., Charles, I., Pickard, D., Levine, M., Hone, D., Posada, M., Strugnell, R.A., and Dougan, G. (1991a) Construction of a genetically defined *Salmonella typhi* Ty2 *aroA aroC* mutant for the engineering of a candidate typhoid - tetanus vaccine. Vaccine, in press.

Chatfield, S.N., Dorman, C.J., Hayward, C., and Dougan, G.D. (1991b) Role of *ompR* dependent genes in *Salmonella typhimurium* virulence: mutants deficient in both OmpC and OmpF are attenuated in vivo. Infect. Immun. 59:449-452.

Chatfield, S.N., Strahan, K., Pickard, D., Charles, I.G., Hormaeche, C.E., and Dougan, G. (1991c) Evaluation of *Salmonella typhimurium* strains harbouring defined mutations in *htrA* and *aroA* in the murine salmonellosis model. Microbiol. Pathogenesis, in press

Clark-Curtiss, J.E., Thole, J.E.R., Sathish, M., Bosecker, B.A., Sela, S., de Carvalho, E.F., and Esser, R.E. (1990) Protein antigens of *Mycobacterium leprae*. Res. Microbiol. 141:859-871.

Clements, J.D., and El-Morshidy, S. (1984) Construction of a potential live oral bivalent vaccine for typhoid fever and cholera-*Escherichia coli* related diarrhoea. Infect Immun 46:564-569.

Clements, J.D., Lyon, F.L., Lowe, K.L., Farrand, A.L., and El-Morshidy, S. (1988) Oral immunization of mice with attenuated *Salmonella enteritidis* containing a recombinant plasmid which codes for production of the B subunit of heat labile enterotoxin of *Escherichia coli*. Infect Immun 53:685-692.

Clements, J.D., and Cárdenas, L. (1990) Vaccines against enterotoxigenic bacterial pathogens based on hybrid *Salmonella* that express heterologous antigens. Res. Microbiol. 141:981-993.

Collins, F.M. (1974) Vaccines and cell-mediated immunity. Bact. Revs. 38:371-402

Cooper, G.A., Nicholas, R.A.J., Cullen, G.A., and Hormaeche, C.E. (1990) Vaccination of chickens with an *S.enteritidis aroA* live oral salmonella vaccine. Microbial Pathogenesis 9:255-265.

Crysz Jr., S.J., Furer, E., Baron, L.S., Noon, K.F., Rubin, F.A., and Kopecko, D.J. (1989) Construction and characterization of a Vi-positive variant of the *Salmonella typhi* live oral vaccine strain Ty21a. Infect. Immun. 57:3863- 3868.

Curtiss III, R. (1990) Antigen delivery systems for analysing host immune responses and for vaccine development. TIBTECH 8:237-240.

Curtiss III R., Goldschmidt R.M., Fletchall N.B. and S.M. Kelly (1988) Avirulent *Salmonella typhimurium* Δcya Δcrp oral vaccine strains expressing a streptococcal colonization and virulence antigen. Vaccine 6:155-160.

Curtiss III, R., Kelly, S.M., Gulig, P.A., and Nakayama, K. (1989) Selective delivery of antigens by recombinant bacteria. Curr. Top. Microbiol. Immunol. 146:35-49.

Dorman, C.J., Chatfield, S., Higgins, C.R.F., Hayward, C., and Dougan, G. (1989) Characterization of porin and *ompR* mutants of *Salmonella typhimurium: ompR* mutants are attenuated *in vivo*. Infect. Immun. 57: 2136-2140.

Dougan, G. (1989) Molecular characterization of bacterial virulence factors and the consequences for vaccine design. J. Gen Microbiol. 135:1397-1406.

Dougan, G., Sellwood, R., Maskell, D., Sweeney, K., Beesley, J., Hormaeche, C.E. (1986) *In vivo* properties of a cloned K88 adherence antigen determinant Infect. Immun. 52:344-347.

Dougan, G., Maskell, D., Pickard, D., Hormaeche, C.E. (1987) Isolation of stable *aroA* mutants of *Salmonella typhi* Ty2: Properties and preliminary characterization in mice. Mol. Gen. Genet. 207:402-405.

Dougan, G., Smith L., and Heffron, F. (1988) Live bacterial vaccines and their application as carriers for foreign antigens. In: Advances in Veterinary Science and Comparative Medicine, (Ed. Bittle, L.J.) Academic, Press, vol 33, p 271-300.

Dougan, G., Chatfield, S., Pickard, D., Bester, J., O'Callaghan, D., and Maskell, D. (1988) Construction and characterization of vaccine strains of *Salmonella* harbouring mutations in two different *aro* genes. J. Infect. Dis. 158:1329-35.

Dougan, G., Hofnung, M., Lanzavecchia, A., and Leclerc, C. (1991) Immune response to proteins with recombinant epitopes, perspectives for vaccines. In: Conferences Philippe Laudat 1990, INSERM, Paris 1991; pp171-220.

Eisenstein, T.K., and Sultzer, B.M. (1983) Immunity to *Salmonella* infections. Adv. Exp. Med. Biol. 162:261-296.

Eisenstein, T.K., Killar, L.M., and Sultzer, B.M. (1984) Immunity and infection in *Salmonella typhimurium*: mouse-strain differences in vaccine and serum-mediated protection. J. Infect. Dis. 150:425-435.

Fahey, K.J., and Cooper, G.N. (1970) Oral immunization against experimental salmonellosis. Infect. Immun. 1:263-270.

Fairweather, N.F., Chatfield, S.N., Charles, I.G., Roberts, M., Lipscombe, M., Jing Li, L., Strugnell, D., Comerford, S., Tite, J., and Dougan, G. (1990*a*) Use of live attenuated bacteria to stimulate immunity. Res. Microbiol. 141: 769-773.

Fairweather, N.F., Chatfield, S.N., Makoff, A., Strugnell, R., Bester, J., Maskell, D.J., and Dougan, G. (1990*b*) Oral vaccination of mice against tetanus by use of a live attenuated *Salmonella* carrier. Infect. Immun. 58:1323-1326.

Fields, P.I., Groisman, E.A., and Heffron, F.A. (1989) A salmonella locus that controls resistance to microbicidal proteins from phagocytic cells. Science 243:1059-1062.

Formal, S.B., Baron, L.S., Kopecko, D.J., Washington, O., Powell, C., and Life, C.A. (1981) Construction of a potential bivalent vaccine strain: introduction of *Shigella sonnei* form I antigen genes into into the *galE Salmonella typhi* Ty21a typhoid vaccine strain. Infect. Immun. 34:746-750.

Forst, S., Conneau, D., Norioka, S., and Inouye, M. (1987) Localization and membrane topology of EnvZ, a protein involved in osmoregulation of OmpF and OmpC in *Escherichia coli*. J. Biol. Chem. 262:16433-16438.

Germanier, R., and Furer, E. (1971) Immunity in experimental salmonellosis. II. Basis of the avirulence and protective capacity of *galE* mutants of *Salmonella typhimurium*. Infect. Immun. 4:663-673.

Germanier, R., and Furer, E. (1975) Isolation and characterization of *galE* mutant Ty21a of *Salmonella typhi*: a candidate strain for a live oral typhoid vaccine. J. Infect Dis. 131:553-558

Gilman, R.H., Hornick, R.B., Woodward, W.E., DuPont, H.L., Snyder, M.J., Levine, M.M., and Libonati, J.P. (1977) Evaluation of a UDP galactose-4-epimeraseless mutant of *Salmonella typhi* as a live oral vaccine. J. Infect. Dis.136:717-723.

Groisman, E.A., and Saier, M.H. (1990) *Salmonella* virulence: new clues to intramacrophage survival. TIBS 15:30-33.

Hackett, J. (1990) Salmonella based vaccines. Vaccine 8:5-11.

Heltvig,T.B., and Nau, H.H. (1984) Analysis of the immune response to papain digestion products of tetanus toxin. Acta Pathol. Microbiol. Scand. Sect. C, 92:59-63.

Herrington, D.A., van de Verg, L., Formal, S.B., Hale, T.L., Tall, B.D., Crysz, S.J., Tramont, E.C., and Levine, M.M. (1990) Studies in volunteers to evaluate candidate *Shigella* vaccines: further experience with a bivalent *Salmonella typhi - Shigella sonnei* vaccine and protection conferred by previous *Shigella sonnei* disease. Vaccine 8:353-357.

Hoiseth, S.K., and Stocker, B.A.D. (1981) Aromatic dependent *Salmonella typhimurium* are non-virulent and effective as live vaccines. Nature 291:238-239.

Hone, D., and Hackett, J. (1989) Vaccination against enteric bacterial diseases. J. Infect. Dis. 11:853-877.

Hone, D., Morona, R., Attridge, S., and Hackett, J. (1987) Construction of defined *galE* mutants of *Salmonella* for use as vaccines. J. Infect Dis. 156:167-174.

Hone, D., Attridge, S., van den Bosch, L., and Hackett, J. (1988) A chromosomal integration system for stabilization of heterologous genes in *Salmonella* based vaccine strains. Microbial Pathogenesis 5:407-418.

Hone, D.M., Attridge, S.R., Forrest, B., Morona, R., Daniels, D., LaBrooy, J.T., Bartholomeusz, RCA, Shearman, D.J.C., and Hackett, J. (1988) A *galE* (Vi antigen negative) mutant of *Salmonella typhimurium* Ty2 retains virulence in humans.Infect. Immun. 56:1326-1333.

Hormaeche, C.E., Pettifor, R., and Brock, R.J. (1981) The fate of temperature sensitive salmonella mutants *in vivo* in naturally resistant and susceptible mice. Immunology 42:569-576.

Hormaeche, C.E., Fahrenkrog, M.C., Pettifor, R.A., and Brock, J. (1981) Acquired immunity to *Salmonella typhimurium* and delayed (footpad) hypersensitivity in BALB/c mice. Immunology 43:547-554.

Hormaeche, C.E., Joysey, H.S., Desilva, L., Izhar, M, and Stocker, B.A.D. (1991) Immunity conferred by Aro⁻ salmonella live vaccines. Microbial Pathogenesis, 10:149-158;1991.

Hsu, H.S. (1989) Pathogenesis and immunity in murine salmonellosis. Microb. Revs. 53:390-409.

Izhar, M,. Desilva, L., Joysey, H.S., and Hormaeche, C.E. (1990) Moderate immune suppression does not increase susceptibility to *aroA* salmonella vaccine strains. Infect Immun 58:2258-2261.

Johnson, K.S., Charles, I.G., Dougan, G., Miller, I.A., Pickard, D., O'Goara, P., Costa, G., Ali, T., and Hormaeche, C.E. The role of a stress-response protein in bacterial virulence. Molecular Microbiology, in press, 1990.

Jones, P.W., Dougan, G., Hayward, C., Mackensie, N., Colins, P., and Chatfield, S.N. (1991) Oral vaccination of calves against experimental salmonellosis using a double *aro* mutant of *Salmonella typhimurium*. Vaccine 9:29-34.

Killar, L.M., and Eisenstein, T.K. (1986) Delayed-type hypersensitivity and immunity to *Salmonella typhimurium*. Infect. Immun. 52:504-508.

Leclerc, C., Charbit, A., Molla, and A., Hofnung, M. (1989) Antibody response to a foreign epitope expressed at the surface of recombinant bacteria: importance of the route of immunization. Vaccine 7:242-8.

Levine, M.M., DuPont, H.L., Hornick, R.B., Snyder, M.J., Woodward, W., Gilman, R.H., and Libonati, J.P. (1976) Attenuated, streptomycin-depndent *Salmonella typhi* oral vaccine: potential deleterious effects of lyophilization. J. Infect. Dis. 133:424-429.

Levine, M.M., Kaper, J.B., Black, R.E., and Clements, M.L. (1983) New knowledge on pathogenesis of bacterial enteric infections as applied to vaccine development. Microbiological Reviews 47:510-50.

Levine, M.M., Herrington, D., Murphy, J.R., Morris, J.G., Losonsky, G., Tall, B., Lindberg, A.A., Svenson, S., Baqar, S., Edwards, M.F., and Stocker, B.A.D. (1987) Safety, infectivity, immunogenicity and *in vivo* stability of two attenuated auxotrophic mutants strains of *Salmonella typhi*, 541Ty and 543Ty, as live oral vaccines in humans. J. Clin Invest. 79:888-902.

Maskell, D.J., Hormaeche, C.E., Harrington, K.A., Joysey, H.S., and Liew, F.Y. (1987a) The initial suppression of bacterial growth in a salmonella infection is mediated by a local rather than a systemic response. Microbial Pathogenesis 2:295-305.

Maskell, D.J., Sweeney, K.J., O'Callaghan, D., Hormaeche, C.E., Liew, F.Y., Dougan, G. (1987b) *Salmonella typhimurium aroA* mutants as carriers of the *Escherichia coli* heat-labile enterotoxin B subunit to the murine secretory and systemic immune systems. Microbial Pathogenesis 2:211-221.

McFarland, W.C., and Stocker, B.A.D.(1987) Effect of different purine auxotrophic mutations on mouse-virulence of a Vi-positive strain of *Salmonella dublin* and of two strains of *Salmonella typhimurium*. Microbial Pathogenesis 3:129-141.

Marmion, D.E., Naylor, G.R.E., and Stewart, I.O. (1953) Second attacks of typhoid fever. J. Hyg. Camb. 51:260-267.

Mastroeni, Pi., Arena, A., Costa, G.B., Liberto, M.C., Bonina, L., and Hormaeche, C.E. (1991) Serum TNFα in mouse typhoid and enhancement of a salmonella infection by anti-TNFα antibodies. Microbial Pathogenesis, in press.

Miller, I.A., Chatfield, S., Dougan, G., Desilva, L., Joysey, H.S., and Hormaeche, C.E., (1989) Bacteriophage P22 as a vehicle for transducing cosmid gene banks between smooth strains of *Salmonella typhimurium*: use in identifying a role for *aroD* in attenuating virulent *Salmonella* strains. Mol. Gen. Genet. 215:312-316.

Miller, S.I., Mekalanos, J.J., and Pulkkinen, W.S. (1990) Salmonella vaccines with mutations in the *phoP* virulence regulon. Res. Microbiol. 141:817-821.

Molina, N.C., and Parker, C.D. (1990) Murine antibody response to oral infection with live *aroA* recombinant *Salmonella dublin* vaccine strains expressing filamentous haemagglutinin antigen from *Bordetella pertussis*. Infect. Immun. 58:2523-2528.

Mukkur, T.K.S., McDowell, G.H., Stocker, B.A.D., and Lascelles, A.K. (1987) Protection in salmonellosis in mice and sheep by immunization with aromatic dependent *Salmonella typhimurium*. J. Med. Microbiol. 24:11-19.

Murray, P.K. (1989) Molecular vaccines against animal parasites. Vaccine 7:291-299.

Nakayama, K., Kelly, S.M., Curtiss III, R. (1988) Construction of an Asd+ expression - cloning vector: stable maintenance and high level expression of cloned genes in a salmonella vaccine strain. Bio/Technology 6:693-697.

Nauciel, C. (1990) Role of CD4+ T cells and T-independent mechanisms in acquired resistance to *Salmonella typhimurium* infection. J. Immunol. 145,1265-1269.

Newton, S.M.C., Jacob, C.O., and Stocker, B.A.D. (1989) Immune response to cholera toxin epitope inserted in *Salmonella* flagellin. Science 244:70-72.

Newton, S.M.C., Kotlo, M., Poirier, T.P., Stocker, B.A.D., and Beavchey E.H. (1991) Expression and immunogenicity of a streptococcal M protein epitope inserted into salmonella flagellin. Infect Immun. 59:21589-2165.

Nnalue, N.A., and Stocker, B.A.D. (1986) Some *galE* mutants of *Salmonella choleraesuis* retain virulence. Infect. Immun. 54:635-640.

O'Callaghan, D., Maskell, D., Liew, F.Y., Easmon, C.S.F., and Dougan, G. (1988a) Characterization of aromatic- and purine-dependent *Salmonella typhimurium*: attenuation, persistence, and ability to induce protective immunity in BALB/c mice. Infect. Immun 56:419-423.

O'Callaghan, D., Maskell, D.J., Beesley, J.E., Lifely, M.R., Roberts, I., Boulnois, G., and Dougan, G. (1988b) Characterization and *in vivo* behaviour of a *Salmonella typhimurium aroA* strain expressing *Escherichia coli* K1 polysaccharide. FEMS Microbiology Letters 52:269-274.

O'Callaghan, D, Maskell, D.J., Tite, J, and Dougan, G. (1990*a*) Immune responses in BALB/c mice following immunization with aromatic compound or purine-dependent *Salmonella typhimurium* strains. Immunology 69:184-189.

O'Callaghan, D., Charbit, A., Martineau, P., Leclerc, C., van der Werf, S., Nauciel, C., and Hofnung, M. (1990*b*) Immunogenicity of foreign peptide epitopes expressed in bacterial envelope proteins. Res. Microbiol. 141:963-969.

O'Callaghan, D., Maskell, D.J., Tite, J., and Dougan, G. (1990) Immune responses in BALB/c mice following immunization with aromatic compound or purine dependent *Salmonella typhimurium* strains. Immunology 69:184-189.

Playfair, J.H.L., Blackwell, J.M., and Miller, H.R.P. (1990) Lancet 335:1263-1266.

Poirier, T.P., Kehoe, M.A., and Beachey, E.H. (1988) Protective immunity evoked by oral administration of attenuated *aroA Salmonella typhimurium* expressing cloned streptococcal M protein. J. Exp. Med. 168:25-32.

Reitman, M. (1967) Infectivity and antigenicity of streptomycin dependent *Salmonella typhosa*. J. Infect. Dis. 117:101-107.

Robbins, A. (1990) Progress towards vaccines we need and do not have. Lancet 335:1436-1438.

Robbins, J.B., Schneerson, R., Acharya, I.L., Lowe, C.U., Szu, S.C., Daniels, E., Yang, Y., and Trollfors, B. (1988) Protective roles of mucosal and serum immunity against typhoid fever. Monogr. Allergy 24:315-320

Sadoff, J.C., Ballou, W.R., Baron, L.S., Majarian, W.R., Brey, R.N., Hockmeyer, W.T., Young, J.F., Crysz, J.J., Ou, J., Lowell, G.J., and Chulay, J.D. (1988) Oral *Salmonella typhimurium* vaccine expressing circumsporozoite protein protects against malaria. Science 240:336-338.

Sanderson, K.E., and Roth, J.R. (1988) Linkage map of *Salmonella typhimurium* edition VII. Microbiol. Revs. 52:485-532.

Schödel, F., and Will, H. (1990) Expression of hepatitis B virus antigens in attenuated salmonellae for oral immunization. Res. Microbiol. 141:831- 837.

Schodel, F., Milich, D.R., and Will, H. (1990) Hepatitis B virus nucleocapsid / pre-S2 fusion proteins expressed in attenuated salmonellae for oral vaccination. J. Immunol 145: 4317-4321.

Sigwart, D.F., Stocker, B.A.D., and Clements, J.D. (1989) Effect of a *purA* mutation on efficacy of *Salmonella* live vaccine vectors. Infect Immun 57:1858-1861.

Silva-Salinas, B.A., Rodríguez-Aguayo, L., Maldonado-Ballesteros, A., Valenzuela-Montero, M.E., and Seoane-Montecinos, M.(1985) (Properties of two *gal*+ derivatives from vaccine strain *S.typhi* mutant Ty21a. Bol. Hosp. Infant. Mex. 42:234-239.

Sjösted, A., Sandström, G., and Tärnvik, A. (1990). Immunization of mice with an attenuated *Salmonella typhimurium* strain expressing a membrane protein of *Francisella tularensis*: a model for identification of bacterial determinants relevant to the host defence against tularemia. Res. Microbiol. 141:887-891.

Smith, B.P., Reina-Guerra, M., Hoiseth, S.K., Stocker, B.A.D., Habasha, F., Johnson, E., and Merritt, F. (1984) Aromatic-dependent *Salmonella typhimurium* as modified live vaccines for calves. Am. J. Vet. Res. 45:59-66.

Smith, H.W., and Linggood, M.A. (1971) Observations on the pathogenic properties of the K88, HLY and ENT plasmids of *Escherichia coli* with particular reference to porcine diarrhoea. J. Med. Microbiol. 4:467-485.

Stabel, T.J., Mayfield, J.E., Tabatabai, L.B., and Wannemuehler, M.J. (1990) Oral immunization of mice with attenuated *Salmonella typhimurium* containing a recombinant plasmid which codes for production of a 31-kilodalton protein of *Brucella abortus*. Infect. Immun. 58:2048-2055.

Stevenson, G., and Manning, P. (1985) Galactose epimeraseless (*GalE*) mutant G30 of *Salmonella typhimurium* is a good potential live oral vaccine carrier for fimbrial antigens. FEMS Microbiology Letters, 28:317-321.

Stocker, B.A.D. (1988) Auxotrophic *Salmonella typhi* as live vaccines. Vaccine 6:141-145.

Stocker, B.A.D. (1990) Aromatic-dependent *Salmonella* as live vaccine presenters of foreign inserts in flagellin. Res. Microbiol. 141:787-796.

Strahan, K.M., Kitchin, P., and Hormaeche, C.E. (1992) SIV-*Salmonella* constructs and their potential as vaccine candidates. Vaccine Research, in press.

Strugnell, R.A., Maskell, D., Fairweather, N.F., Pickard, D., Cockayne, A., Penn, C., and Dougan, G. (1990) Stable expression of foreign antigens from the chromosome of *Salmonella typhimurium* vaccine strains. Gene 88:57-63

Tacket, C.O., Forrest, B., Morona, R., Attridge, S.R., LaBrooy, J., Tall, B.D., Reymann, M., Rowley, D., and Levine, M.M. (1990) Safety, immunogenicity and efficacy against cholera challenge in humans of a typhoid - cholera hybrid vaccine derived from *Salmonella typhi* Ty21a. Infect. Immun. 58:1620-1627.

Tacket, C.O., Hone, D.M., Curtiss III R., Kelly, S.M., Losonsky, G., Guers, L., Harris, A.M., Edelman, R., and Levine, M.M. (1992a) Comparison of the safety and immunogenicity of Δ*aroC* and Δ*aroD* and Δ*cya* Δ*crp* *Salmonella typhi* strains in adult volunteers. Infect.Immun 60: 536-541.

Tacket, C.O., Hone, D.M., Losonsky, G., Guers, L., Edelman, R., and Levine, M.M. (1992b) Clinical acceptability and immunogenicity of CVD908 *Salmonella typhi* strain.Vaccine, in press.

Tarkka, E., Muotiala, A., Karvonen, M., Saukkonen - Laitinen, K., and Sarvas, M. (1989) Antibody production to a meningococcal outer membrane protein cloned into a live *Salmonella typhimurium aroA* vaccine strain. Microbial Pathogenesis 6:327-335.

Taylor, D.W., Cordingley, J.S., Dunne, D.W., Johnson, K.S., Haddow, W.J., Hormaeche, C.E., Nene, V., Butterworth, A.E. (1986) Molecular cloning of schistosome genes. Parasitology 91:s73-s81.

Tite, J.P., Dougan, G., and Chatfield, S.N. (1991) The involvement of tumour necrosis factor in immunity to *Salmonella typhimurium*. J. Immunol., in press.

Tite, J.P., Gao, X-M., Hughes - Jenkins, C.M., Lipscombe, M., O'Callaghan, D., Dougan, G., and Liew, F-Y. (1990) Anti - viral immunity induced by recombinant nucleoprotein of influenza A virus. III. Delivery of recombinant nucleoprotein to the immune system using attenuated *Salmonella typhimurium* as a live carrier. Immunology 70:540-546; 1990.

Wu, J.Y., Newton, S.M.C., Judd, A., Stocker, B.A.D., and Robinson, W.S. (1989) Expression of immunogenic epitopes of hepatitis B surface antigen with hybrid flagellin proteins by a vaccine strain of *Salmonella*. Proc. Natl. Acad. Sci. USA 86:4726-4730.

Yan, Z.X., Reuss, F., and Meyer, T.F. (1990) Construction of an invertible DNA segment for improved antigen expression by a hybrid *Salmonella* vaccine strain. Res. Microbiol. 141:1003-1004.

COMMERCIAL USE OF MICROBIAL INOCULA CONTAINING LIVE GENETICALLY MODIFIED MICROORGANISMS (GEMMOs)

William J Harris

Department of Molecular and Cell Biology
University of Aberdeen
Aberdeen, AB9 1AS, Scotland

INTRODUCTION

A wave of technologies developed in the 1970's formed a new Biotechnology industry which is expected to mature in the 1990's. The first commercially successful products in human health-care have already made their impact and a track record of success against prediction can now be examined. New waves of technology are rapidly appearing and offer product opportunities to compete with products still in the developmental pipeline. A good example is the predicted *in vivo* use of rodent monoclonal antibodies. Despite early data that rodent antibodies have very short half lives and induce an immune response in humans several hundred have entered clinical trials on the basis that their value overrides the difficulties. Now, however, we have a protein engineering technology to convert a rodent monoclonal into a human one. Suddenly, the commercial future of these rodent antibodies is very limited. It is against this background that we can appraise the likely commercial use of GEMMOs and, the use of live inocula in the market-place.

TABLE 1   FACTORS FOR COMMERCIALIZATION

TECHNICAL

RECOMBINANT   V   NATURAL
LIVE INOCULA   V   KILLED CULTURE
COMPETING TECHNOLOGIES

ECONOMIC

WILL THE CUSTOMER BUY IT?
WILL THE LEGISLATORS PERMIT ITS USE?
WILL IT MAKE MONEY?
COMPETING TECHNOLOGIES

The Release of Genetically Modified Microorganisms
Edited by D.E.S. Stewart-Tull and M. Sussman, Plenum Press, New York, 1992

85

The factors to be considered are described in Table 1, with competing technologies requiring both technical and economic assessment. One must not only consider the scientific merits of technologies, but also address who will finance their use. On the one hand, the top-tier multinational companies have the resources to bring products to market and can select the technology which will provide them with the highest profitability, while the small- and medium-sized companies (SMEs) and new Biotechnology companies are based upon more limited expertise and must be careful to match products to a discernible edge for their technology. In the early 1980's a plethora of new companies were set up with the same lead product targets - human growth hormone, albumin, tissue plasminogen activator - and clearly only one or two could finally succeed. The recent history of the human health-care market is littered with failed companies which did not appreciate this, and the number of companies currently committed to products based upon *Bacillus thuringiensis* toxin suggests that this error is about to be repeated. In this chapter proposed commercial opportunities for GEMMOs will be revised and their progress to date summarized in a search for clues to defining those likely to be commercial successes.

In the context of the new Biotechnology industry as a whole, it is clear that GEMMOs, as final products, can only contribute to minority markets. The combined plant biotechnology/algae, aquaculture, metal-mining, animal health and bioremediation markets only represent about 17% of the 1989 industries (Burrill, 1989). It should be emphasized that these are personal views, from one not directly involved in any of the discussed industrial sectors, and drawn mainly from the experiences of the impact of biotechnology in the health-care market. One aspect of the use of GEMMOs, as vaccines in animal health-care is not considered since this will mature as a legitimate business sector and parallel the development of vaccines in human health-care.

POLLUTION CONTROL AND BIOREMEDIATION

This is perhaps the area of most obvious use of living GEMMOs and is proposed as a lucrative market. Water pollution is the most damaging and widespread environmental hazard for farming, costing in excess of $2 billion per year in the USA. Industrial use of microbial inocula has been the treatment of choice in the USA for the remediation of land based oil-spills for 25 years. Successful acceleration of mineralization has been achieved at sites including farm tanks, coal-tar pits, ship's bilges etc. Also of course, municipal water treatment has relied on the use of microorganisms for decades. The use of microbial inocula is therefore, accepted practice.

There are now in our environment pollutants representing a broad range of chemical entities and since microorganisms can be found to biodegrade these (Alexander, 1980), there would seem to be considerable scope for expansion of this market sector. However, without biological contractors, trickling filters and various kinds of digesters, it would be prohibitively expensive to purify municipal waters. Inoculation with microorganisms to clean up groundwater is an attractive proposition and would be cost-effective particularly if such organisms could survive and grow for some time from starting inocula. While less inoculum means lower cost, it also lowers the profit to the supplier. There is a balance to be struck, further complicated by the fact that the driving-force for clean up is legislation or public demand. Biotreatment methods must be flexible, that is, technology must be geared to each specific application and is determined by local situation, contamination type, concentration and location of contaminant, and ability to use any complementary physical or chemical treatments. All of these factors reduce potential profitability to the supplier.

There are three common methods for bioremediation (Table 2) (Devine, 1990). In biostimulation, contaminated groundwater is pumped to the surface and treated above the ground in a reactor and then the effluent from the reactors, containing oxygen and acclimatized

TABLE 2

BIOREMEDIATION : EXAMPLES

METHOD	POLLUTANT	EFFECTIVENESS
BIOSTIMULATION		
GROUNDWATER STORAGE SITE	TOLUENE/ZYLENES/PHENOLS	100-200MG/LITRE DECONTAMINATED IN 2 YEARS
SLURRY TANKS		
SLUDGES FROM PLANT WOOD TREATMENT	PHENANTHRENE/ANTHRACENE, BENZPYRENES	300,000 PPM TO 65PPM 1100PPM TO 3PPM
COMPOSTING		
CONTAMINATED SOIL	HYDROCARBONS	31G/KG TO 14G/KG: in 9 MONTHS (NATURALLY); in 20 DAYS (HYDROLIC COMPOSTER)

microorganisms, is percolated or injected back into the ground. This has been used to clean up a site of stored solvents where the groundwater concentration of toluene, zylenes and phenol was in the 100-200 mg/litre range. Groundwater was treated with a submerged, fixed-film aerobic biotreatment system and a portion of the bioreactor effluent containing active biomass was sprayed on the ground and allowed to percolate through the soil. After two years of operation the site was removed from the list of contaminated soils.

Bioslurry treatment is a modified version of the activated sludge process and is used to treat highly-contaminated (greater than 1%) soil. A 10-20% solution of sludge is prepared with a simple reactor built on site by digging a hole and installing a liner. Aeration and mixing is supplied with surface aerators. A wood-treatment plant which produced creosote contaminated sludges managed to decontaminate about 100 tons of soil per week with 4 semi-batch reactors.

Bioenrichment is a version of composting where soils are manipulated to enhance the breakdown of target organics. Nutrients and surfactants are added, the pH adjusted and the soil turned over *in situ* or gathered up in windrows and tilled periodically. Alko Biotechnology, Finland, has carried out studies since 1984 and has shown that in 200m^3 of soil, chlorophenol diminution of 90-99% is achievable. This includes the slow-down of biological activity in Finnish winters.

Similarly, a windrow of 8000m^3 of oil-containing sludge had its hydrocarbon content reduced from 31g/kg to 14g/kg in 9 months including winter months and hydrolic composter treatment lowered a concentration of 320g/kg to 130g/kg in just 20 days. Bioaugmentation, the addition of exogenous microbial inocula can be applied to all these methods.

What is the remedy when physical movement and containment are not possible, as with oil spills? The Valdez oil spill in 1989 represented the first major use by the USA government of biological methods for site remediation. In this case, bioenrichment was used through the application of nitrogen and phosphate to contaminated shore-lines and natural biodegradation rates were significantly increased. The use of microbial inocula was rejected because it was thought that the Alaskan Shoreline Commission would oppose the introduction of non-indigenous bacteria (Moreland, 1989). In the more recent spills in the Gulf of Mexico, Texas and New Jersey the development of a generalized procedure has started. Ten to thirty percent of the spilled oil was rapidly removed by manual methods - booms, dykes, vacuum barges, skimmers. Natural processes of evaporation, photo-oxidation and dilution by water eventually eliminated 30% of the oil. Chemical dispersants will effectively break up the oil but their use remains controversial because of potential toxicity. Last year, the USA Environmental Protection Agency conducted field trials of the ability of two commercial bacterial cultures to biodegrade weathered crude oil in Prince William Sound in Alaska; preliminary data seemed to suggest some enhancement of degradation. An open ocean spill in the Gulf of Mexico was treated by applying commercial microbial cultures directly to the floating slick with inconclusive results. The treatment of an estuary in Texas with a proprietary mixture of nutrients and non-indigenous organisms was successful though it cannot be concluded that the microbial inocula played a significant role.

What then is the future for GEMMOs in these processes? All of these examples use the endogenous microflora or commercial cocktails of naturally-occuring organisms. Naturally occuring organisms are accepted for practical use and a GEMMO will always have to prove that it offers considerably more benefit before it is accepted by regulatory authorities. Arguments that the use of GEMMOs will reduce the time required to clean up contaminated land from say three to two years is unlikely to exert much weight in the argument. It is a fact that environmental clean-up, though it provides good election fodder for politicians and is liked by the general public, does not have the cumulative driving forces of immediate impact upon society, significant profit generation or job creation; the factors required to drive the use of new biotechnologies on to the market.

One strong driving force would be legislation which could be met effectively by the use of GEMMOs. In many countries the legislation defining the acceptable level of organic pollution is unclear. The EEC guidelines for organics in drinking-water, for example, suggest levels of the order of 0.2µg litre or 'absent'. The technologies and knowledge base required to provide GEMMOs to meet this market need do not seem to be near to exploitation. A GEMMO must be able to demonstrate that it can remove organics present in low concentrations more effectively than natural isolates but this will be no easy task. A naturally-occuring *Pseudomonas cepacia* can reduce groundwater levels of trichloroethylene from 3000ppb to 100ppb when inoculated along with a substrate to induce the enzyme system (Devine, 1990). A basic commitment to the study of the kinetics of removal of trace concentrations of a particular pollutant would seem appropriate.

One must also be aware of the potential of competing technologies. Removal of polychlorinated biphenyls (PCBs) from groundwater for example would be better and more effectively achieved by chelation with high-affinity adsorbents (Srinivasan and Fogler, 1989) and one can expect transgenic plants to be developed which will extract organics from the soil and chelate them to peptides within the plant.

As illustrated by the examples above microbial inocula are only a small part of an environmental management package. Microbial inocula do not demand a high price, and if used as live material to regenerate *in situ*, profit to a supplier is very low indeed. In this respect, inocula are similar to vaccines where one treatment provides life-long protection - an excellent product but of such low profitability that they are nowadays provided by only a few select companies, or their production is subsidized by Government. The commercial future of GEMMOs is likely to lie in the same direction. Most industrial R&D in GEMMO development is carried out in a few new biotechnology companies such as Biotechnika, Envirogen, Celgene, Ecova Corporation but their lead products are naturally-occuring microorganisms (Wick, 1990).

Market forecasts predict a healthy future for environmental biotechnology business (B.C.C., 1989). The fastest growing area is the bioremediation of hazardous wastes. The

USA market was $34 million in 1989 and may be $153 million by 1995. The total environmental Management through Biotechnology business in 1990 is estimated at $1 billion. However, the proprietary microbial cultures market share in the USA was $24 million in 1989 and is expected to grow to $60 million by 1995 - not a sufficiently lucrative business opportunity to support one embarking upon the additional rigors of R&D, field-trials, and regulatory approval required for GEMMOs. Rather, GEMMOs and microbial inocula are likely to be subsumed as a small part of the business of Companies offering full-service remediation, currently engineer driven.

BIOPESTICIDE INDUSTRY

Chemical pesticides have been the mainstay of agriculture for most of this century but escalating environmental, human and animal health problems have caused a shift in attitudes toward the use of biocontrol. Biocontrol methods include the use of natural predators exemplified by the earliest recorded application, the use of ant nests in barns in China in 1200 AD, and the recent use of *Encarsia* wasps which are now used to control whitefly in 95% of the crops in Dutch glasshouses.

The introduction of selected microbes for agricultural purposes has also been carried out for nearly a century for a variety of purposes; these have included the use of rhizobacteria to increase nitrogen utilization and baculoviruses to control insects. The biological insecticide based upon *Bacillus thuringiensis* (BT) toxin has been available for more than 30 years. There would seem again then to be clear opportunities for GEMMOs in this market.

However, the current market share of biopesticides is very small ($30 million out of a total of $21.6bn, less than 1%) and in 1988 only 58 products were marketed world-wide (Zechendorf, 1990). This is partly due to their reputation of being "notoriously unreliable". Despite the lack of any evidence that this reputation is unfounded, or that recent technology has provided a solution to this problem, analysts repeatedly predict a huge increase in biopesticide utilization. Market projections of a total of $6-8 billion in the USA by the year 2000 have been promoted. These are fuelled by expectations that a general market demand will be created by legislative measures banning synthetic pesticides. Abbott, the current market leader, is accompanied by at least 23 separate companies who claim to be already engaged in the field (Zenchendorf, 1990).

Biocontrol products are derived from insects, micro-organisms (fungi, bacteria viruses), other animals and plants, and include BT toxin, spider or ant venom, lectins, pyrethrins, photo-sensitive chemicals, hybrid biotoxins. Genetic modification offers the possibility of a variety of delivery systems for these agents, and the use of microbial inocula containing live GEMMOs will need to compete with the application of killed micro-organisms, purified chemicals (produced both synthetically and from GEMMOs), and transgenic plants and insects. One of the earliest methods, the release of sterilized insects, has already been updated through the use of genetically modified blow-flies and fruit-flies.

The release of ice-minus *P. syringae* was the first example of deliberate release of a GEMMO and it was initially envisaged that a live inoculum would have a lucrative market. Field-tests of the organism (Forstban) in 1987 were thwarted by public opposition with the result that a naturally-occurring form of ice-minus *P. syringae* is likely to be commercialized (McCormick, 1985). In addition, the EPA recently approved the first recombinant BT products, namely M-Trak and MVP manufactured by Mycogen Corp. Both products contain killed *Ps. fluorescens* carrying the delta toxin. With these early products then the option to use live GEMMOs was rejected.

At the same time transgenic plant technology has moved forward relentlessly (Brunke & Meeusen, 1991). By 1987, the BT gene was introduced into tobacco plants by 1987 and subsequently into potato and cotton plants. Analogous transgenic plants containing proteinase inhibitor genes, and anti-fungal chitinase were shown to display field resistance at levels equivalent to those obtained with chemical pesticides. By June 1990, the USA had issued 48 permits for field-trials of engineered plants compared to only a few for microbial inocula. Proponents of transgenic plant technology point to the potential advantages of plant expression as control throughout the growing season and control extended to roots, shaded leaves and new growth, while supporters of the use of a microbial inoculum raise the spectre of higher human exposure through the ingestion of plant material containing toxin. This argument remains to be won. It is clear however that the multinational commercial commitment is to transgenic plants with their considerably higher profit potential. Permits for their use have been granted for herbicide tolerance and insect and disease resistance in tomato, tobacco, rice, corn, potato, soy bean, alfalfa, cotton and cucumber plants, poplar and walnut trees.

It seems likely then that GEMMOs may not find a use in the treatment of the major world crops and this is increasingly more likely as toxins with very broad anti-insecticide activity and antifeedants are transferred into plants. This scenario is supported by evidence that the majority spending on biological pesticides is concentrated in newly-formed companies. Of the three new biopesticides registered in USA in 1988, all were from new biotechnology companies : Ecogen's biofungicide, DaggerG; Mycogen's bioinsecticide, M-ONE; and Igene's nematocide CLANDOSAN (Davies, 1989). The major producers of chemical pesticides continue to devote R&D to new chemical products, at a rate of $432 million in 1985.

One must evaluate the commercial potential for engineered viral insecticides in the same way. Companies must look upon a microbial inoculum as just one component to add value to a package in the market-place. As an example, the company Crop Genetics International is developing INCIDE Vaccines based upon genetically-modified endophytes carrying selected genes but used in combination with seed inoculation technology which allows the company to market genetically modified seed directly to the farmer (Burrill, 1989). A few biocontrol agents are near to market exploitation: AGC. Ltd have a nematode to control black vine weevil (NEMASYS), Bayer have a fungus for the same purpose, and Hoechst a viral insecticide against codling-moth, an apple pest (GRANU-POM) (Dunn, 1990). However, it is salutory to note that there have been more field trials with recombinant plants, insects and animals than with micro-organisms, even though technically, the molecular biology is simpler and more understood in the latter.

MICROBIAL METAL MINING

Microbial metal mining is an industry currently worth about $350 million per annum for copper and $20 million per annum for uranium but suggestions of $90 billion markets by 2000 have been suggested (Munroe, 1985). The last 20 years has seen the identification of a broad range of microorganism with the ability to extract metals. Commercial interest is driven by the continued depletion of high-grade mineral resources leaving low-grade sources which require increasingly expensive mining operations. There are many attractions of microbial mining; it requires less energy, is pollution free and lower in cost. Existing commercial operations certainly show that suitable scales of operation are possible. Microbial mining in Ontario recovers about 5,000kg of uranium per month from low-grade ores. This mechanism of extraction is based upon the use of thiobacilli to produce ferric sulphate and sulphuric acid. The ferric iron readily oxidizes uranium-containing ores which are then easily dissolved in dilute sulphuric acid. Current methods are however generally uncontrollable and any application of GEMMOs would seem highly unlikely until there is a better understanding of microbial/mineral interactions in these complex systems. It must also be open to question whether GEMMOs would be an improvement of natural isolates. For example, *Pseudomonas aeruginosa* rapidly accumulate uranium which can account for 50% of the total dry weight (Strandberg *et al*, 1981). Extracellular biosorption or surface ion-exchange is another way in which many microorganisms accumulate metals; yeast cells will bind up to 15% of the total dry weight as uranium.

CONCLUSIONS

The use of microbial inocula is a well-established but small market. Undoubtedly the market will expand and diversify through the use of both bacterial and viral inocula. Such inocula whether derived from organisms indigenous to the treatment site, exogenously derived natural microorganisms or GEMMOs share a number of barriers to extensive commercial use:

- Viable cultures are difficult to reproduce to satisfactory quality assurance standards.
- Many bacteria or viruses are slow-growing. Scale-up of manufacture and concentration will present downstream developmental problems.
- Liquid concentrates are expensive to transport and spray-dried material often loses viability. A 10-million barrel oil spill would require thousands of tons of inoculum quickly (Simpson, 1991).
- Survival varies considerably in different environments ruling out the likelihood that the same inoculum would be suitable for Nottingham, Nebraska and Nagasaki. Each GEMMO will then have a limited geographical market.

These considerations argue against the massive use of microbial inocula whether natural or genetically-modified, hence the lack of interest from multinational industries. There are as yet few data that genetically-modified bacteria for soil/water remediation have significant increased value over the use of natural isolates or even simpler stimulation of the endogenous microflora. As discussed earlier, since the inoculum is only a small part of a remediation exercise other factors will determine the nature of treatment. There are specialized niches, such as treatment of industrial effluents at source or clean-up of storage sites, but assuming the continued trend away from the manufacture and use of such chemicals, the market size for any one product is small and short term. It is difficult to escape the conclusion that the level of profitability from a GEMMO for bioremediation is small and exploitation will require direct financial subsidy from Government for R&D and very simple regulatory requirements. Since it would be legitimate to argue that a detoxifying GEMMO is in the public good and treating an "environmental illness", legislative fast-track procedures equivalent to Orphan Drug status could be used to distinguish such a GEMMO from say genetically-modified Rhizobia which increase nitrogen utilization.

In plant agriculture, the limitation on the commercial use of GEMMOs, both bacterial and viral, is competition from transgenic plants, the provision of chemical pesticides based upon natural toxins, and the farming of natural predators of disease. The world's major crops will be genetically modified to protect them against the major pests. The chemical pesticide industry is large and well-structured with defined distribution routes. Methods for the effective delivery to the field are well-defined. The large-scale spraying of microbial inocula is a new technology to be defined. There are though clear niche opportunities where viral insecticides are the only effective treatment other than non-specific chemical pesticides.

Commercial drive to exploit GEMMOs seems to be stalled at the moment. The prime driving force -profit- seems inadequate. Legislative demand to clean-up the environment is a strong driving force for bioremediation and would force the multinationals to develop products for in-house use, but it must be accompanied by simple legislation to allow their use. The recent EPA prepared guidelines seem to meet this requirement but EEC legislation is far from clear. As scientists, we can see the value of GEMMOs, but to champion their commercial use we need to demonstrate their superiority over natural isolates under field conditions. In general, it seems that GEMMOs will find their place as a legitimate range of products servicing small but important markets. They will probably be brought to market by one or two small companies. Analysts predict large markets in Environmental and Agricultural Biotechnology but only a small percentage can be attributed to GEMMOs. The profitability from such products is insufficient to sustain investment from venture capital as stand-alone business opportunities, and top-tier companies will only consider those specific products which impinge directly upon their core business. In some cases competitive technologies have appeared and now offer better and more competitive product opportunities than those envisaged

10 years ago. Despite my rather conservative conclusions, it would not be surprising if the ultimate major commercial use of GEMMOs has yet to be realized.

REFERENCES

Alexander, M. (1980) Biodegradation of chemicals of environmental concern. Science 211,132-138.

Brunke, K.J. and Meeusen, R.L. (1991) Insect control with genetically engineered crops. TibTech. 9, 197-200.

Burrill,G.S. (1989) Biotech 90: Into the next decade, Mary Ann Liebert Inc., New York.

Business Communications Co. (1989) Market Study: Environmental Management through Biotechnology.

Davies, J.E. (1989) Young entreprenerial firms will serve as trendsetters for the biopesticide industry. Genet. Eng. News, March, 4-5.

Devine, K. (1990) Environmental factors dictate choice of specific bioremediation method. Genet. Eng. News, May 1990. 3-5.

Dunn, N. (1990) Natures prescriptions for plant protection. The Furrow, 95, issue 3, 8-10.

McCormick, D. (1985) No escaping free release. Biotechnology 3, 1065-1071.

Moreland, S.M. (1989) Bioremediation. Genet. Eng. News, December. 66.

Munroe, D. (1985) Microbial metal mining. Int. Biotech. Lab. June, 19-29.

Simpson, K. (1991) Can biotechnology react to emergencies? Biotech. Forum Europe 8, 149-150.

Srinivasan, K.R. and Fogler, H.S. (1989) Use of modified clays for the removal and disposal of chlorinated dioxins and other priority pollutants from industrial wastewaters. Chemosphere. 18, 333-342.

Strandberg, G.W., Starling, E.S. & Parrott, J.R. (1981) Microbial cells as biosorbents for heavy metals. Appl. Rev. Microbiol. 41, 237-245.

Wick, C.B. (1990) Biodegradation will play key roles in hazardous waste treatment in '90's. Genet. Eng. News, May, 5-7

Zechendorf, B. (1990) Biocontrol - Market Study: A review article on biological pest control. Biotech. Forum. Europe 7, 212-214.

RISK ASSESSMENT

H John Dunster

Consultant in Radiation Protection
19 Diamond Court, 153 Banbury Road
Oxford OX2 7AA

INTRODUCTION

There is confusion and controversy about the meaning of the term "risk". Its common usage is descriptive, as in the phrase "the risks of rock climbing". In a quantitative sense, it has sometimes been used to mean the probability of occurrence of an unwelcome event. For accidents, it has also been used to mean a combination of probability and consequences. This combination is intended to represent the expectation of harm arising from a specified event or situation. I prefer the term "detriment" for that combination. Risk can then be reserved for use in its common, descriptive sense and in well-established technical senses such as "excess relative risk".

This chapter will outline how risk has been used in setting standards for protection against ionizing radiation in both normal and accident situations. The procedures call for both scientific and social judgments, but are fairly straightforward. Ionizing radiation has been recognized as hazardous for nearly a century, but is applications in science, medicine and industry have been too important for it to be abandoned. The risks have therefore been assessed and controlled. Some of the experience gained in this process is specific to radiation, but the approach and many of the resulting policies are relevant to other areas, including microbiology. Corresponding procedures for toxic chemicals and pathogenic organisms are more complex, if only because the types of harm and the pathways from the source of risk to its expression are much more diverse and complicated.

THE CONCEPT OF DETRIMENT

In 1977, the International Commission on Radiological Protection (ICRP, 1977b) introduced the concept of detriment. This was a measure of the total harm that would eventually be experienced by an exposed group and its descendants as a result of the group's exposure to a radiation source. In practice, the only harm considered was danger to health. The original definition was the expectation value of the number of health effects, weighted for severity, that will be experienced by the exposed group and its descendants. In respect of an individual, the detriment could also be expressed as the product of the probability of a deleterious effect and a measure of the severity of that effect. Only death and severe hereditary conditions due to the exposures were dealt with quantitatively, although the contribution from less severe conditions was implicit in the definition.

The Release of Genetically Modified Microorganisms
Edited by D.E.S. Stewart-Tull and M. Sussman, Plenum Press, New York, 1992

This approach to detriment was useful, but too limited. A serious problem is posed by the multifarious nature of the possible outcomes, so that probability and severity can be combined in many different ways to represent detriment. Thus, even in the simple case of exposure to low doses of ionizing radiation, the outcome may be fatal cancer, with a delay of anything from a few years to many decades, non-fatal cancer, or a very wide range of hereditary disorders that occur in later generations but mainly in the first two. The early discussions used only the probability of attributable death or of severe hereditary disorders. The time of occurrence and the effect of competing causes of death were recognized, but largely ignored.

Since mortality is necessarily 100%, the probability of an attributable death is a poor measure of detriment. The date and cause of death also matter. Non-fatal effects may be less important individually, but if their frequency is much higher than that of attributable death, they may represent more total harm. Hereditary conditions range from the trivial to the fatal and, perhaps worse, may be severely crippling but not life-threatening.

There are recognized methods for combining wide-ranging consequences by using weighting factors leading to an aggregated index of harm. Alternatively, the different consequences can be kept separate to give a multi-dimensional representation of detriment. There is no universal choice and, for many purposes, it is best to use a multi-attribute presentation with some aggregation to simplify the presentation. This is the approach now used in radiological protection (ICRP, 1991).

THE QUANTIFICATION OF RADIATION RISKS

The most useful information for quantifying the risks following exposure to radiation comes from epidemiological studies. The most effective study remains that of a group of nearly 100,000 of the survivors of the nuclear weapon explosions at Hiroshima and Nagasaki. This is supported by studies on radiotherapy patients. The effects of occupational exposure and of variations in the natural background of radiation are being examined, but the studies have not yet provided consistent information.

The epidemiology has to be supported by biological work on cells *in vitro* and on experimental animals because the human data do not yet cover the lifetime of the study groups and usually relate to radiation doses and dose rates much higher than those for which risk estimates are needed. The risk estimates for hereditary effects are based solely on animal data. All the risk estimates are thus subject to uncertainties resulting from the application of scientific judgment. The characteristics used to specify the effects attributable to a low dose of radiation include the probability of fatal cancer, the mean time lost if a fatal cancer occurs, the probability of non-fatal cancer, and the probability of hereditary effects crudely weighted for severity relative to death. For these effects, there are no grounds for expecting a threshold below which the probabilities will be zero and the relationship between probability and dose is broadly proportional, at least at low doses. The detriment due to a unit dose can then be expressed either by aggregating these attributes, relative to a specified baseline, or by considering all the attributes separately. The choice of method depends on the purpose for which the detriment is intended. In both cases, the result is a probabilistic statement, not an absolute one.

The next stage in estimating risks is that of assessing the doses associated with the situation under review. In normal operations, the doses are those received routinely. They are actual doses, either measured or forecast from previous experience. There will be some measure of uncertainty, but the probability of receiving the doses is essentially unity. The only probabilistic component is that in the detriment per unit dose. In assessing the importance of possible accidents, there is a further probabilistic stage, that of assessing the likelihood of the doses being received, more precisely the assessment of the distribution of the probabilities of the doses in relation to their magnitudes.

The final stage of risk assessment is that of judging the social implications of the risk estimates and the use of this judgment in taking decisions.

THE BASIS OF DECISIONS IN NORMAL OPERATIONS

The first clear statement of radiological protection policy based on risk was in the 1965 recommendations of the International Commission on Radiological Protection (ICRP, 1966).

With subsequent developments and clarification, this policy still underlies the 1990 recommendations (ICRP, 1991). It contains three interrelated components that can be paraphrased as follows.

(a) No practice that causes radiation exposures should be adopted unless it produces enough benefit to offset, and thus justify, the radiation detriment it causes.
(b) Within a justified practice, the protection of each source of exposure should be adjusted so that all reasonable steps are taken to reduce the exposures.
(c) The total exposure of each individual caused by the relevant practices should not be permitted to exceed specified limits.

Component (a) calls for an estimate of the total radiation detriment and of the conventional costs and benefits of the project. It is rare for the radiation detriment to play a decisive role in the decision to proceed with a practice. It is usually overwhelmed by the conventional aspects.

Component (b) is both the most complex and the most effective part of the system of protection. It imposes a duty on all concerned to consider the available options and to choose those that most effectively reduce the dose, provided that they are not unreasonable. This judgment calls for a comparison of the differential costs and dose reductions for the available options and for a basis for judging dose reductions against costs. The costs include the deployment of resources, the social implications of the precautions and any repercussions on the effectiveness of the practice. This process tends to emphasize the costs and benefits to society or to a project as a whole. The individual may not be adequately considered. Furthermore, the procedure is often somewhat intuitive, and always judgmental. It needs a safety net to ensure that no gross errors occur. It should be supplemented by a constraint on the dose to the most highly exposed individual.

Although expressed in more specific terms, component (b) differs only in emphasis from the long established requirement of British safety legislation to do all that is reasonably practicable to prevent hazards.

The final component, the dose limit, may be used to provide the constraint needed in component (b), or it may be used to provide a limit on the total dose if an individual is exposed as the result of several practices.

ACCIDENT PREVENTION AND MITIGATION

No operation can be totally free from the risks of accidents. There is always a spectrum of accidents each with a probability of occurrence and a severity of the consequences. In the classical approach to safety, the aim was simply to prevent accidents. The occurrence of an accident was regarded as a culpable failure. Although this attitude persists, especially, and reasonably, among the victims of accidents, the practice of accident prevention now recognizes that accidents *will* occur. The emphasis is now on the reduction of their probability and of their consequences - prevention and mitigation.

In the nuclear industry, and increasingly in other industrial contexts, one basic safety tool is probabilistic safety analysis. This complex procedure considers all the identifiable causes of error or failure, links them to the engineering and operational features that are intended to intercept the progress towards an accident, and then assesses the probability of the initial cause and the probabilities of failure of the subsequent chain of protective features. The overall assessment then leads to an estimate of the spectrum of probabilities and consequences of all the foreseeable sequences.

There are many uncertainties, some in the initial data about failure rates of components or items of plant, some due to conceptual difficulties in achieving a complete inventory of possible sequences and in judging whether the successive stages of prevention are genuinely independent or may suffer from common mode failures. Despite the uncertainties, probabilistic safety analysis is increasingly being used as an input to decisions about the acceptability of new plants or practices. Probabilistic safety analysis also provides insights into the relative significance of different sequences. Design and operational resources can then be concentrated on the problems where solutions will do most good.

THE BASIS OF SOCIAL JUDGMENT

Although the concept of risk is clearly understood in general terms by members of the public, it has not proved easy to involve them in direct discussions about the tolerability of industrial risks. In practice, the initial discussions have taken place between specialists. Lay members of advisory committees, lawyers, and elected politicians, including government ministers, have progressively become more involved as the importance and prominence of the relevant cases increase.

In the context of ionizing radiation, ICRP has used three words to indicate the degree of tolerability of an exposure, and thus of a risk. **Unacceptable** is used to indicate that the exposure would not be regarded as acceptable on any reasonable basis in the normal operation of any practice of which the use was a matter of choice. Such exposures might have to be accepted in abnormal situations, such as those during accidents. Exposures that are not unacceptable are then subdivided into those that are **tolerable**, meaning that they are not welcome but can reasonably be tolerated, and **acceptable**, meaning that they can be accepted without further improvement, provided that they are justified by the benefits of the practice that causes them.

To provide a quantitative basis for judging the tolerability of an exposure, it is necessary to quantify a range of attributes of the detriment. For none of them is it possible to establish a categorical criterion against which to define unacceptable and tolerable, but, taken together, they provide a basis for judgment.

The attributes typically associated with the death of one individual are as follows. Additional allowance is made for non-fatal conditions:

- The attributable probability of death summed over the whole lifetime.
- The time lost if the attributable death occurs.
- The reduction of life expectancy (a combination of the first two attributes).
- The annual distribution of the attributable probability of death.
- The increase in the age-specific mortality rate, i.e. in the probability of dying in a year at any age, conditional on reaching that age.

It is also necessary to consider the collective detriment due to the exposure of all the relevant individuals, sometimes called the societal risk. There seems to be no satisfactory set of attributes for this situation, partly because of the way in which society's attitude to the risk when it is expressed depends dramatically on the way in which the consequences are distributed. If many people are affected by a single event, the consequences are seen as being much more serious than if the same total number of people are affected by several separate events. Even if death is the only significant endpoint, which is rarely true, the expected (or even the observed) number of deaths is a very poor indicator of society's reaction.

Even more difficulties arise when the source of exposure is itself probabilistic, i.e. if it may or may not occur, as is the case with accidents. The "risk" is then often expressed as the expectation value of the number of deaths. This is a very unsatisfactory simplification. One death resulting from a type of accident with an annual probability of occurrence of 0.1 gives an expectation death rate of 0.1 deaths per year. So does a type of accident with a death toll of 1000 and an annual probability of 1 in 10 000, but the two situations are certainly not equivalent. A multi-attribute approach allows these problems to be discussed openly, but does not provide easy answers.

RISK ASSESSMENT AS A DIRECT INPUT TO DESIGN AND OPERATION

There are examples in radiation protection of a simple form of risk assessment being used to give direct guidance to designers and operators. One, dating back to 1954 (Dunster, 1954), used a toxicity classification of radionuclides, coupled with a grouping of operations based on the likelihood of dispersing the radioactive material being handled, to give a link between the amount of radioactivity in use and the necessary quality of the laboratory and the operating procedures. Similar systems are still recommended both nationally and internationally (NRPB, 1988; ICRP, 1977a; ICRP,1989).

Another example concerns the transportation of radioactive materials and has been recently updated (IAEA, 1990). The international requirements call for well-constructed packaging, adequate to survive the normal stresses of transport. Since the packaging will not survive a major accident, the activity of the contents is limited on a scale based on the radiotoxicity of the nuclide and its physical form. Larger activities have to be transported in specially constructed and tested containers.

RISK ASSESSMENT FOR TOXIC AND PATHOGENIC MATERIALS

It is tempting to suggest that the techniques developed for ionizing radiation could be modified and applied to toxic, ecotoxic, and pathogenic materials. At the fundamental level that is true. Upper limits can be set, at least conceptually, for the risks to man and the environment. Above this limit, the project is not permitted. Below the limit, all reasonable steps should be taken to reduce the risks, even if they are already small. In my view, the more recent environmental requirement to use the "best available technology not entailing excessive cost" should not be interpreted as calling for unreasonable steps and thus should lead to decisions not very different from those called for by the older requirement to do all that is "reasonably practicable" to reduce risks. Beyond the level of this general statement of policy, the problems look very intractable. The pathways between an exposure to the harmful agents and the actual expression of the harm are much more complex and less well defined than those for radiation. Often they are less well quantified. Inevitably, the process of risk assessment depends heavily on professional judgment and, at least initially, on arbitrary decisions that may later appear to have been unduly cautious.

At a more practical level, the systems of classification used in the handling and transportation of radioactive materials can be seen as the forerunners of the procedures required for the handling of genetically modified microorganisms. That system, set out by the Advisory Committee on Genetic Manipulation (ACGM, 1988), categorizes genetic manipulation experiments and gives guidance on the necessary standard of containment.

Several properties of radiation and radioactive materials make it difficult to transfer the techniques of radiation risk assessment to the field of microbiology, particularly to the release of genetically modified microorganism. In practice, the only significant risks from radiation are those to health. It is very rare for radiation to damage an environment. The activity of radioactive material decreases with time. This is also true of microorganisms that have been deliberately disabled. Other microorganisms, unlike radioisotopes, are modified with the aim of achieving multiplication in the environment. There is a universal background of radiation from natural sources. This can be used as a basis of comparison. Nevertheless, some of the principles and policies developed for radiation can be applied to the management of the risks from genetically modified microorganisms.

Probably the most important lesson to be transferred is the importance of a structured and transparent approach. This approach is also well established by the Advisory Committee on Genetic Manipulation (ACGM, 1990). This publication provides a very detailed structure for reviewing the dangers of releasing genetically modified microorganisms into the environment. Quantification does not yet seem to be possible, but this only adds to the importance of structure and transparency. Work sponsored by the Laboratory of the Government Chemist on the project called Prosamo (Planned Release of Selected and Modified Organisms) should be able to make a substantial contribution to this difficult field. Most social judgments are arbitrary to some extent and they all depend on who is making the judgments and on whose behalf. Clarity of thought and of presentation may not be enough to carry conviction, but their absence will guarantee failure.

REFERENCES

Advisory Committee on Genetic Manipulation (1988) Guidelines for the categorisation of genetic manipulation experiments, ACGM/HSE/NOTE 7. Health and Safety Executive, London, 1988.
Advisory Committee on Genetic Manipulation (1990) The intentional introduction of genetically manipulated organisms into the environment: Guidelines for risk assessment and for the notification of proposals for such work, ACGM/HSE/NOTE 3 (revised). Health and safety Executive, London, 1990.

Dunster, H.J. (1954) The protection of personnel working with radioactive materials and the disposal of radioactive waste. Medicine Illustrated 8 No 11.

International Atomic Energy Agency (1990) Regulations for the safe transport of radioactive material (As amended 1990), Safety series No 6. IAEA, Vienna, 1990.

International Commission on Radiological Protection (1966) Recommendations of the International Commission on Radiological Protection, ICRP Publication 9. Pergammon Press, Oxford, 1966.

International Commission on Radiological Protection (1977a) The handling, storage, use and disposal of unsealed radionuclides in hospitals and medical research establishments, ICRP Publication 25. Annals of the ICRP, 1 No 2.

International Commission on Radiological Protection (1977b) Recommendations of the International Commission on Radiological Protection, ICRP Publication 26. Annals of the ICRP, 1 No 3.

International Commission on Radiological Protection (1989) Radiological protection of the worker in medicine and dentistry, ICRP Publication 57. Annals of the ICRP, 20 No 3.

International Commission on Radiological Protection (1991) 1990 Recommendations of the International Commission on Radiological Protection, ICRP Publication 60. Annals of the ICRP, 21 No 1-3.

National Radiological Protection Board (1988) Guidance notes for the protection of persons against ionising radiations arising from medical and dental use. HMSO, London, 1988.

SOCIAL IMPLICATIONS AND PUBLIC CONFIDENCE: RISK PERCEPTION AND COMMUNICATION

Ray Kemp

School of Environmental Sciences
University of East Anglia
Norwich, NR4 7TJ, U.K.

INTRODUCTION

Public attitudes to risk have developed into an important area of concern for both operators and regulators of industrial processes as they face increasing demands to explain the likely consequences of potentially hazardous products and activities to both workers and the general public, and to the environment. There is public concern about the long-term health risks arising from industrial processes and products; risks from accidental releases; the effects of exposure to chemicals at low dose levels; and the inter-generational effects of such exposures. These problems are not only crucial areas of concern for the health and safety of the public, they also raise important questions of equity, distributional effects, efficiency of resource use and economic planning. The *sine qua non* of any action to address such issues has to be a careful assessment of the role and effects of information generation and dissemination - the question of risk communication.

In what follows, the complex problem of public attitudes to the release of genetically modified microorganisms will be addressed in five stages. First, the basic premises and findings from mainstream social psychological investigations of public perceptions of risk will be introduced. The central purpose being to demonstrate that much is already known about the factors which most strongly influence public attitudes to potentially hazardous processes and products. Within this material there is evidence to suggest that public perceptions of the risks associated with genetically-modified organisms (GEMMOs) are volatile. Secondly, the results of a review of expressed public concerns about biotechnology which are to be found in the published literature will be set out. For convenience, these are categorized as socio-economic; environmental; health-related; and ethical concerns. Loss of public trust in science is another important effect. These are not mutually exclusive categories and many of the concerns appear to reflect a lack of confidence in the regulation of science and technology.

Thirdly, two recent and related developments in social scientists' analyses of the process of risk communication will be outlined. On the one hand, there is a model for explaining how risk attitudes to activities and products come to be *amplified*, that is, over-estimated given the technically assessed level of risk. On the other hand, attention is now being drawn to the potential for processes and products to become *stigmatized*, with possible significant costs for regulators and industry alike. Fourthly, the safety assessment, communication of advice, and public reception of a genetically-modified bakers' yeast in 1990 will be examined. Particular attention to this case study will be given as an exemplar of the social amplification of risk.

Finally, some conclusions will be drawn regarding the nature of public concerns about the release of genetically-modified micro-organisms and I shall argue for a broadening of the concept of risk communication. Steps need to be taken to avoid the unnecessary social amplification of perceived risk and its associated costs to industry, the regulatory system and to consumers. At the same time, public perceptions should not be seen to be irrational, rather

The Release of Genetically Modified Microorganisms
Edited by D.E.S. Stewart-Tull and M. Sussman, Plenum Press, New York, 1992

they are the direct response to the forms of technological development, regulatory and advisory system and commercial processes which have emerged over recent years.

RISK PERCEPTION STUDIES

Arguably the most comprehensive approach to understanding public attitudes to risks, has evolved over the past decade or so in the work of Paul Slovic, Baruch Fischhoff and colleagues in the United States (Fischhoff *et al*, 1978; Slovic, 1987, 1990). Known as the "psychometric paradigm", their wide-ranging empirical studies set out to ask people directly about their perceptions of risks and benefits, and their expressed preferences for various risk/benefit trade-offs. "Perception" in this field refers to various kinds of attitudes and judgements about risk. Data, from extensive questionnaire surveys, has been gathered for a wide range of activities, technologies and products, and analysed by multiple regression and related statistical techniques, in order to disentangle multiple influences at work in the results. Further, the studies elicited judgements by people about various qualities or characteristics (such as voluntariness, novelty, controllability, dread) which have been hypothesized to influence public attitudes towards risk and the acceptability of risks. The psychometric paradigm assumes that risk is subjectively defined by individuals who may be influenced by a variety of psychological, social, institutional and cultural factors (Slovic, 1990). It has its detractors and its critics (Wildavsky and Dake, 1990; Wynne, 1984). Nevertheless, it is the most fully documented and widely-replicated approach, and it provides an appropriate framework to address public concerns about the release of genetically-modified organisms (GEMMOs) in particular, and about biotechnology in general. More recent considerations of the interaction between risk perceptions and communication are also addressed.

The approach to risk communication which has evolved out of recent revisions to the psychometric paradigm (Slovic, 1990; Kasperson, *et al*, 1988), draws attention to the horns of a difficult dilemma. Where the risk communication process, through press statements, ministerial announcements and the like, over-emphasizes the potential harm due to a particular hazard, leading to unnecessary public anxiety, the social *amplification* of risk is said to have occured. Such amplification can lead to increased regulation and disproportionate levels of public and private investment in safety. Conversely, the social *attenuation* of risk is to be avoided. Where the potential harm from a hazardous activity or substance is underestimated or given insufficient weight due to ignorance or lack of institutional and/or public concern, then the public and the environment may be exposed to unnecessary and unacceptable risks. It is, therefore, important to set out the potential causes of distortion in the risk communication process which can give rise to the likely alteration of risk perceptions and public attitudes.

SCIENCE: THE DECLINE OF DEFERENCE

There is a growing belief among many scientists that public policy has become hijacked by an irrational public led by environmental pressure groups and fuelled by a hostile press, whose assessment of risks is largely based on fear, not fact. Public pressure leads to the inefficient use of resources; significant delays in the development process due to the Not In My Back Yard (NIMBY) and Not In Anyone's Back Yard (NIABY) syndromes (Tait, 1988); increased costs to the economy exacerbated by inconsistent safety investment requirements across sectors of industry; and a continual challenge to established scientific expertise which undermines its legitimate authority to advise governments on rational safety investment and technology policy (Hadden, 1991; Otway and Peltu, 1985). According to Hadden (1991):

.... the public is no longer willing to defer to institutions and groups that may have commanded more respect in earlier years. Both government and industry are regarded as less than completely trust-worthy. Even scientists, who are still held in some regard, are not as respected as heretofore. Before, scientists who had difficulty in conveying their risk messages to citizens could fall back on appeals to expertise and status. This is less true now;"

Laird (1989) sees this "decline of deference" as something to be acknowledged, if not lamented. Indeed, this is an anxiety which has also been expressed much closer to home. Referring to the vigorous public campaign which has been conducted against food irradiation, the Chairman of the Advisory Committee on Novel Foods and Processes (ACNFP), (Burke, 1991), was moved to soliloquise:

"This is not the language of reasoned argument; this is a Crusade. What went wrong, why are people reacting like this? Did we just fail to explain it clearly enough in scientific language? I don't think so; there is something much more deep-seated behind this resistance which I only partially understand. Certainly the image of the scientist has become tarnished; the man in the white coat has become the enemy, the expert committee is no longer trusted. This troubles me deeply. We must try and tease out what the problems are and how we can explain our position, not that we shall win over the Crusaders, but we must explain our position to the reasonable person in the general public who is so far listening more to the Crusaders than to ourselves."

In order to understand why this is the case, we need to examine carefully why expert and lay assessments of risk differ so markedly.

THE SOCIAL PSYCHOLOGY OF EXPERT AND LAY ASSESSMENTS OF RISK

While the precise definition of different terms encountered in risk assessments will vary from organization to organization and according to context, the *technical* approach to risk, which is in common among risk professionals, defines risk as:

- the likelihood (or expected frequency) of an (adverse) specified consequence.

Expert judgements of risk, therefore, are concerned primarily with risk as *probability x consequences*. Expert assessments rely upon available data, fault-tree analyses, mathematical calculations, and they are prone to uncertainties. Those uncertainties relate to both sides of the equation, that is, both to the calculation of probabilities of occurrence and to the estimation of the consequences (Roberts and Hayns, 1989). Risk experts, generally speaking, are at pains to clarify where their assessments are open to uncertainty. Their assessments are merely an aid to decision-making for investment in safety and to risk management. As such, although exact statements of the *level* of uncertainty to be attached to risk calculations are rare, risk experts take considerable care to ensure the scientific exactitude of their assessments.

The difficulty with this technical approach to assessments often lies in the practical context of the hazard being assessed. Thus, calculating the probabilities associated with the possible exposure of a local environment or population to toxic materials is difficult to make precise. Similarly, toxicological data and their likely effects upon human beings may also be imprecise. Thus, calculating the health consequences is also prone to uncertainty. Nevertheless, expert assessments of risk continue to be a valuable aid to decision-making on important environmental issues: risk professionals feel comfortable in making such assessments, given the qualification that where uncertainties exist, they should be allowed for with caution ("conservatism"), and where expert judgement is involved, it should be made explicit. The Royal Commission on Environmental Pollution's espousal of the GENHAZ approach to assessing the hazards associated with GEMMOs is a good example of this method (RCEP, 1991). Care must be taken, however, not to assume that the provision of a rigorous technical assessment will, of itself, quieten public concerns.

In the USA in 1978, Baruch Fischhoff and his colleagues set out to examine the differences between expert and lay assessments of risk, and to determine those factors which most strongly influenced public assessments of risk associated with some 90 diverse activities/technologies. In general, empirical studies from within the psychometric paradigm pursued by Fischhoff *et al* (1978) demonstrate that perceived risk is both quantifiable and predictable. Moreover, the work has identified systematic differences in risk perception between groups (Slovic *et al*, 1985).

Whereas experts' judgements of risk correlate highly with technical estimates of annual fatalities, lay people can also produce estimates of annual fatalities, but their judgements of risk appear to be more strongly influenced by other hazard characteristics such as catastrophic potential and threat to future generations. Experts have a tendency to under-emphasize risks of low consequence, high probability events, whereas lay people tend to over-emphasize the risk of high consequence, low probability events.

Concerns about these differences in perception, meaning the various attitudes, judgements and predispositions held by people, led Slovic *et al* (1985) to develop questionnaires in order to

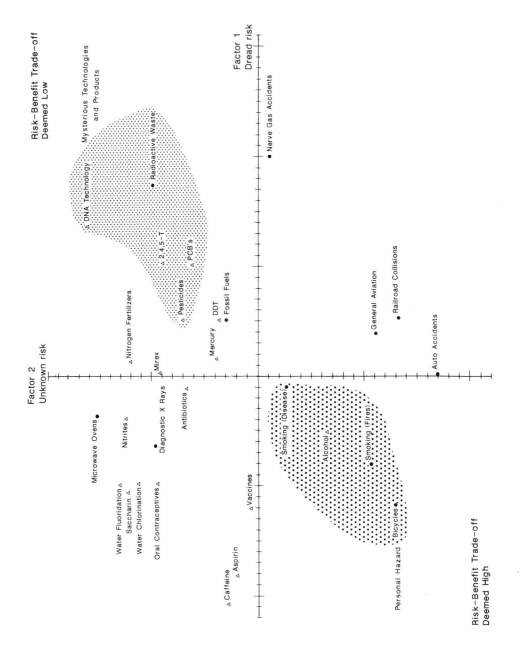

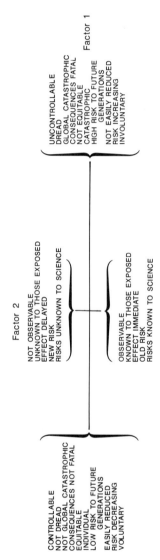

Factor 2

NOT OBSERVABLE
UNKNOWN TO THOSE EXPOSED
EFFECT DELAYED
NEW RISK
RISKS UNKNOWN TO SCIENCE

OBSERVABLE
KNOWN TO THOSE EXPOSED
EFFECT IMMEDIATE
OLD RISK
RISKS KNOWN TO SCIENCE

UNCONTROLLABLE
DREAD
GLOBAL CATASTROPHIC
CONSEQUENCES FATAL
NOT EQUITABLE
CATASTROPHIC
HIGH RISK TO FUTURE
    GENERATIONS
NOT EASILY REDUCED
RISK INCREASING
INVOLUNTARY

Factor 1

CONTROLLABLE
NOT DREAD
NOT GLOBAL CATASTROPHIC
CONSEQUENCES NOT FATAL
EQUITABLE
INDIVIDUAL
LOW RISK TO FUTURE
    GENERATIONS
EASILY REDUCED
RISK DECREASING
VOLUNTARY

Fig.1. Location of hazards on factors 1 and 2 derived from the relationship among 18 risk characteristics. Each factor is made up of a combination of characteristics, as indicated by the lower diagram (after Slovic, 1987).

elicit people's "expressed preferences" in relation to the risks and benefits associated with various hazards and activities. Drawing on personality theory, the psychometric approach characterizes the 'personality of hazards' by rating them in relation to key characteristics or factors which are hypothesized to influence risk perception. Subjecting the results of a national survey of public perceptions of risk to factor analysis, the studies revealed that in addition to the balance of *benefits* judged to result from an activity, three central characteristics associated with the risks themselves, most strongly influenced public perceptions. These were:

- **Dread** - the extent to which the activity/technology induced a sense of fear;
- **Familiarity** - the extent to which the activity/technology was a part of people's everyday lives;
- **Number of people exposed** - the catastrophic potential of the activity/technology for causing harm to humans.

In other words, the emphasis within public assessments of risks lies more in the area of *consequences* of the unfavourable outcome rather than the probabilities of occurrence in the first place.

The results also showed that every hazard had a particular 'personality profile'. By principal components factor analysis, Slovic *et al* (1985) investigated the inter-relationships between hazard characteristics. The upper diagram in Figure 1 presents a spatial representation of selected hazards within a factor space reflecting the perceived risk characteristics set out in the lower part of the figure, that is, the degree to which a risk is understood (Factor 2) and the degree to which it creates a feeling of dread (Factor 1). Thus DNA Technology scores highly in terms of dread, potential for catastrophe and risk to future generations on the one hand, (Factor 1), and in terms of being a new risk, poorly understood by the lay public on the other hand, (Factor 2). It is, therefore, located in the upper right quadrant of the diagram. Research has demonstrated that lay perceptions of risk are closely related to the position of a hazard within this factor space. The following are the most important features of lay perceptions of risk:

1) the most influential factor is dread risk;
2) the further to the right in Figure 1, the higher the perceived risk;
3) the further to the right in Figure 1, the more people wish to see current risk levels reduced and stricter regulation of the hazard imposed.

In 1983, Lindell and Earle undertook a survey to determine what factors affected people's willingness to live near a range of different industrial facilities. A central finding of this study was that the public often did not *trust* the facility operators. The final perceived risk factor which needs to be specified therefore is that of *personal control* of consequences. This is a finding which has been supported by many subsequent studies. It also goes some way to explain why so many people will willingly expose both themselves and their children to high levels of risk in their everyday lives on the roads, at home and at play, if they feel they are exercising some personal choice in driving too fast, in carrying out do-it-yourself repairs in the home, and in choosing to go hang-gliding or mountaineering for pleasure. They value the benefits, exercise choice, and therefore accept the consequences.

In a recent contribution to risk perception research in relation to hazardous waste disposal, Hadden (1991) argues that:

".... public perception is part of a broader perception of the distribution of power in society, so that public reaction to risky activities reflects concerns about power as well as about risk."

That is, we need to address the question of *control* as an element of risk perception. Access to the decision-making process becomes a prime target not only for drawing scientific expertise and public opinion together, but also for generating the conditions necessary for the resolution of conflicts around environmentally impacting technologies.

The policy implications are significant. Should investment decisions be based upon expert assessments of the likely health risks or rather upon the public's perceived risk assessment? American researchers provide an alternative solution, namely, to invest in *risk communication* projects in order to attempt to reconcile public concerns with expert judgements about health impacts. The word *reconcile* is employed cautiously here. For there is no wish to give the impression that either the expert assessment as opposed to the public perception, that is, the "objective risk" as opposed to the "subjective risk", is correct, and that the public view needs correction. Rather, the point is that expert and lay-people's assessments of risk differ markedly, and mutual enlightenment of the alternative perspective is the *sine qua non* for improved environmental risk decisions. In the words of Paul Slovic (1987):

> "Perhaps the most important message from this research is that there is wisdom as well as error in public attitudes and perceptions. Lay people sometimes lack certain information about hazards. However, their basic conceptualization of risk is much richer than that of the experts and reflects legitimate concerns that are typically omitted from expert risk assessments. As a result, risk communication and risk management efforts are destined to fail unless they are structured in a two-way process. Each side, expert and public, has something valid to contribute. Each side must respect the insights and intelligence of the other."

A closer look at the selected hazards portrayed in Figure 1 reveals that biotechnology-related risks appear to have a high "dread factor". They tend to group towards the upper right-hand quadrant of the factor space. The public appears to judge such risks as holding potential for future harm and would support investment to reduce the likelihood of such harm occurring. Hypothetically, in order to move the perception of biotechnological risks towards the bottom left-hand quadrant, questions of control over the technology and its use, greater understanding of the products and processes involved, and improved communication of advice would appear to be necessary.

PUBLIC CONCERNS ABOUT BIOTECHNOLOGY

The growth of environmental awareness in the late 1980s and early 1990s cuts across a wide range of industrial processes and products. The *bête noire* of many environmental pressure group campaigns has largely been the nuclear industry, but the *bête naissante* may be biotechnology. According to Collins (1990):

> "In a survey of German public opinion on current technology in which the likelihood of a mishap/misuse and the extent of a resultant catastrophe arising from such a mishap were evaluated, gene technology ended up in a 'danger class' of its own, far above nuclear energy, lasers, chemistry, and microelectronics."

Clearly, we need to understand the types of issues in relation to biotechnology which may be driving public opinion and attitudes in this direction. A systematic review of the published literature suggests that public attitudes are underpinned by five main areas of concern. These are socio-economic implications, environmental effects, health concerns, ethical problems, and last, but not least, loss of trust in science. Table 1 outlines the main issues which are to be found in published literature on biotechnology and genetic engineering. These are not mutually exclusive categories. Questions which have been classified here, say, under the socio-economic and environmental heads, often have an ethical dimension also. A good example is the concern expressed about the extent of particular commercial interests in the development of genetically-modified agricultural products and foodstuffs, and the potential conflict of interest between the necessary commercial confidentiality on the one hand and the need to protect more universal interests in protecting the ecosystem, on the other.

A thoroughgoing review of this material is in its own right a substantial subject for research which lies beyond the scope of this chapter. Table 1, however, does provide a relatively comprehensive overview of the topics of concern. It will suffice here to make two central points. First, it is an interesting empirical question whether such concerns are *real* concerns of substance or whether they are "vocabularies of motive and of justification" (Kemp, 1977, 1990). In other words, they may be expressions of concern which are being raised as a means of creating public access to an area of applied scientific and public policy which many feel would otherwise be closed to them. Secondly, and flowing from this, we can interpret these publicly-voiced concerns as expressions of the classic factor 1 and factor 2 dimensions of risk perception outlined above. Namely, they reflect judgements that biotechnology has

Table 1   Public Concerns over Biotechnology

SOCIO-ECONOMIC	• Commercial Exploitation • Role of Multi-nationals	• Patenting Rights • Effects on Third World Agriculture and Profitability

ENVIRONMENTAL	• Threat to Ecosystems and
	  Indigenous Species
	  - Competition
	  - Predation
	  - Parasitism
	  - Food Webs
	  - Habitat Destruction
		• Pollution
		• Loss of Biodiversity
		• Transfer of DNA

HEALTH	• Resistance Transfer
	• Long-term Effects
	
	  - Gene Therapy
	  - Genetic Drift
		• Toxins in Novel Foods
		• Creation of Uncontrolled
		  Organisms

ETHICAL	• Man playing God
	
	• Human/Animal Rights
	• Links to Biological Warfare
	
	• Who Decides?
		• Effects on Evolutionary
		  Process
		• Equality and the Third World
		• Secrecy/Commercial
		  Confidentiality
		• Malthusianism

TRUST IN SCIENCE	• Human Error • Commercial Science • Mitigation Measures	•Dose-Response Relationships • Relevance of Animal and Tissue Studies • Predictability, Reliability

Sources: Ashwell, 1990; Brill, 1988; Collins, 1990; Evers-Kiebooms *et al*, 1987; Fiksel and Covello, 1986; Flanagan, 1990; Hicks, 1990; Kramer and Penner, 1990; Motulsky, 1983; Murray, 1985; Pimentel *et al*, 1989; Straughan, 1990; Tait, 1988; Watts, 1990; Yoxen and Di Mento, 1989.

catastrophic potential for the environment; there is a lack of trust in those responsible for creating and regulating the new products and processes; there is a lack of basic understanding of the technology; and there is a sense of loss of personal control over the consequences. One way of attempting to alleviate these public concerns is to see them as a consequence of ineffective communication about the risks and the benefits of biotechnology.

RISK COMMUNICATION

Risk communication is a collective noun for a variety of procedures expressing quite different attitudes towards the relationship between the general public and the technical-managerial elite (Covello *et al*, 1986; US National Research Council, 1989). The orthodox approach to risk communication, which developed in part out of the psychometric studies of risk perception, conceptualized the risk communication process as involving an information *source*, a *channel* for the transmission of a message or *signal*, and a *receiver*.

An important element in this framework is the *signal value* by which Slovic *et al* (1984) meant the extent to which an event or activity provides new information about the likelihood of similar or more hazardous future such events. The *signal potential* of an event reflects its potential social impact, and there are some interesting features here. For example, an accident that kills a large number of people may produce very little social disturbance, if it occurs as part of a familiar, well understood system - such as in the case of an air crash due to pilot error. Conversely, a small accident in an unfamiliar or mistrusted system or product such as with biotechnology may have immense social and economic consequences if it has high *signal potential*, that is, if it is seen as a harbinger of worse things to come. Within the factor space shown in Figure 1, events with high signal potential were found to predominate in the top right-hand corner of the diagram (Slovic *et al*, 1987). According to Slovic, (1990):

> One implication of the signal concept is that effort and expense beyond that indicated by a cost/benefit analysis might be warranted to reduce the possibility of 'high-signal accidents'."

Thus our attention is drawn to the types of information which, when communicated to the general public, may create even wider gulfs between expert, "objective", technical assessments of the risks associated with an industrial operation and the perceptions of the general public and even political decision-makers.

The earliest solution to this problem was to develop a set of guidelines and manuals for various sectors of industry to create "effective risk communication" in order to calm public "outrage" (Covello *et al*, 1988; Hance *et al*, 1987; Sandman, 1986). Typical rules for the effective communication of risk information included:

- Accept and involve the public as a legitimate partner
- Plan carefully and evaluate performance
- Listen to your audience
- Be honest, frank and open
- Co-ordinate and collaborate with other credible sources
- Meet the needs of the media
- Speak clearly and with compassion.

These "seven cardinal rules" were posited to absolve the industrialist of the sin of communicating ineffectively with American communities armed with the 'The Emergency Planning and Community Right-to-Know Act' (Title III of Superfund) 1987. Although such orthodox approaches recognize that risk communication must be two-way to be effective (Hance *et al*, 1987), the general sense is nevertheless one of aiming risk communication at the concerns and information needs of specific target audiences. This is patently sensible. But what of those situations where the concerns and information needs of particular sections of society are poorly understood?

The US National Research Council, Committee on Risk Perception and Communication, published *Improving Risk Communication* in 1989. The following definitions were offered:

> "We see risk communication as an interactive process of exchange of information and opinion among individuals, groups, institutions." "We construe risk communication to be successful to the extent that it raises the level of understanding of relevant issues or actions for those involved and satisfies them that they are adequately informed within the limits of available knowledge."

The point to note about these definitions is that they admit to the complex and interactive nature of the process, (Kasperson *et al*, 1988), the focus, nevertheless, is on the *transmission of information*. Further, the implication is that effective risk communication must send signals to a public which may well be deficient in terms of its ability to comprehend those signals or messages. Fischhoff (US National Research Council, 1989) summarizes a section on "strategies for risk communication" thus:

> "... communicators (should) avoid simplistic strategies that leave recipients, at best, unsatisfied and, at worst, offended by the failure to address their perceived needs. In some cases this will be for better information; in other cases, they will be for better protection. Only after communication programs are recipient centred in this respect can they productively begin to be recipient centred in the sense of .... considering laypeople's strengths and weaknesses in understanding risk information".

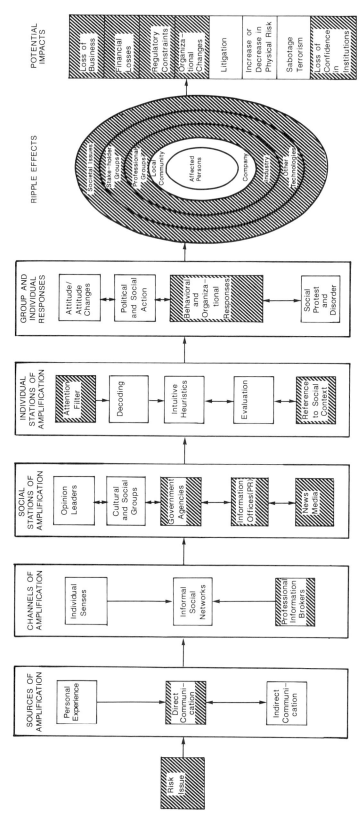

Fig. 2. A framework of social amplification of risk. (after Kasperson *et al*. 1988)

SOURCES OF AMPLIFICATION

Risk Issue

Personal Experience → Direct Communication ↔ Indirect Communication

CHANNELS OF AMPLIFICATION

Individual Senses → Informal Social Networks ← Professional Information Brokers

SOCIAL STATIONS OF AMPLIFICATION

Opinion Leaders ↔ Cultural and Social Groups ↔ Government Agencies ↔ Information Offices (PR) ↔ News Media

INDIVIDUAL STATIONS OF AMPLIFICATION

Attention Filter → Decoding → Intuitive Heuristics ↔ Evaluation ↔ Reference to Social Context

GROUP AND INDIVIDUAL RESPONSES

Attitude/Attitude Changes ↔ Political and Social Action ↔ Behavioral and Organizational Responses ↔ Social Protest and Disorder

RIPPLE EFFECTS

Affected Persons — Company — Local Community — Professional Groups — Stake-holder Groups — Societal Issues — Industry — Other Technologies

POTENTIAL IMPACTS

Loss of Business
Financial Losses
Regulatory Constraints
Organizational Changes
Litigation
Increase or Decrease in Physical Risk
Sabotage Terrorism
Loss of Confidence in Institutions

108

The argument here appears to suggest that the source of the problem is one of presentation, that the message is not "getting across" because of certain identifiable characteristics of the recipient. The orthodox perspective still implicitly places the scientist, technologist, regulator or bureaucrat in an implied position of superiority and control. While acknowledging the need to respond to people's express needs and interests, and also the importance of the findings of the psychometric approach, we rather need to shift the focus of attention away from the "recipients" and *back on to the responsible organizations*. A concern for avoiding distortions or alterations in public perceptions of risk, means that we need to pay attention to two further central considerations:

a) the organizational structure in operation; and
b) *the process* of risk communication.

RISK COMMUNICATION: THE CRITICAL APPROACH

A way out of the impasse between "expert" and "lay" assessments of risk is to create legitimate decision-making mechanisms in which all the relevant social and environmental issues can be fairly addressed. This is also the position taken by Hadden (1991) who argues for new forms of risk communication processes which focus on the need for equity and control. These, she argues, must include:
- early participation in the decision-making process;
- attention to issues of compensation for loss, and for increased risk;
- the careful use of independent experts to aid the decision process;
- progress towards new participatory institutions, including citizens' advisory panels;
- continual review to assess whether alternative methods/products offer the best practicable solution.

There is a further point. The Critical Approach cautions against potential *distortions* in the risk communication process. Distorted communication not only creates inequalities and unfairness in the decision-making process, it also creates the possibility of human beings and the environment being placed at greater risk, because the controlling decision was made either without full or with false information or without proper participation.

RISK AMPLIFICATION, BAKERS' YEAST, AND STIGMA

In setting out the main factors which influence individual attitudes and societal judgements about 'risk events' and which bring about the divergence of perceptions between technical expertise and the general public of the level of risk associated with a particular activity, the psychometric paradigm provides the basis for a model of the *social amplification of risk* (Kasperson *et al*, 1988) (Figure 2). Simply put, the model draws on communications theory to represent a one-way transmission of a message: signal - transmission - receiver - response. The authors' chief contribution is to point out that the risk message is prone to alteration or distortion during transmission, with a range of potentially costly outcomes or "second order impacts". According to the model, amplification occurs at two stages; in the transfer of information about the risk and in the response mechanisms of society. Signals about risk are processed by individual and by social amplification 'stations', including the scientific expert or committee which communicates the risk assessment; the news media; environmental pressure groups; and interpersonal networks. The response mechanisms reflect the extent to which 'ripple effects' occur - the greater the amplification, the greater the spread of effects. Individuals, communities, companies, professional groups, whole industries and related technologies may incur costs as a result of the risk amplification process. The classic example of this process is the Three Mile Island accident in 1979, but we can also point to public concerns about health risks from asbestos in buildings, bovine spongiform encephalopathy (BSE), and Perrier water. The second order effects listed by Kasperson *et al* (1988) include financial costs to business, increased regulatory activity and associated costs, loss of confidence in institutions, and litigation.

Confirmatory work by Burns *et al* (1990) showed that analysis of 108 accident events was consistent with the social amplification of risk hypothesis. A particularly important *signal* was the perception that an event was caused by managerial incompetence.

One salutary aspect of the discussion of societal responses to risk information in this model deserves some further consideration. Kasperson *et al* (1988) suggest that *stigma*, the negative imagery associated with an undesirable victim as judged by some observer, may also come to be attached to polluted environments or an hazardous technology (Kasperson *et al*, 1988). According to Slovic (1990) stigmatization may be generalized from persons to environments, technologies and products, and since the normal response to stigmatized objects is avoidance, risk-induced stigma may have significant social and policy implications.

The chief characteristics or dimensions of social stigma have been elucidated by Jones *et al* (1984) as follows:

1. *Concealability*   Is the condition hidden or obvious? To what extent is its visibility controllable?

2. *Course*   What pattern of change-over time is usually shown by the condition? What is its ultimate course?

3. *Disruptiveness*   Does the condition block or hamper interaction and communication?

4. *Aesthetic*   To what extent does the condition make
   Qualities   the possessor repellent?

5. *Origin*   How did the condition originate? Was anyone responsible for it, and what was he or she trying to do?

6. *Peril*   What kind of danger is posed by the condition and how imminent and serious is it?

Slovic (1990) examines proposals for high-level radioactive waste disposal in Nevada and suggests that dimension 6, *Peril* is the key link between stigma and perceived risk. It is, however, possible to see strong potential connections between genetically modified organisms and *all* of the dimensions of stigma outlined above. Indeed, there are strong correspondences with the central themes of expressed public concerns about biotechnology discussed earlier (Table 1). The point, therefore, is that the release of genetically modified microorganisms into the environment displays the potential for a significant increase in public opposition due to processes of social amplification of risk, reinforced by the possibility of social stigma driven by expressed public concerns and attitudes.

It is, therefore, instructive briefly to examine in terms of the social amplification of risk model the recent decision by the Advisory Committee on Novel Foods and Processes (ACNFP) to approve the release of a genetically-modified bakers' yeast. This is an important case study and six points should be made which indicate that we should attempt to understand why the social amplification of risk occurred in this instance. The yeast, which was developed by Gist-Brocades by intraspecies modification, is the first example of a GEMMO to be assessed and to be approved by the ACNFP. The modification involved the alteration of the ability of the yeasst to utilize maltose from an inducible to a continuously expressed function. It was not considered to raise any serious environmental problems by the International Releases Sub-Committee which advises the Health and Safety Commission (Beringer, 1991). Secondly, the United Kingdom is said to be ahead of the rest of Europe and the US in regulating genetic modification in food (Watts, 1990). Thirdly, release was sanctioned on the ground that it "posed no unacceptable risk to human consumers" (ACNFP, 1989). Fourthly, the public furore which ensued, involved national press coverage which was largely antipathetic. Fifthly, subsequent commercial exploitation of the yeast appears to have been slowed. Sixthly, the ACNFP has now revised its procedures for publicising its decisions, and finally, questions of consumer attitudes and perceptions of GEMMOs, particularly in relation to novel foods and processes, are now being given much more serious consideration in the determination of acceptable risks of environmental release (Beringer, 1991; Burke, 1991).

The chronology of events in relation to the bakers' yeast episode is outlined in Table 2. Careful steps to assess the risks associated with the yeast were taken by the appropriate advisory committees in the months leading up to the press announcement of 1 March 1990. Given the range of potentially much more complex processes which are likely to arise, a

Table 2   Genetically-modifed Bakers' Yeast: Chronology of Events

December 1988	ACNFP - asked to examine safety in food use of a genetically manipulated bakers' yeast developed by Gist-Brocades. First example of a GMO for food to be assessed by the ACNFP.
January 1989	ACNFP meeting - considers safety of bakers' yeast for use in food.
June 1989	H.S.C's Intentional Introductions Sub-Committee cleared yeast on the grounds that it posed no significant environmental impact risk or risk under the Health and Safety at Work Etc. Act.
July 1989 (6th)	ACNFP meeting - approved release to interested parties. Ministers advised to grant formal approach.
1989	Issue is now passed by all concerned and is finalised. Food Advisory Committee advises no special labels necessary on packaging. Reason: genetic manipulation has occurred intraspecies.
March 1990 (1st)	ACNFP/MAFF/DOH Press Release: ".... the product may be used safety." Press responses: "Bionic bread sales wrapped in secrecy" (Today) "Mutant yeast is half baked way to slice up Nature" (Today) "Are the boffins taking the rise out of bread?" (The Star) "Man-made yeast raises temperatures" (The Independent) "Genetic yeast passed for use" (The Times) Consumers' Association: "We think all genetically altered foods should be labelled." ACNFP decides to review procedures to ensure press releases include adequate information on the product and the approval process.
October 1990	ACNFP Workshop on Consumer Concerns. Proposals passed to Ministers include: • broadening membership of ACNFP; • allowing observers in to ACNFP meetings; • improved access to information; • increased consultation; • provision of educational material; • research into consumer perceptions and food choice; • guidance notes on commercial confidentiality • FAC to publish guidelines on food labelling.

bakers' yeast involving intraspecies modification, albeit the first-of-a-kind decision, was not expected by Committee members and civil servants alike to cause a public reaction in the way that it did. The two main areas of public concern which arose were:

• the lack of information about the assessment process - the decision appeared to have been taken "behind closed doors"; and
• the decision to grant use of the yeast without any requirement for product labelling.

The baldness of the official press release led to a large and weighty rock being dropped into the tabloid press news pool. The model of Kasperson *et al* (1988) highlights the news media as a significant station of risk amplification - here we have a classic example of the amplification process at work (see Figure 2). The shaded areas in Figure 2 show the course of social amplification for the Bakers' Yeast example. The Figure emphasizes the public and press response to official sources and channels of risk information. In this instance, there were no "affected persons" or victims who were exposed to any physical harm. Nevertheless, the ripple effects and potential impacts associated with the incident were extensive, as shown. Clearly, in terms of the public risk perception framework discussed above, the lack of available knowledge and personal control for the general public are important factors. The potential for generating stigma in both the product and the approval process arises through the dimensions of concealability, of origin, and of peril mentioned above. These are worrying insights, given the scientifically assessed level of risk.

CONCLUSIONS

It would appear that the 1990 ACNFP workshop, though restricted to 16 members drawn from various Advisory Committees, 15 Civil Servants and 10 others, including two professors and four PhDs, addressed two of the salient issues of public concern, namely, the need to increase the transparency of the approval process and the representativeness of committee membership. It would appear that further attention might profitably be paid to questions of risk communication. As noted above, the key influences upon public attitudes to risk and the processes which create social amplification of risk are now well understood by social psychologists. The potential for social stigma to be attached to the products of biotechnology has been discussed. The resulting second-order costs associated with amplification and stigma may be considerable. If the potential benefits associated with genetic modification are to be judged as socially acceptable, then it appears to be all the more important to recall both the recommendations and the advice of the Royal Commission on Environmental Pollution's Thirteenth Report (1989, para 5.47)

"We propose a precautionary but realistic system of regulation. This should allow safety issues to become part of the development of the technology rather than having to be introduced following problems .... which caused serious damage to human health or to the environment and destroyed public confidence in both the science and the scientists."

Studies of the public perception of risk demonstrate that public confidence may also be harmed by paying inadequate attention to the nature of public attitudes to innovation and public concerns about GEMMOs. Risk communication and public participation in generating acceptable risk also deserve careful consideration before, not after the event.

ACKNOWLEDGEMENTS

My thanks to Sam Jones and Nia Blank for assistance in the preparation of this paper. Thanks also to Professor Derek Burke for informal discussions on the work of the ACNFP. The views expressed in this paper are the sole responsibility of the author.

REFERENCES

ACNFP 1989 Advisory Committee on Novel Foods and Processes. *Annual Report 1989.* London, Department of Health; Ministry of Agriculture, Fisheries and Food.
ACNFP 1990 Advisory Committee on Novel Foods and Processes. *Annual Report 1990.* London, Department of Health; Ministry of Agriculture, Fisheries and Food.
Ashwell, M. 1990 Consumer perception of food related issues. *ACNFP Workshop*, Paper 19.
Beringer, J. 1991 The release of genetically modified organisms. In: Roberts, L.E.J. and Weale, A. (Editors) *Innovation and Environmental Risk.* London, Belhaven Press.
Brill, W.J. 1988 Why Engineered Organisms are Safe. *Issues in Science and Technology,* 44-50.
Burke, D. 1991 Public acceptance of innovation. In: Roberts, L.E.J. and Weale, A. (Editors) *Innovation and Environmental Risk.* London, Belhaven Press.

Burns, W., Kasperson, R., Kasperson, J., Renn, O., Emani, S. and Slovic, P. 1990 Social amplification of risk: an empirical study. Unpublished. Oregon, Decision Research.

Collins, J. 1990 A little knowledge is dangerous. *Biotechnology*, 8, 6.

Covello, V.T., Sandman, P.M. and Slovic, P. 1988 *Risk communication, risk statistics and risk comparisons: a manual for plant managers*. Washington DC, Chemical Manufacturers Association.

Covello, V.T., von Winterfeldt, D. and Slovic, P. 1986 Risk communication: a review of the literature. *Risk Abstracts*, 3, 4, 171-182.

Douglas, M. 1985 *Risk acceptability according to the social sciences*. New York, Russell Sage.

Douglas, M. and Wildavsky, A. 1982 *Risk and Culture: an essay on the selection of technical and environmental dangers*. Berkeley, University of California Press.

Evers-Kiebooms, G., Cassiman, J-J, Van den Berghe, H. and d'Ydewalle, G. 1987 (Editors) *Genetic Risk, Risk Perception, and Decision Making*. New York, Alan R. Liss, Inc.

Fiksel, J. and Covello, V.T. 1986 (Editors) *Biotechnology Risk Assessment. Issues and Methods for Environmental Introductions*. Elmsford NY, Pergamon Press.

Fischhoff, B., Lichtenstein, S., Slovic, P., Derby, S.L. and Keeney, R.L. 1978 *Acceptable Risk*. Cambridge, Cambridge University Press.

Flanagan, P.W. 1990 Do genetically engineered micro-organisms pose risks to ecosystems? In: Cox, L.A. and Ricci, P.F. (Editors) *New Risks*. New York, Plenum Press.

Goffman, E. 1970 *Stigma*. Harmondsworth, Penguin.

Hadden, S.G. 1991 Public perception of hazardous waste. *Risk Analysis*, 11, 47-57.

Hance, B.J., Chess, C. and Sandman, P.M. 1987 *Improving dialogue with communities: a risk communication manual for government*. New Brunswick, N.J., Rutgers University.

Hicks, R.M. 1990 Novel Foods; Problems of Safety and Acceptability. *ACNFP Workshop*, Paper 3.

Jones, E.E. *et al* 1984 *Social stigma: the psychology of marked relationships*. New York, W.H. Freeman.

Kasperson, R.E., Renn, O., Slovic, P. *et al* 1988 The social amplification of risk: a conceptual framework. *Risk Analysis*, 8, 177-187.

Kemp, R. 1977 Controversy in scientific research and tactics of communication. *Sociological Review*, 25, 515-534.

Kemp, R. 1990 Why Not In My Back Yard? A radical interpretation of public opposition to the deep disposal of radioactive waste in the United Kingdom. *Environmental Planning*, A, 22, 1239-1258.

Kramer, C.S. and Penner, K.P. 1990 Consumer Response to Food Risk Information. In: Cox, L.A. and Ricci, P.F. (Editors) *New Risks*. New York, Plenum Press.

Kraus, N., Malmfors, T. and Slovic, P. 1990 Intuitive toxicology: expert and lay judgements of chemical risks. [Manuscript in preparation, referenced in Slovic, P. 1990.]

Laird, F. 1989 The decline of deference: the political context of risk communication. *Risk Analysis*, 9, 543-550.

Lindell, M.K. and Earle, T.C. 1983 How close is close enough: public perceptions of the risks of industrial facilities. *Risk Analysis*, 4, 245-253.

Motulsky, A.G. 1983 Impacts of genetic manipulation on society and medicine. *Science*, 219, 135-140.

Murray, T.H. 1985 Ethical issues in genetic engineering. *Social Research*, 471-489.

Nighswonger Kraus N. and Slovic, P. 1988 Taxonomic analysis of perceived risk: modelling individual and group perceptions within homogenous hazard domains. *Risk Analysis*, 8, 435-455.

Otway, H. and Peltu, M. 1985 (Editors) *Regulating Industrial Risks. Science, hazards and public protection*. London, Butterworths.

Pimentel, D., Hunter, M.S., LaGro, J.A., Efroymson, R.A., Landers, J.C., Mervis, F.T., McCarthy, C.A. and Boyd, H.E. 1989 Benefits and risks of genetic engineering in agriculture. *Bioscience*, 39, 606-614.

Roberts, L.E.J. and Hayns, M. 1989 Limitations on the usefulness of Risk Assessment. *Risk Analysis*, 9, 483-494.

R.C.E.P. 1989 Royal Commission on Environmental Pollution, *Thirteenth Report. The Release of Genetically Engineered Organisms to the Environment*. London, H.M.S.O.

R.C.E.P. 1991 *Fourteenth Report. GENHAZ* London, HMSO.

Sandman, P.M. 1986 *Explaining environmental risk*. Washington DC, TSCA Assistance

Office, US Environmental Protection Agency.

Shepherd, R. 1990 Consumer perceptions of potential risks from the application of biotechnology to food production. *ACNFP Workshop*, Paper 18.

Slovic, P. 1987 Perception of risk. *Science*, 236, 280-285.

Slovic, P. 1990 Perceptions of risk: reflections on the psychometric paradigm. Mimeo 54 pp. Forthcoming in Golding, D. and Krimsky, S. (editors) *Theories of Risk*. Dover, Mass. Auburn House.

Slovic, P., Lichtensten, S. and Fischhoff, B. 1984 Modelling the societal impact of fatal accidents. *Management Sciences*, 30, 464-474.

Slovic, P., Fischhoff, B. and Lichtenstein, S. 1985 Regulation of risk. A psychological perspective. In: R.G. Noll (Editor) *Regulatory Policy in the Social Sciences*. Berkeley, University of California Press.

Slovic, P. *et al* 1987 Risk perception, risk-induced behavior, and potential adverse impacts from a repository at Yucca Mountain, Nevada. Unpublished manuscript, Decision Research, Oregon.

Straughan, R. 1990 The genetic manipulation of plants, animals and microbes - the social and ethical issues for consumers: a discussion paper. *ACNFP Workshop*, Paper 13.

Tait, J. 1988 NIMBY and NIABY : Public perception of biotechnology. *International Industrial Biotechnology, 8, 6,*pp5-9.

US National Research Council 1989 *Improving Risk Communication*. Washington DC, National Academy Press.

Watts, S. 1990 Have we the stomach for engineered food? *New Scientist*, 3 November 1990, 24-25.

Wildavsky, A. and Dake, K. 1990 Theories of risk perception: who fears what and why? *Daedalus*, 119, 4, 41-60.

Wynne, B. 1984 Perceptions of risk. In: Surrey, J. (Editor) *The Urban Transportation of Irradiated Fuel*. London, Macmillan.

Yoxen, E. and Di Martino, V. 1989 (Editors) *Biotechnology in Future Society. Scenarios and Options for Europe*. Aldershot, Dartmouth Publishing.

WORKSHOP 1

PERSISTENCE AND SURVIVAL OF GENETICALLY-MODIFIED MICROORGANISMS RELEASED INTO THE ENVIRONMENT

Chairman and Rapporteur:

Ronald M. Atlas
Department of Biology
University Louisville
Louisville Kentucky USA

Participants:

Allan M. Bennett
Division of Biologics
PHLS CAMR
Salisbury, Wilts, United Kingdom

Rita Colwell
Department of Microbiology
University of Maryland
College Park Maryland USA

Jan van Elsas
Institute for Soil Fertility Research
Wageningen Netherlands

Steffan Kjelleberg
Department of Marine Microbiology
University of Goteborg
Goteborg Sweden

Jens Pedersen
Division of Marine Ecology and Microbiology
National Environmental Research Institute
Soborg Denmark

Walter Wackernagel
Department of Genetics
University of Oldenburg
Oldenburg Germany

When recombinant DNA technology first permitted the creation of genetically-modified microorganisms (GEMMOs), efforts were made to prevent the release of these GEMMOs into the environment. A decision was made at the Asilomar Conference to establish appropriate containment facilities and to develop GEMMOs that could not survive outside the laboratory. Gradually as confidence built regarding the safety of GEMMOs and consideration was given

The Release of Genetically Modified Microorganisms
Edited by D.E.S. Stewart-Tull and M. Sussman, Plenum Press, New York, 1992

115

to the benefits that could be achieved through the deliberate release of GEMMOs into the environment, questions arose about the survival and persistence of GEMMOs released into the environment. The ability of a GEMMO to survive is central to its usefulness (functionality) and the persistence of a GEMMO is critical in assessing risk.

The ability of any introduced organism to survive and to persist in a particular ecosystem - regardless of whether it has been genetically-modified - depends upon the physiology of that organism, its ability to tolerate environmental conditions, and its capacity to compete with the indigenous microorganisms. GEMMOs can be formed with physiological properties that enhance their survival capabilities. Alternatively, GEMMOs can be created with debilitated survival properties and even can be made suicidal to limit their potential to persist and spread should they produce any untoward effects after their introduction into the environment.

MEASURING SURVIVAL AND PERSISTENCE OF INTRODUCED MICROORGANISMS

The ability to monitor the survival and persistence of introduced organisms depends upon the suitability of the detection methodologies. One method can indicate persistence and another, lack of survival. The various methods that can be employed for detecting microorganisms differ in their sensitivities and capabilities of distinguishing live from dead cells. Without a marker system, the detection system often is not sensitive enough. Therefore, emphasis often is placed on marked organisms that can be distinguished from background organisms. This may involve the use of selective media, such as media containing antimicrobics to which the strain is resistant or the use of genetic (hybridizable) markers. For nucleic acid detection, probes are based upon markers or upon random sequences. The type of probe, e.g., size and target, is important and will give varied specificity. Methods for detecting viable cells by direct microscopic observation of metabolically active cells - such as incubation with nalidixic acid or INT - allows three classes of microorganisms to be detected: viable culturable cells, viable non-culturable cells, and dead cells.

Major differences are often found in numbers as determined by immunofluorescence and by plate counts (Fig.1). While the difference in some cases may be due to dead cells that are detected by immunofluorescence and not by plating methods, at least in the case of *Rhizobium*, it often is possible to revive up to 100% of the cells by culture methods. The ability to detect microorganisms by direct immunofluorescence and not by plating applies to a variety of microorganisms and will apply to GEMMOs introduced into the environment. At least 1-2 orders of magnitude difference are often observed between direct detection and viable plating.

The suitability of a detection method may vary between environments. For example, viable selective plating and immunofluorescence microscopy show differences when used to

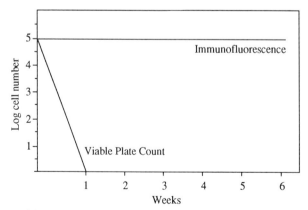

Fig.1 Persistent of introduced microorganisms as detected by viable-plate count versus direct immunofluorescence microscopy.

monitor the survival and persistence of *Enterobacter cloacae* on leaves and in soil. The fate of *E. cloacae* inoculated into soil near water-holding capacity or air-dried was of the same magnitude (10^7 per gram) in the wet soil, but in the dry soil selective plating dropped to about 10^3 cells/gram, while total immunofluorescence remained about 10^7 cells/gram. Immunofluorescence microscopy counts of barley and bean leaves were equivalent, but selective plating showed more culturable *E. cloacae* on barley compared to bean leaves. Immunofluorescence microscopy and selective plating results of top soil were comparable, with the exception of a period in a field experiment in which the top soil was extremely dry and hot and during which the numbers of colony-forming units decreased dramatically while the number of cells enumerated by direct microscopy was unchanged.

It is not clear how to treat the different results that are obtained with varying methodologies. To monitor reliably survival and persistence it may well be necessary to use a combination of techniques which may include direct microscopic detection by immunofluorescence, plating techniques, and gene probe hybridization.

PERSISTENCE OF INTRODUCED DNA

When considering the environmental introduction of GEMMOs it is important to consider not only the survival of the GEMMO itself but also the persistence of the recombinant DNA. DNA often persists longer than cells. It is important, therefore, to consider the persistence of DNA in the environment and whether free DNA can transform cells. It is possible that high molecular weight DNA that is released, including plasmids, may persist for hours. Such DNA may remain undergraded even in the presence of cells. DNA that remains stable is available to transform competent cells.

DNA released from cells into groundwater and soils rapidly adsorbs to surfaces, e.g., minerals and clays. This occurs rapidly as long as cations are present, with binding at pH 5-9 occuring in seconds. DNA associated with particles can still transform cells. Once adsorbed, the DNA becomes very resistant to nuclease digestion, showing a 100-fold greater resistance to nuclease digestion than unbound DNA. Starvation produced greatly enhanced competence and starved cells are better able to take up DNA than cells in non-stressed soils.

PERSISTENCE OF VIABLE NON-CULTURABLE AND STARVATION-INDUCED DIFFERENTIATED MICROORGANISMS

Bacteria introduced to the environment may enter non-culturable state. The concept of viable non-culturable bacteria originated from outbreaks of cholera in the United States with the finding that *Vibrio cholerae* is widely distributed in estuaries. These *V. cholerae* usually are not detected by plating on selective media. However, when incubated in the presence of nalidixic acid, their morphologies change and elongated cells form, demonstrating that these cells are alive and metabolically active. This response is not seen in all cells and only 1-10% of the cells observed microscopically could be confirmed to be viable. These viable cells are still pathogenic. These observations with *V. cholerae* also have been found with other bacteria including *Salmonella* and enteropathogenic *E. coli*. In cases where no viable cells were detected by plating on selective media, 10^4-10^6 cells were observed by direct microscopy and 1-10 percent of these could be shown to be viable to incubation with nalidixic acid. It has been possible to maintain viable non-culturable bacteria for several years and later to revive them. It also has been possible to demonstrate that these viable non-culturable bacteria remain pathogenic.

Non-sporulating bacteria can undergo differentiation that involves specific gene induction events over a period of time. Starvation will induce differentiation that often results in the formation of small cells. The formation of such microcells is the result of the expression of multiple genes. Starved cells are not the same as viable non-culturable cells. They are the result of different differentiation programs. Viable non-culturable cells can be recovered by a temperature shift that induces a resuscitation phase. Starved cells respond rapidly to nutrients. Both viable non-culturable or starved, that is differentiated cells, have mechanisms that enhance survival. During adaptation, differentiation, multiple stress tolerances arise. These multiple tolerances are of ecological importance since cross protection occurs, for example against predation, when viable non-culturable cells form.

SURVIVAL OF GEMMOs IN AIR

Many GEMMOs intended for deliberate release will be released as aerosols. The ability to survive in such aerosols is important to the success of such introductions. Additionally, the spread of introduced microorganisms may occur through the air. Hence, the survival of GEMMOs in air is an important consideration in assessing the risk of dissemination and persistence of introduced GEMMOs.

Water activity is a critical factor in determining the stability of GEMMOs in aerosols that affect how long an organism can survive and spread. When an aerosol is sprayed outside, it is very difficult to capture organisms. Aerosols can be captured on spider webs which are normally sterile. Gram-negative bacteria tend to survive less well than Gram-positive bacteria but the variation is strain specific. A difference of a factor of 10 can readily be seen between very closely-related organisms. Survival during stationary phase is better than during growth phase. Humidity enhances survival. There is an important difference in survival between indoor and outdoor air. Ozone and car exhaust products are quite toxic. This can be seen by differences in survival based upon wind direction.

DETECTION METHODS FOR MODIFIED ORGANISMS IN THE ENVIRONMENT

Chairman and Rapporteur:

Gary S. Sayler
The Centre for Environmental Biotechnology
The University of Tennessee
10515 Research Drive, Suite 100
Knoxville, Tennessee 37932

Participants:

Richard Burns
Biological Laboratory
University of Kent
Canterbury Kent CT2 7NJ

J.E. Cooper
Department of Food and Agriculture Microbiology
The Queens University of Belfast
Newforge Lane
Belfast BT9 5PX
Northern Ireland UK

Tim Gray
Department of Biology
University of Essex
Colchester Essex CO4 3SQ

Jens Pederson
Ministry of the Environment
National Environmental Research Institute
Division of Marine Ecology and Microbiology
Soborg, Denmark

Jim Prosser
Department of Molecular and Cell Biology
University of Aberdeen
Marischal College
Aberdeen Scotland AB9 1AS

Sonja Selenska
Department of Genetics
University of Bayreuth
Universitaetsstr. 30
W-8580 Bayreuth Germany

The Release of Genetically Modified Microorganisms
Edited by D.E.S. Stewart-Tull and M. Sussman, Plenum Press, New York, 1992

Methods for detection of genetically-modified microbes in the environment continue to evolve in terms of improved sensitivity and specificity, applications in varying environmental matrices, and use in fundamental studies of microbial ecology and environmental microbiology. There appears to be three discrete strategies currently under investigation, summarized as follows:

(1) Molecular detection of a specific recombinant (modified) DNA sequence within a host microorganism.
(2) Detection of marker traits or gene cassettes introduced into host organisms to specifically improve monitoring of microorganisms.
(3) Detection instrumentation for analytical monitoring of phenotypic characteristics of host or indigenous microorganisms.

These detection strategies are being applied to a diversity of genetically-modified as well as environmentally relevant organisms. Currently, virtually every environmental matrix; including soils, sediments, water, waste-waters, air and plant surfaces (both below and above ground), are being examined by one on more cellular or DNA specific detection methods. From these investigations it is becoming clear that no one method is universally suited to all types of environmental sample (without some degree of modification).

A recent advance in detection is the development of methods to detect the activity of specific genes or organisms in the environment. This development complements the existing capacity of microbial ecologists to determine the presence or absence of specific genes in a recombinant or indigenous host in the environment. As a consequence detection methods at various stages of use can provide quantitative information of the prevalence and activity of recombinant genes in the environment.

NUCLEIC ACID ANALYSIS AND HYBRIDIZATION

DNA/RNA probe technology and polymerase chain reaction (PCR) amplification of nucleic acid sequences from environmental samples continue the expansion of sensitive and diverse applications. A subtraction hybridization procedure coupled with PCR complication has been developed for selective detection of strain-specific organisms in mixed microbial cultures and for the preparation of highly discriminatory DNA probes. In this procedure target, single-stranded DNA is sonicated and the mixture is mercurated and biotinylated. The modified DNA is then allowed to hybridize with the sample DNA under solution phase conditions. The non-hybridizing single-stranded probe DNA is recovered following separation from the doubl-stranded hybrids by Sepharase and Strepavidin column chromatography .

Because this procedure used an undesirable reagent (Hg) and had a low recovery of the probe, the procedure was modified. Mercuration of the DNA was eliminated and the DNA sample was subjected to *Sau* 3A digestion rather than sonication followed by addition of PCR primers. The biotinylation was retained and the hybridized molecules were removed by avidin treatment and the single-stranded material was extracted with phenol/chloroform and precipitated with ethanol. The resulting product of the procedure can be directly subjected to PCR for probe preparation. Conversely, the sample can be subjected to gel electrophoresis and followed with PCR for probe preparation and sequence analysis. At an early stage of restriction enzyme treatment, appropriate selection of primers can permit cloning in vectors such as pUC18 and amplification of the fragments in the vector. These techniques have been successful in developing probes for analyzing soil microorganisms and are capable of discrimination among very closely-related strains as demonstrated for *Rhizobium loti*.

The extraction of DNA from soil samples, including methods to improve the quantity and quality of DNA recovered, permits discrimination of chromosome or plasmid-maintained genes and detection of genes harbored in a single organism. Direct cell lysis, together with polyethylene glycol concentration, $CsCl_2$ centrifugation, and phenol extraction, can routinely recover DNA from soils in a size class of approximately 25 kb. This DNA has been used as a target for Tn5 probing for an introduced *E. agglomerans* strain in soil. The procedure was highly-sensitive for up to 35 days following soil inoculation with 10^7 cells per gram of soil. It was noted that Tn5 was detected in Southern hybridizations in a new DNA fragment size class suggesting either a gene transfer event or molecular rearrangement in the host DNA.

The yield of DNA in the extraction procedure increased with time beyond the theoretical population density in soil. These results suggest that either many of the soil organisms were in a non-culturable state after prolonged incubation or that there was production of significant amounts of cell-free DNA in the soil. These hypotheses are in part based on a two-order of magnitude (from 10^9 to 10^7 cells/g) loss of the initial inoculum to the soil.

While $Tn5$ detection sensitivity by probing in DNA blot hybridizations or Southern transfers was adequate, it was demonstrated that the extracted DNA could be amplified (for $Tn5$) by PCR technology. However, this may not be universally true for soil-extracted DNA. There are questions concerning inhibition of the polymerase, the concentration of DNA, and optimized conditions in the PCR procedures. In the case of $Tn5$ there is also the question of a stem-loop structure of IS50 that may inhibit primed amplification.

Inclusion of diethyl pyrocarbonate (DEPC), as an RNase inhibitor in the extraction, permitted simultaneous recovery of total RNA. However, it was not clear whether this was predominantly a tRNA or rRNA. However, quantitative mRNA analysis methods for DEPC or hot phenol soil-extracted RNA are becoming available. Such methods have demonstrated *in situ* expression of indigenous biodegradative genes in soil.

FLUORESCENT ANTIBODY AND FLOW CYTOMETERY

Detection of bacteria by immunofluorescence microscopy (IF) has two major advantages: It is a direct method, and information on cell morphology and size can be obtained. Both steps, the immunological staining and the microscopic examination, are useful in other microbial ecological techniques: flow cytometry, scanning EM. It is apparent that IF methods are well-suited for detection of genetically-modified microorganisms when the host is relatively unique or when highly discriminatory mono- or poly-clonal antibodies are available.

Four major technical problems persist in the use of immunofluorescent methods:

(1) The threshold of detection is rather high (about $5x10^3$ and $5x10^5$ cells/g in soil and on leaves respectively. In our current procedure for leaf samples, stomacher extracts are stained without any separation step. A separation of bacteria and leaf debris, to improve the threshold of detection, might be achieved by filtration through Mirocloth™, which has been shown to give satisfactory recovery of bacterial DNA from stomacher extracts.

(2) Over-estimation: Standard IF does not distinguish between living and dead cells. This can be achieved by incubation with nalidixic acid. This technique, however, is not unproblematic; the distinction of elongated/swollen ("viable") cells is very subjective, and one still cannot say, that cells which are not counted as viable by this technique are really dead.

(3) Under-estimation: Algae and soil are separated from bacterial cells before enumeration in standard procedures. This separation may lead to loss of bacteria. Soil samples are usually flocculated. Flocculation must however be optimized with each soil, and bacterial species, and recovery may change through time.

(4) IF is a very laborious technique. The counting could be automated, and made more objective, with digitalized image analysis.

While still expensive in routine application, laser-based fluorescence flow cytometery has an important developmental application for automating detection methods for environmental and ecological use. In applications for cell detection, discrimination of bacteria and submicron sized soil particles can be achieved. Detection and discrimination of organisms based on forward light scattering is used to achieve data on abundance relative to organismal size. Analysis of light reflected as 90° angles provides information on topography (granularity) of the organisms or particle surfaces. Laser-induced fluorescence measured at 90° is appropriate for single cell detection of organisms; organisms targeted with fluorescent antibodies, 16S rRNA probes, or organisms containing intrinsically fluorescent cellular components.

For *Bacillus subtilis*, fluorescent antiserum stained preparations can distinguish vegetative cells or spores at 5000 cells per second under flow operations. Coupled with cell-sorting capability, it is also possible to isolate and/or encapsulate individual cells of a target organism. Practical applications currently exist for water and groundwater monitoring. However, extensive developmental research is still required for routine application to soil samples. For *B. subtilis* spiked soil samples, homogenization of the soil and passage through a 200 µm sieve still results in poor detection of the bacteria even with fluorescent antibodies. This is due to interfering particles. Blocking steps to reduce non-specific fluorescence and background subtraction of signals from soil particles has however achieved detection of *B. subtilis* inoculated into soil. This development is essential for routine use of the technology.

BIOLUMINESCENT MARKERS AND REPORTERS

The past three years has seen considerable developments in the use of bioluminescent (*lux*) gene cassettes for marking and detecting organisms to be released to the environment. Strategies used for developing such marker systems include:

(1) Incorporation of two structural genes for luciferase on a vector plasmid into the genome of the host organism,
(2) plasmid introduction of constitutively expressed *lux* genes containing all the structural genes for the fatty acid synthetase, reductase and the luciferase and
(3) introduction of heterologously regulated *lux* operons.

The attractiveness of this genetic marking system is that light sensing can be potentially visualized for single cells and that the light itself may be detected *in situ*. Beyond the issue of simple detection, the light production itself can be coupled to generalized or specific measures of cellular activity *in situ*.

A variety of organisms such as *E. coli*, *Erwinia carotovora*, *B. subtilis* and *P. fluorescens* have been examined for their ability to act as a suitable host for light production with applicability for environmental analysis. The *lux* molecular constructions in these organisms include chromosomal, plasmid or transposon maintained genes. Some constructions are auto-inducible, others constitutive, and others substrate inducible. Light production and sensitivity is virtually strain specific. The *lux* genes themselves are also a useful target for DNA probing since they are essentially absent from terrestrial and freshwater environments.

Light production can be correlated with physiological growth as a measure of cellular activity and has been found to be more sensitive than measures of dehydrogenase activity. Practical detection limits using luminometers is approximately 10^2 *E. coli* per ml or 10^3 per gram soil. However, charged coupled detectors (CCD) interfaced with direct microscopy can permit single cell detection. Photon-counting methods can quantify individual cell bioluminescence. Remote-sensing with fiber optic or liquid light pipes coupled to sensitive photomultipliers has been applied for *in situ* detection of naphthalene induced bioluminescence in soil. In addition, scintillation counting, photography and exposing X-ray film have all been used successfully for quantitatively monitoring bioluminescence. Consequently, bioluminescent monitoring costs can range from $100K for intensified CCD/microscopic photon-counting to a few dollars for photographic or X-ray film.

It is evident that there has been great activity in the development of detection systems for organisms in the environment. Many of these systems have important environmental and ecological applications beyond the issue of genetically-modified organisms in the environment. However, some of the detection methods themselves such as *lux* engineered marker strains are in their own right genetically-modified microorganisms with applications and potential for release to the environment.

WORKSHOP 3

ENVIRONMENTAL IMPACT

Chairman and Rapporteur:

Prof. Rita R. Colwell
President Maryland Biotechnology Institute
1123 Microbiology Building
College Park
Maryland 20742 USA

Participants:

Dr. Peter B. Baker
Laboratory of the Government Chemist
Queen's Road
Teddington
Middlesex
TW11 OLY UK

Dr. Martin A. Collins
Dept. of Food Microbiology
Agriculture and Food Science Centre
Queen's University of Belfast
Newforge Lane
Belfast   BT9 5PX
Ireland

Prof. G. Hamer
ETH/EAWAG
CH-8600
Dubendorf
Switzerland

Dr. Dorothy Jones
Dept. of Microbiology
Medical Sciences Building
University of Leicester
Leicester
LE1 9HN UK

Prof Jim Lynch
AFRC Institute of Horticultural Research
Worthing Road
Littlehampton
West Sussex
BN17 6LP UK

The Release of Genetically Modified Microorganisms
Edited by D.E.S. Stewart-Tull and M. Sussman, Plenum Press, New York, 1992

Dr. Ian P. Thompson
NERC Institute of Virology and
Environmental Microbiology
Mansfield Road
Oxford
OX1 3SR UK

The workshop on environmental impact was well attended, with approximately 70 participants. The session opened with a presentation by Dr. Lynch, who discussed exchange processes. He stated his lack of conviction that a principal ecological issue is at stake. That introduction of genetically-modified microorganisms into the environment is damaging environmentally has not been definitively demonstrated. An impetus for new methodologies in microbial ecology is plant genetic modification. In trying to determine the impact of genetically-modified plants on microbial communities, the British Department of the Environment sought to determine critical baseline factors. However, the dominant bacterial members of a community are not clearly understood. The Oxford Experimental Virology Group has carried out releases, e.g. studies of leaf and root surfaces in soil have been done to determine the dominant microbial members. The dominant versus genetically-modified microorganisms are appropriately marked and have been studied to determine any impact. Bacteria do not dominate soil; it is mainly the fungal community. Thus, it is necessary to determine the baseline fungal community.

Another critical component is protozoa-what is the protozoan role in turnover or other activities in soil. Elizabeth Scott has constructed a mathematical model of protozoan interaction, and has shown it is, in fact, chaotic, providing interesting information that, within chaos, one can analyze the community. She has determined that from factors defined in chaos, one can predict predator/prey relationships.

On leaf surfaces, it is crucial to look at those fungal communities which are dominant in the phylloplane microflora. It is now possible to study yeasts, especially since fatty-acid profiles aid in identification. However, it is impossible to recover all organisms in an ecosystem into culture. One can culture only some of the microorganisms using a range of markers to assess potential hazards in disturbing the community.

One has to start using glass houses (greenhouses) from microcosms, because one can end up with a different phylloplane ecology, compared with that which is observed in the field. A wheat crop in a sunny climate may be quite different from a sugar-beet field at Oxford, but it is useful to determine common parameters. By studying a vast number of isolates, by the fluorescent antibody technique one can obtain instructive information. In the early stages of such studies, it has been possible to construct suitable markers, especially with the *lux* insertion marker.

By the addition of a genetically-modified organism to the environment, the intent in fact is to alter the environment. The intent is to make a profitable product to help society. Scientists have studied single organisms in different environments, in soil, on a leaf, in air. The question of whether organisms are viable or nonviable in the environment, of course, is of great interest with a lot of evidence now accumulating, showing that in the natural environment, a majority of the organisms may be in a non-culturable state, either dormant, oligotropic or starved. The evidence for the viable but non-culturable condition is fairly extensive.

The impetus is to determine environmental impact. The technique of fluorescent antibody marker tracking opens up new areas which can be used effectively. A question was raised by Dr. Kalakoutski about the hypothesis of "threshold", which speaks in favor of a requirement for a certain concentration of microorganisms in a given space, such as soil. Aside from physical/chemical factors framing a biological interaction, newcomers, in concentrations in lower numbers can persist for long periods of time, or even increase. On the other hand, if the concentrations are higher, there would be a tendency for a decline in given microbial populations, thus, the hypothesis of threshold concentrations, that is, if one thinks in terms of perturbations. Microorganisms introduced at numbers a level lower than the threshold level would, at lower concentrations show a decline in interaction. Microorganisms introduced at lower concentrations should not influence community structure. Whereas others introduced at concentrations greater, would influence community structure. Dr. Lynch responded that it is totally depended on the mode of interaction and the ecological fitness of the organism that is introduced. That is, if it becomes a competitor, these factors are important. What one generally tries to do in introducing an organism into soil, is to use other attributes such as antibiosis, introduction of lytic enzymes, and therefore, the introduced organism is introduced in rather low concentrations with ecological success.

For example, Agrobacterium, has beneficial effects in low doses. Generally speaking, if the competition is the single mechanism involved, these low concentration have a significant effect. In soil, in the crop protection sense, genetic manipulation of properties, such as pesticide properties, can be a concern if secondary metabolites accumulate. There can be a much greater barrier in ultimately using organisms if secondary metabolites are produced. However, it depends on the phenotypic property of the introduced organism.

Dr. Hamer queried whether, as in enrichments done in the laboratory, one gets primary and a number of ancillary strains supporting the growth of the primary utiliser in the environment, since the usual procedure is to use pure cultures. Is it a better policy to create "niches", with an ancillary strain. Dr. Lynch agreed that this is a useful approach and, Dr. Alan Bull and Dr. Howard Slater have studied communities, finding that there was greater stability in communities than when microcosms contained a single component. Until now one could cope with only one organism. As we move into more biotechnological studies, the concept of the industrialists of mixed fermentations is being more widely viewed.

Quite a small community can yield good information. For example, *Clostridium butyricum*, *Trichoderma* sp., and an obligate anaerobe can be very effective as cellulose degraders. The *Clostridium* fixes nitrogen and, within a range of oxygen concentration, the *Trichoderma* utilizes the oxygen, taking it out of the system, protecting the *Clostridium*.

Another example is *Enterobacter* yielding a carbohydrate (polysaccharide), with a good community established. In the case of the *Trichoderma*, it is a good disease control agent. Dr. Kalakoutski again inquired about the threshold dose hypothesis. There are a number of individual microorganisms, under natural conditions, in a soil sample. With chlorophenol-degraders in pure culture, apparently at least 10^3 cells, in a sterile medium are needed.

Dr. Hamer also suggested remembering the work of Kluyver, Gottschalt at Grönigen, who studied "specialists"and "generalists" and unavailability of nutrients. Dominance could be determined, thiobacilli are, more generally, specialists.

Persistence can be viewed in several ways and can be significant as a genetic novelty. There is an accumulating body of evidence that genetic recombinants will persist until they find some opportunity to reach ascendancy. So genetic novelty will be available and retained amongst the generalists. What does that tell us about natural communities? Where do niches come from? Are niches "pre-destined" by an environment or are there an infinite number? Is there evidence from natural history, of mutation, or other factors, to substantiate the idea of the niche. Dr. Beringer suggested that a number of clonal types in Rhizobia exist. The American experience of the Chestnut blight, gypsy moth, etc., need to be kept in mind. The question is what is common? The list of things that are not a problem is much greater. What makes the gypsy moth successful and not other moths cannot be predicted. Dr. McCartney mentioned the basic ecology of soil and the phylloplane (the area on plant leaves, the potential to move from where you put bacteria to other environments. These studies have been instructive. The mechanism that makes the organism move, that is, dispersal, such as of fungal pathogens, needs to be studied.

Wind or rain splash is effective in dispersal. Fungal spores can be distributed by the wind. The potential of rain to remove bacteria from leaf surface and to extend the range of dispersal has been reported. Three species of bacteria were studied - *Bacillus*, *Klebsiella*, and *Pseudomonas syringiae* - in different crop plants. The findings showed that artificial rain studies resulted in changes in the populations of bacteria on the leaves - they decreased by one to two orders of magnitude (10^{-1} to 10^{-2}). Bacteria ran off the leaves in large numbers. Dispersal by splash, that is, distribution of the cells in a splash, could be over quite substantial distances, as was shown with fungal pathogens. Bacteria can be carried in droplets greater than 100 microns in diameter. A small proportion of the bacteria are dispersed by aerosols - about 1% to 5%. The initial population is dispersed more than a few centimeters from the leaf. During a rain storm, multiple splash effects occur. A major finding is that the large number of cells in water run down the stems and into the soil. Dispersal over short distances is caused by rain. Also, there is a problem in transporting bacteria via leaves. Leaf microbiology is a very useful subject for study. Also, soil groundwater distribution can be important.

Dr. Thompson discussed the sugar beet study done at Oxford University, which is in the second year of survey. Problems have been determined, showing that organisms in the indigenous community decomposing the sugar beet, are present at 10^9 population numbers.

One can detect changes in the numbers, with respect to marker organisms comprising the structure of the community.

It is difficult to identify individual community components. It is possible to identify 60 or 70 bacterial species per day and about 300 in a week, by traditional methods. Another approach is to use selective plates and monitor certain organisms, such as *Pseudomonas*. Hopefully the numbers on selective plates can be interpreted appropriately, but one may be missing an important component by virtue of selection. How to detect impact on a leaf? How to develop methods to identify changes? These are questions that need answering.

With genetically-modified microorganisms in the environment, the purpose is to have them "do something" which is why they are put in the environment. You look at what they do, such as, adding bacteria to sugar-beet leaves for a purpose.

Warren, in the 1970s, showed that when one introduced plant leaf pathogens, one could get a mechanism to study such changes. But, one needed to go to process effect, as well as diversity of the community effect. The Shannon Index is acceptable, but it is still necessary to study effect(s) on the processes.

The application of biocontrol agents, where bacteria are controlled, such as root rot conditions, one looks at treated and untreated plots that the biocontrol bacterium "died out" was determined by screening randomly detected bacteria, with 40 different bacteria and doing cluster analysis. The communities could be shown to be different.

The use of metabolic properties can be more effective. In general, lists of species names are not useful. Properties of the organisms are better indices. The point is to induce some impact by addition of a genetically-modified organism. Negative environmental impact, that is, the organism will persist or spread, may not necessarily be selected for.

In the United States, the National Academy of Sciences report on release of genetically-modified organisms into the environment is based on familiarity. In the United Kingdom, the ground rule is not to harm the environment. We will, in effect, have to generate a history of cases, as introductions are made.

It is important to point out that, if the organism "works", it has an impact. That is an underlying principal. If the introduced organism works, it *must* have an impact. Therefore, we are trying to produce data to satisfy a legal definition. Definitions should be read, understood, and communicated. Legal definition of procedures should be reviewed by scientists when they are promulgated. Laws require certain procedures to be followed. In the United Kingdom, the E.C. Directives are being implemented. Definitions are the most argued points. Ambiguity in scientific and legal terms requires discussion to resolve. The courts, in the end, will decide, but, scientists' confusion on these issues does not help.

Monitoring is labor-intensive. In the case of pollen transfer in plants, manpower is needed to monitor microbial field additions. If it is easy to monitor, fine, if not, then the addition will be less likely to be allowed.

We need to study predation versus competition, stable versus unstable environments. Invasions by fungi have been studied, but none have been recorded for bacteria.

The familiarity rule is useful in doing releases. There has been no measured impact, but several possible impacts have been suggested by Dr. Seidler, who summarized from the literature five examples of effects. Briefly stated, they are:

(1) Interactions of bacterial and fungal test control agents that inhibit growth/function of mycorrhizal fungi with host plants. This work was conducted by Linderman and colleagues at the USDA laboratories in Corvallus, Oregon.

(2) Inhibition of predatory protozoa exposed to recombinant strains of *Pseudomonas solanacearum*. This work was conducted by Peter Hartel and colleagues at the University of Georgia in the United States.

(3) Expression of recombinant lignin peroxidase by *Streptomyces* in soil. This work was conducted by Dr. Crawford at the University of Idaho. The activity greatly stimulated production of carbon dioxide from soil. The authors hypothesized that the action of the recombinant peroxidase increased the rate of oxidation of relatively unavailable, lignin-derived soil organic matter. This oxidation partially opened up the lignin molecule and made lignified carbon more available to the soil microflora, that were then able to respire the carbon more rapidly. No studies were conducted to determine subsequent effects, if any of lignin removal on other soil processes, including general soil fertility, moisture holding capacity, etc.

(4) Eco-system structural and functional changes due to exposure to the root colonizing fluorescent *Pseudomonad* strain RC1. This work was done by Fredrickson and

colleagues at the Battelle Laboratories in Richland, Washington. The percentage of total fluorescent *Pseudomonads* on wheat plant roots was lower in treatments that received strain RC1. Thus, RC1 out-competed a significant proportion of the native soil *Pseudomonads* for colonization of the root surface. This altered the bacterial composition of the rhizoplane by decreasing the percentage of fluorescent *Pseudomonads* from 24% in the control, to 1% in the RC1-treated systems.

(5) Ecological effects from 2,4-Dichlorophenol, a metabolite of 2,4D biodegradation, on soil populations of fungi. This work was conducted by Drs. Seidler and Stotzky. The soil fungi were eliminated from microcosms due to the accumulation of dichlorophenol caused by mutant strain that biodegraded 2,4D. Fungi did not regrow in soil following extended incubations after additions of the 2,4D degrader.

However, introductions of genetically engineered corn, pines, and other plant species, have shown no effect.

In summary, the discerned concerns arising from introduction of genetically engineered organisms into the environment associated with persistence, transfer of genetic material, etc. are socioeconomic, environmental (ecological), health (antibiotic resistance, for example) ethical - humans playing god or third world agriculture being used as a guinea pig system, and reduced trust in scientists (human error, secrecy, and commercial confidentiality being part of this factor).

When a legislature gets involved in the release issue, as in the case of baker's yeast, it is long-winded in dialogue and costly in the legal paths followed. The tendency is to deal with very formal processes in a very long-winded way. We need to do research to determine what the effects will be, if any.

The press is not comprised of people waiting for stories to come to them. The press, in fact, seeks out stories. The relationship between science and the press has changed. Scientists should be more willing to give public lectures, to talk with journalists, rather than allow journalists to take on the role of giving public lectures on science.

Media reaction to risk information in the U.S.A. seems to have been much less aggravated compared, for example, to the nuclear industry, which was the "bête noir" in the 1970s. Is biotechnology to be the "bête noir" of the 1990s? The fact remains that, to date, there has been no documented adverse effect on the environment arising from the introduction of a genetically-modified microorganism. They hypothesized disasters have not occured and no adverse effect can be documented. Of those measurable "effects", cited above, adversity has not been documented. Clearly, progress has been made but much more work needs to be done.

CHARACTERIZATION OF MICROBIAL EMISSIONS FROM A FERMENTATION PLANT USING A GENETICALLY-MODIFIED BACILLUS STRAIN

K. Smalla[1], M. Isemann[2] and K.H. Weege[2]

Institut fur Biochemie und Pflanzenvirologie[1]
Biologische Bundesanstalt für Land und Forstwirtschaft
W-3300 Braunschweig

Hygieneinstitut[2]
O-3010 Magdeburg
Germany

INTRODUCTION

Unintended releases of genetically-modified microorganisms from contained facilities are to be expected for safety level 1 organisms because the inactivation of biomass and waste-water, as well as the installation of sterile filters for exhaust air, is not provided by law (EC regulation, GILSP concept of OECD) for low-risk organisms.

In a case study for industrial alpha amylase production with a rDNA *B. subtilis* we investigated biomass and waste-water from a fermentation plant. The rDNA *B. subtilis* carried 4-6 copies of an erythromycin resistance plasmid with a duplicated amylase gene insert. The plasmid used, pDB101, was a deletion derivative of the natural streptococcal resistance plasmid pSM19035.

Even though the rDNA *B. subtilis* is expected to be a low-risk microorganism, a critical point was seen to be a potential selective advantage of the production strain, or a dissemination of the MLS resistance gene, amongst indigenous microbes due to the presence of the recombinant MLS resistance plasmid. For a reasonable evaluation of potential adverse effects due to the released production strain, the normal microbial burden of bioprocessing emissions should be taken into account.

MATERIALS AND METHODS

Samples of waste-water and biomass from an enzyme factory were taken weekly and spread onto different selective media. A phenylethylethanol (0.25%), erythromycin (5 µg/ml) and starch (1%) containing selective medium was used to detect the production strain amongst Gram-negative bacteria which are mostly intrinsically resistant to MLS-antibiotics (Smalla *et al*, 1991). Gram-negative bacteria were selectively cultivated on a vancomycin (10 µg/ml) nutrient agar plate (Weege and Thriene,1990). About 100 Gram-negative bacteria from waste-water and from biomass were presumptively identified according to a scheme described by Aislabie and

The Release of Genetically Modified Microorganisms
Edited by D.E.S. Stewart-Tull and M. Sussman, Plenum Press, New York, 1992

129

Table 1.Frequency of antibiotic resistance and heavy metal tolerance in Gram-negative bacteria of different origin (% resistant isolates) * from the fermentation plant

sampling sites	n	Hg	Cd	Sm	Cm	Tc	Tp	Ap	Gm	Km
waste water*	104	29	8	1	6	9	0	57	0	3
biomass*	101	25	27	1	8	36	2	65	1	3
sewage water	103	13	30	21	5	5	12	63	17	7
river water	133	29	3	2	4	5	3	69	1	2
storage reservoirs	99	3	4	2	1	2	16	31	1	1

Loutit (1984). The antibiotic resistance pattern determined by the agar diffusion method and heavy metal tolerance were studied as described by Weege and Thriene (1990). Plasmid screening was performed according to Kado and Liu (1981).

RESULTS AND DISCUSSION

The investigation of biomass and waste-water by selective cultivation showed a high proportion of Gram-negative bacteria. When the rDNA *B. subtilis* was detected the titre was mostly at the level of the sporulation rate (Smalla *et al*, 1991). About 100 Gram-negative bacteria, randomly collected from biomass and waste-water were presumptively identified and assigned to the groups Enterobacteriaceae, *Pseudomonas/Alcaligenes*, *Alcalifaciens* and Vibrionaceae. Most isolates were members of the Enterobacteriaceae (80% for biomass, 70% for waste-water). The frequency of antibiotic-resistant bacteria from biomass and waste-water and their tolerance towards mercury and cadmium are summarized in Table 1 and compared with Gram-negative bacteria of different origins. A high proportion of isolates from biomass was resistant to tetracycline (36%). More than 30% of the tested isolates from waste-water showed multiple resistance against the antibiotics tested, whereas the proportion of multiple resistant isolates from biomass surprisingly was found to be 10%. Furthermore, we observed a high percentage of bacteria tolerating mercury and cadmium. Contrary to relevant literature on the frequent association of bacterial resistance to antibiotics and heavy metal resistance in clinical isolates, we found a higher proportion of antibiotic-sensitive bacteria tolerating heavy metals. Plasmid screening indicated that 44% of tested biomass isolates and 64% of waste-water isolates harbored plasmids. In contrast to municipal waste-water or polluted river-water most isolates harbored small plasmids.

Biomass and waste-water of the fermentation plant were shown to be a pool of plasmid carrying, antibiotic-resistant and/or heavy metal-tolerant Gram-negative bacteria. Relative to the background burden of microbial emission from the enzyme plant it seems likely that the effects of released spores of the production strain are negligible. However, it should be emphasized that the microbial composition and the prevalence of antibiotic-resistance and heavy metal-tolerance found for Gram-negative isolates of biomass and waste-water resembled that of municipal waste-water or polluted river-water and is typical not only of biotechnological waste.

REFERENCES

Aislabie, J. and M.W. Loutit (1984) The effect of effluent high in chromium on marine sediment aerobic heterotrophic bacteria. Marine Environ. Res. 13:69-79.
Kado, C.I. and Liu (1981) Rapid procedure for detection and isolation of large and small plasmids. S.T. J. Bacteriol. 145:1365-1373.

Smalla, K., Isemann, M., John, G., Weege, K.H., Wendt, K. and Backhaus, H. (1991) In: Biological monitoring of genetically engineered plants and microbes. Ed.: MacKenzie, D.R. and Henry, S.C. Proceedings of the Kiawah Island Conference, 205-220.

Weege, K.H. and Thriene, B. (1990) Die bakterielle resistenz als bioindikator für die schwermetallbelastung der umwelt. VDI Kolloquium "Wirkungen von Luftverunreinigungen auf Böden," Lindan In: VDI-Berichte 837, 1009-1028.

TRANSFER OF INC.P AND INC.Q PLASMIDS IN SOIL AND FILTER MATINGS FROM *PSEUDOMONAS FLUORESCENS* TO A RECIPIENT STRAIN AND TO INDIGENOUS BACTERIA

E.Smit and J.D.van Elsas

Institute for Soil Fertility Research
Wageningen
The Netherlands

Heterologous DNA of GEMMOs released into soil can potentially be transferred to members of the indigenous microbial community.However,so far few studies have shown experimentally genetic transfer from introduced bacteria to indigenous bacteria in soil (Henschke and Schmidt, 1990; Brokamp and Schmidt, 1991). Until now, the lack of a suitable donor counter-selection technique has hindered the detection of indigenous transconjugants (Van Elsas and Trevors, 1991). A method is described for donor counter-selection and two specifically marked plasmids which enabled us to study plasmid transfer and mobilization,in the soil and rhizosphere,to indigenous microorganisms.

DEVELOPMENT OF A DONOR COUNTER-SELECTION METHOD

Plasmid transfer in soil can only be adequately studied by a donor counter-selection method to prevent growth of relatively high numbers of donor cells on low-dilution transconjugant selective plates. This is usually achieved with an antibiotic resistant recipient (Smit and van Elsas, 1992). Since plasmid transfer to indigenous bacteria cannot be studied in this way, other solutions have been proposed such as: an inducible host- killing gene on the donor chromosome, a bacteriophage that specifically lyses the donor, an auxotrophic mutant as donor, a donor which dies rapidly in soil and a donor that cannot express the selective marker, as described in more detail by Smit and van Elsas (1992).

Bacteriophage R2f was isolated from agricultural drainage- water, and was shown to specifically lyse the donor strain *P. fluorescens* R2f. R2f proved to be suitable for donor counter-selection in soil plasmid transfer experiments, since it reduced the donor cell number with 10^{-5} without any major effects on the indigenous (potential recipient) soil bacterial population (Smit *et al.*, 1991). In addition, plate matings could thus be avoided.

USE OF MARKED PLASMIDS AND PROBES

The self-transmissible broad-host-range (Inc. P1) plasmid RP4 was provided with a eukaryotic hybridization marker to permit the detection of the plasmid by colony filter hybridization. The molecular marker, which is a part (0.69 kb) of the patatin gene from potato, was inserted into the kanamycin resistance gene of RP4,rendering it inactive and yielding

The Release of Genetically Modified Microorganisms
Edited by D.E.S. Stewart-Tull and M. Sussman, Plenum Press, New York, 1992

133

RP4p. Bacteria possessing RP4p were selected by adding ampicillin and tetracycline to the medium.

A mobilizable plasmid was constructed on the basis of the replication functions and *mob* site of the Inc. Q plasmid RSF1010. Two antibiotic resistance genes, *npt* II, conferring resistance to kanamycin, and *aad*B, conferring resistance to gentamycin and part of the cryIVB gene from *B. thuringiensis* were ligated into HindIII/Eco RI-cut pSUP104 to produce pSKTG.

The use of different antibiotic resistance genes in combination with different unique hybridization markers enabled us to detect low cell numbers possessing either of the plasmids in the soil environment. Four strains were used in the transfer experiments: 1] *P. fluorescens* R2f (RP4p); 2] *P. fluorescens* R2f (pSKTG); 3] *P. fluorescens* R2f (RP4p, pSKTG) and 4] *P. fluorescens* R2f resistant to rifampicin and nalidixic acid (recipient strain). Transfer of both plasmids to a co-introduced recipient was studied on filters and in sterile soil using different donor strain combinations. Transfer to indigenous soil bacteria was studied with the same donor combinations in soil planted with wheat plants.

PLASMID TRANSFER ON FILTERS AND IN SOIL

In two- and tri-parental filter-mating studies between *P. fluorescens* donor and recipient strains with different chromosomal markers, both RP4p and pSKTG were transfered at frequencies of $2x10^{-2}$, regardless of whether the plasmids were jointly present in one, or, separate in two donor strains. Apparently pSKTG is very efficiently mobilized and cell to cell contact does not limit the transfer in the triparental mating on filters. However, the result was different when this experiment was performed in sterile soil; when both plasmids were present in the same donor strain they were transferred to the recipient with equal frequencies (about $2x10^{-4}$), but when the plasmids were present in different, separately-inoculated donor strains, the numbers of recipients harboring pSKTG were 600-fold lower than those exclusively harbouring RP4. This suggests that in contrast to the triparental matings on filters, in sterile soil only part of the cells containing pSKTG receives RP4p, which results in a reduction of the subsequent mobilization of pSKTG. The lower probability of cell-to-cell contact in sterile soil limits the mobilization of pSKTG when the Tra functions are present in other cells.

Table 1. Transfer of RP4p and mobilization of pSKTG from *P. fluorescens* R2f to indigenous bacteria in Ede loamy sand planted to wheat in soil microcosms (R = Rhizosphere soil, B = bulk soil)

Treat-ment**		Donor		Indig. transconjugants		
		RP4p	pSKTG	both	RP4p	pSKTG
A	R	$7x10^4$	$4x10^5$	-	$2x10^2$	$5x10^2$
	B	$7x10^4$	2.10^5	-	[0]	$1.7x10^2$
B	R	-	-	$6x10^5$	$1x10^3$	$2x10^4$
	B	-	-	$1x10^6$	$1.5x10^2$	$4.8x10^3$
C	R	-	$6x10^5$	-	-	[0]
	B	-	$7x10^5$	-	-	[0]
D	R	-	-	-	[0]	[0]
	B	-	-	-	[0]	[0]

* Cfu counts per g of dry soil after 7 days are presented. Initial inoculum levels (per g dry soil) were: *P. fluorescens* R2f (RP4P): 10^6; R2f (pSKTG): 10^6; R2f (RP4p; pSKTG): 10^7

** Treatments: A: Donor R2f (RP4p) and donor R2f (pSKTG) added; B: Donor R2f (RP4p; pSKTG) added; C: Donor R2f (pSKTG) added; D: no donor added.

[0] = below detection limit (10^2 cfu/g soil).

Plasmid transfer to indigenous soil bacteria was also studied by introducing different combinations of donor strains in soil microcosms planted with wheat. On day 7, rhizosphere and bulk soil samples were taken, treated with R2f and plated on selective media. Indigenous transconjugants were detected by colony filter hybridization. The results indicate (Table 1) that when both plasmids are present in the same donor strain, pSKTG is transferred with higher frequencies than RP4p itself. Several factors, such as differences in host-range, size and sensitivity to restriction might be responsible for this difference. Mergeay (pers. comm.) also found that Inc.P plasmids were not always transferred themselves when mobilizing other plasmids. Surprisingly pSKTG was, albeit at much lower frequencies, also mobilized when RP4p was initially present in a different, co-introduced strain (Table 1). Considering the low mobilization frequency of pSKTG in the triparental mating in sterile soil, one would not expect to detect mobilization to indigenous bacteria with RP4p present in a co-introduced donor strain. Although the transconjugant numbers were just above the detection limit, the fact that a small Inc.Q plasmid was still mobilized when the transfer functions are present in other cells clearly shows the putative hazards associated with the use of such plasmids as vectors, since it is likely that tra^+ plasmids occur naturally in bacteria in the soil environment (Diels et al., 1989; Sayler et al., 1990).

REFERENCES

Bropkamp, A. and Schmidt, F.R.J. (1991). Survival of *Alcaligenes xylosoxidans* degrading 2,2-dichloropropionate and horizontal transfer of its halidohydrilase gene in a soil microcosm. Curr. Microbiol. 22, 299-306.

Diels, L., Sadouk, A. and Mergeay, M. (1989). Large plasmids governing multiple resistances to heavy metals: a genetic approach. Tox. Environ. Chem. 23, 79-89.

Henschke, R.B. and Schmidt, F.R.J. (1990). Plasmid mobilization from genetically engineered bacteria to members of the indigenous microflora. Curr. Microbiol. 20, 100-105.

Sayler, G.S., Hooper, S.W., Layton, A.C. and King, J.M. (1990). Catabolic plasmids of environmental and ecological significance. Microb. Ecol. 19, 1-20.

Smit, E. and Van Elsas, J.D. (1990). Determination of plasmid transfer frequency in soil: consequences of bacterial mating on selective plates. Curr. Microbiol. 21, 151-157.

Smit, E. and Van Elsas, J.D. (1992). Methods for studying conjugative plasmid transfer in soil. In: E.M.H. Wellington and J.D. van Elsas (eds.). Genetic interactions between microorganisms in the environment. p. 112-126. Manchester University Press, Manchester.

Smit, E., Van Elsas, J.D., Van Veen, J.A. and De Vos, W.M. (1991). Plasmid transfer from *Pseudomonas fluorescens* to indigenous bacteria in soil using phage R2f donor counterselection. Appl. Environ. Microbiol. 57, 3482-3488.

Van Elsas, J.D. and Trevors, J.T. (1990). Plasmid transfer to indigenous bacteria in soil and rhizosphere: problems and perspectives. In: J.C. Fry and M.C. Day (eds.). Bacterial genetics in natural environemts. p. 188-199. Chapman and Hall, London.

IMPACT OF INTRODUCING GENETICALLY-MODIFIED MICROORGANISMS ON SOIL MICROBIAL COMMUNITY DIVERSITY

Asim K. Bej, Michael Perlin and Ronald M. Atlas

Department of Biology
University of Louisville
KY 40292 USA

This study examined the impact of introducing a genetically modified herbicide-degrading bacterium on soil community diversity. As a model the genetically modified bacterium *Pseudomonas cepacia* AC1100 was used, which is a 2,4,5-trichlorophenoxyacetic acid (2,4,5-T) degrader, formed through molecular breeding (Karns *et al.*, 1981; Kilbane *et al.*, 1982). Both taxonomic diversity and genetic diversity were measured following introduction of *P. cepacia* AC1100 into soil microcosms alone or in combination with the herbicide 2,4,5-T.

PERSISTENCE OF *P.CEPACIA* AC1100

Studies were conducted in soil microcosms containing an Oregon silt loam soil. Replicate microcosms received either no treatment (control microcosms), treatment with 2,4,5-T to achieve a final concentration of 100 ppm, inoculation with *P. cepacia* AC1100 at a concentration of 10^5/g soil, or treatment with 10^5 *P. cepacia* /g soil + 100 ppm 2,4,5-T. Bacterial populations were enumerated on trypticase soy agar. *P. cepacia* AC1100 was enumerated by colony hybridization with a gene probe for a 1.3 kb repeat sequence that occurs in the chromosome and on plasmids of this bacterium (Steffan and Atlas, 1988).

P. cepacia persisted in soil microcosms with and without the addition of 2,4,5-T for the 6-week experimental period (Fig.1). Concentrations of *P. cepacia* were somewhat higher in the microcosms where 2,4,5-T was present than in those where it was absent. *P. cepacia* AC1100 was not detected in microcosms which had not been inoculated.

TAXONOMIC DIVERSITY

Isolates of randomly selected colonies from trypticase soy agar plates were cultured for taxonomic studies. Approximately 40 phenotypic characteristics - API 20E test supplemented by cellular and colonial morphologies, temperature growth ranges, and abilities to utilize additional substrates - were determined for each isolate. Data were subjected to cluster analysis to determine taxonomic groupings and the number of taxonomic groups and the number of individuals within each group were used to calculate the Shannon diversity index (H') by the method of Kaneko *et al.*, (1977).

The introduction of the genetically modified 2,4,5-T-degrading microorganism *P. cepacia* into soil appears to have caused an imbalance within the microbial community. The addition of *P. cepacia* resulted in a slight increase in taxonomic diversity of culturable populations,

The Release of Genetically Modified Microorganisms
Edited by D.E.S. Stewart-Tull and M. Sussman, Plenum Press, New York, 1992

137

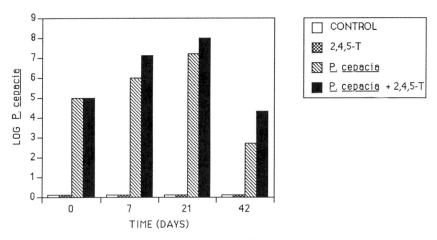

Figure 1. Enumeration of *Pseudomonas cepacia*

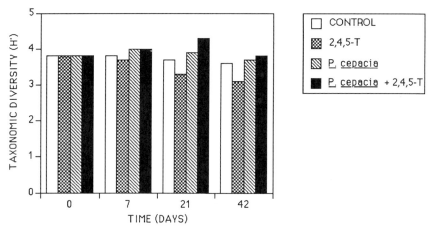

Figure 2. Taxonomic diversity (Shannon index)

particularly when 2,4,5-T was added along with this herbicide-degrading microorganism (Fig.2). The highest diversity occured when the maximal number of *P. cepacia* AC1100 was found.

GENETIC DIVERSITY

DNA was extracted from 100 g soil samples by the method of Ogram *et al.* (1988) as modified by Steffan *et al.* (1988). The DNA was sheared by sonication to approximately 0.5 kb. Thermal denaturation and reassociation was measured in a Lambda III spectrophotometer with an automatic temperature controller (Perkin-Elmer Inc., Norwalk, CT) following the basic procedure of Torsvid *et al.*, (1990). Thermal melting was carried out with a total of 280-300 mg of DNA per ml of 1X SSC for each sample for 10 min to melt the DNA completely. The reassociation of DNA was initiated by lowering the temperature rapidly (>5°C/min) to 60°C. A_{260}nm measurements were made, initially every minute and subsequently as the absorbance became constant every day or week. Percent reassociation was calculated at time t (in seconds) = A_o - A_t/A x 100, where A_t = absorbance at 260 nm at time t, A_o = absorbance at 260 nm when DNA was completely melted, and A= [absorbance at 260 nm for 100% melted DNA] - [absorbance at 260 nm for 0% DNA melted]. C_ot values were calculated as moles / 1 x sec.

The introduction of *P. cepacia* AC1100 to soil resulted in increased C_ot values, reflective of an increase in diversity within the gene pool of the soil microbial community (Fig. 3). The

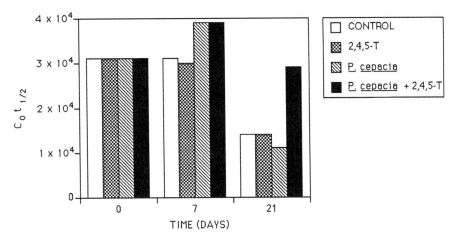

Figure 3. Genetic diversity $(C_0t_{1/2})$

metabolic activities of *P. cepacia* AC1100 may have contributed substrates for other populations leading to a transient imbalance within the community. The increased genetic diversity may have resulted from genetic transfer and/or recombination. *P. cepacia* AC1100 has plasmids and multicopy 1.3 kb sequence, shown to be a transposon. The increased diversity was transient, indicating that if genetic recombination was contributing to the increased genetic diversity most or all of the recombinants were nonfunctional or noncompetitive.

REFERENCES

Kaneko, T., Atlas, R.M. and Krichevsky, M. (1977) Diversity of bacterial populations in the Beaufort Sea. Nature 270, 596-599.

Karns, J.S., Kilbane, J.J., Chatterjee, D.K. and Chakrabarty, A.M. (1981) Laboratory breeding of a bacterium for enhanced degradation of 2,4,5-T, in "Genetic Engineering for Biotechnology" (Crocomo, J., Tavares, F.C.A. and Sodrzeieski, D., Eds.), pp. 37-40. Promocet, Sao Paulo, Brazil, p. 37-40.

Kilbane, J.J., Chatterjee, D.K., Karns, J.S., Kellogg, S.T. and Chakrabarty, A.M. (1982) Biodegradation of 2,4,5-trichlorophenoxyacetic acid by a pure culture of *Pseudomonas cepacia*. Appl. Environ. Microbiol. 44, 72-78.

Ogram, A., Sayler, G.S. and Barkay, T. (1988) DNA extraction and purification from sediments. J. Microbiol. Methods 7,57-66.

Steffan, R.J., Goksyr, J., Bej, A.K. and Atlas, R.M. (1988) Recovery of DNA from soils and sediments, Appl. Environ. Microbiol. 54, 2908-2915.

Steffan, R.J. and Atlas, R.M. (1988) DNA amplification to enhance detection of genetically engineered bacteria in environmental samples. Appl. Environ. Microbiol. 54, 2185-2191.

Torsvik, V., Salte, K., Srheim, R. and Goksyr, J. (1990) Comparison of phenotypic diversity and DNA heterogeneity in a population of soil bacteria. Appl. Environ. Microbiol. 56, 776-781.

THE DETECTION AND SURVIVAL OF RECOMBINANT

PSEUDOMONAS PUTIDA POPULATIONS IN LAKE WATER

J.A.W. Morgan[1], C. Winstanley[2], R.W. Pickup[1] and J.R. Saunders[2]

The Freshwater Biological Association[1]
Institute of Freshwater Ecology
Ambleside, Cumbria
LA22 OLP

The Department of Genetics and Microbiology[2]
University of Liverpool, Liverpool
L69 3BX

Marker plasmids have been constructed to enable the detection and survival of populations of Gram-negative bacteria and to assess their ability to survive after release into lakewater (Winstanley *et al.*, 1989; 1991). On these constructs the marker gene, *xylE* is expressed from the lambda promoters p_L or p_R controlled by the temperature-sensitive lambda repressor cI_{857}. A range of culture and direct detection methods were developed for tracing released hosts in lakewater (Morgan *et al*; 1989; 1991). As a result of using both direct detection and culture methods, the *xylE* gene carried by marker plasmids was considered to be a valid indicator for detecting and studying the survival of *Pseudomonas putida* populations after release into sterile or untreated lake-water. (Winstanley *et al.*, 1991). In these studies the effects of inoculum size , auxotrophic mutation carried by the host strain, metabolic burden imposed on cells by unregulated expression of *xylE*, and the reproducibility of the experiments was investigated.

MARKER SYSTEMS

The following marker plasmids were used:
pLV1016 an Inc P conjugative plasmid with Ap^r Km^r and Tc^r (p_R *xylE* cI_{857})
pLV1017 an Inc P conjugative plasmid with Ap^r Km^r and Tc^r (p_L *xylE*)
pLV1013 an Inc Q conjugative plasmid with Sm^r and Km^r (p_R *xylE* cI_{857})

In the plasmids pLV1016 and ;LV1013 the *xylE* gene is under the control of the lambda promoter p_R and its temperature sensitive repressor cI_{857}. When cells harboring these plasmids are grown below 28°C very little expression of the *xylE* gene occurs, at temperatures above this derepression of the lambda p_R promoter occurs and high level expression of the *xylE* gene is achieved. Therefore the metabolic burden from over expression of the *xylE* gene is minimal when the cells are growing below 28°C, a situation that is probably present in lake-water experiments which are conducted at 10°C. The plasmid pLV1017 is unregulated and high level expression of *xylE* gene occurs continually. The two strains used were *Pseudomonas putida* PaW8, and an auxotrophic mutant of this strain, PaW340 which requires tryptophan for growth.

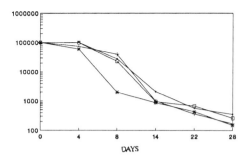

Fig. 1 Survival of *Pseudomonas putida* in lake-water.

+ Paw8 (pLV11016)   □ Paw8 (pLV1017)

● Paw340 (pLV1016)   ✳ Paw340 (pLV1017)

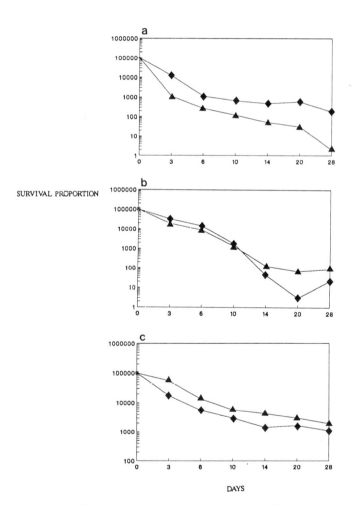

Fig. 2 Survival of *Pseudomonas putida* in lake water collected in a September, b, October; c, November.

▲ Paw8 (pLV1013)

◆ Paw340 (pLV1013)

SURVIVAL OF *P. PUTIDA* IN LAKEWATER

Figure 1 illustrates the recovery of *P. putida* PaW8 (pLV1017), PaW340 (pLV1017, PaW8 (pLV1016) and PaW340 (pLV1016) in untreated lakewater model systems. When the hosts were introduced at approximately 10^5 cfu/ml the populations initially declined within 20 days to 10^3 cfu/ml, they then became stable at 10^3 cfu/ml for up to 28 days. All releases were carried out in triplicate and counts from three agar plates at appropriate dilutions were recorded. An analysis was developed for the data that assumed that the counts were distributed in a way proportional to a Poisson distribution and that the average count arose as the product of the initial inoculum and a survival proportion. The results of the analyses were not significant ($P>0.05$) and suggested that there were no differences between the strains and marker systems. Therefore, the auxotrophic nature of PaW340 did not significantly affect its survival under these conditions. In addition, the metabolic burden from over expression of the *xylE* gene in strains harbouring pLV1017 had no significant effect on the survival of the host in lakewater.

VARIABILITY IN SURVIVAL PATTERNS IN WATER SAMPLES TAKEN ON CONSECUTIVE MONTHS

The survival of *P. putida* PaW8 and PaW340 (both harboring the plasmid pLV1013) in lakewater collected from the same location in September, October and November 1989 can be seen in Figure 2. Statistical analysis of these results showed that they differed significantly ($p<0.001$) between months for both strains. These results suggest that the survival of a strain released into the natural environment could be highly variable and dependent on the time and place of release. it may well be that the natural population varies to such an extent over short distances or short periods of time in natural habitats that it will be impossible to predict accurately the survival of a released organism (Pickup *et al.*, 1991).

ACKNOWLEDGEMENTS

The support of the Freshwater Biological Association, the Natural Environmental Research Council and the Department of the Environment is gratefully acknowledged.

REFERENCES

Morgan, J.A.W., Winstanley, C., Pickup, R.W., Jones, J.G. and Saunders, J.R. (1989). Direct phenotypic and genotypic detection of a recombinant pseudomonad population released into lake-water. Applied and Environmental Microbiology 55, 2537-2544.

Morgan, J.A.W., Winstanley, C., Pickup, R.W., Jones, J.G. and Saunders, J.R. (1991). Rapid immunocapture of *Pseudomonas putida* cells from lake-water by using bacterial flagella. Applied and Environmental Microbiology 57, 503-509.

Pickup, R.W., Morgan, J.A.W. Winstanley and Saunders, R.J. (1991). Implications for the release of genetically engineered organisms. J. Applied Bacteriology Symposium Supplement 70, 195-305.

Winstanley, C., Morgan, J.A.W., Pickup, R.W., Jones, J.G. and Saunders, J.R. (1989). Differential regulation of Lambda p_L and p_R promoters by a cI repressor in a broad-host-range thermoregulated plasmid marker system. Applied and Environmental Microbiology 55, 771-777.

Winstanley, C., Morgan, J.A.W., Pickup, R.W. and Saunders, J.R. (1991). Use of a *xylE* marker gene to monitor survival of recombinant *Pseudomonas putida* populations in lake-water by culture on nonselective media. Applied and Environmental Microbiology 57, 1905-1913.

SURVIVAL OF INTRODUCED BACTERIA IN RHIZOSPHERE AND NON-RHIZOSPHERE SOILS

C. S. Young[1], K. A. Cook[2], G. Lethbridge[2] and R. G. Burns[1]

Biological Laboratory[1]
University of Kent
Canterbury
Kent, CT2 7NJ

Shell Research Limited[2]
Sittingbourne Research Centre
Sittingbourne
Kent ME9 8AG

INTRODUCTION

Bacteria have often been introduced into soil to promote agriculturally beneficial activities. However, the inconsistent results of many field-trials may be due to differences in survival of allochthonous inoculants that are not well-adapted to the soil environment. A bacterial inoculant will be more likely to survive and express its properties in soil if it can compete effectively with the indigenous microorganisms. Thus, it may be appropriate to develop beneficial bacteria by genetic manipulation of naturally-occuring bacteria isolated from the target soil.

The objectives of the present study were to examine and compare the survival of two ecologically-distinct bacteria, *Arthrobacter globiformis* (A109) and *Flavobacterium balustinum* (P25), in field soils from which both bacteria were isolated originally. Survival was monitored in both planted (wheat) and non-planted soil and interactions between A109 and P25 were examined by introducing them simultaneously and sequentially. The results of the field experiments were also compared with the results of the equivalent laboratory experiments (reported in Thompson *et al* (1990)). Furthermore, the potential for horizontal spread of A109 and P25 from plant to plant was investigated.

METHODS

In April 1990, the field site was divided into 1 m² plots (soil type: silt loam; sand 34%, silt 41%, clay 20%, pH 7.0). The inoculum of A109 and P25 was produced by washing and resuspending exponential phase cultures. A 1 m² quadrat, with 81 regularly spaced holes, was placed over each plot and 1 ml of inoculum was deposited in the soil in a 4 cm deep hole. For planted plots, imbibed wheat seeds were placed in each hole before the inoculum was applied. Plots were raked so as to cover each hole but not disturb the inoculum. On each sampling date the quadrat was placed over each plot to locate the inoculation points, and three soil cores or three wheat plants were randomly selected. Soil cores were dispersed in buffer by shaking; excised wheat roots were homogenized in buffer. Suspensions were serially diluted and six 20 µl drops from each dilution were placed on the appropriate selective medium (A109 was

The Release of Genetically Modified Microorganisms
Edited by D.E.S. Stewart-Tull and M. Sussman, Plenum Press, New York, 1992

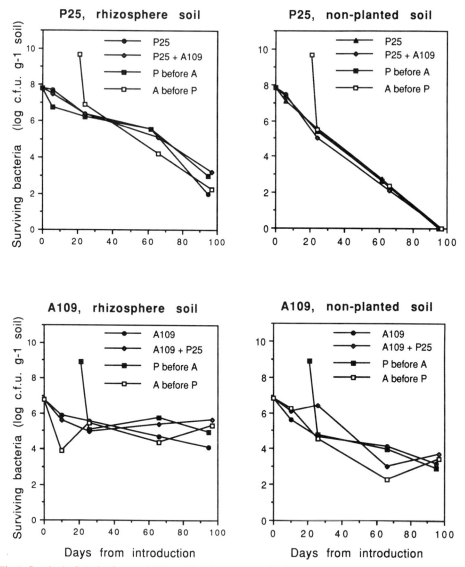

Fig.1 Survival of *Arthrobacter* A109 or *Flavobacterium* P25, introduced (11-5-90) to unplanted and planted field soil. The mean standard deviation was ± 1.4.

resistant to streptomycin and P25 was resistant to streptomycin, kanamycin and rifampicin). A109 and P25 colonies were counted within two days and four days, respectively.

RESULTS AND DISCUSSION

In plots with separate inoculations of A109 or P25, the numbers of A109 or P25 were always higher in planted than in non-planted soils. Greater numbers of A109 survived than P25 at most sampling times, irrespective of the presence of wheat (Thompson *et al* 1992). These same trends in population levels were observed in the equivalent laboratory experiments (Thompson *et al*, 1990). Furthermore, for most comparisons of the rate of decrease in laboratory soil with the corresponding rate of decline in the 1988, 1989 and 1990 field experiments, no significant differences ($p>0.5$) were found. However, the rate of decrease in field soils was usually slower, which may be partly explained by the lower average temperatures in the field soil.

The rate of decline of inoculum of A109 or P25 in separate inoculations was also compared with the decline of the same inoculum species when introduced simultaneously or sequentially (21 days before or after the other species). The decrease of inoculum after day 21 was similar for A109 or P25, in planted or non-planted soil, regardless of whether inocula were combined, separate or at different densities. The final numbers of P25 inocula in planted soil were not significantly different ($p>0.05$); the final numbers in non-planted soil were zero. Similarly, the final numbers of A109 in planted or non-planted soil were not significantly different ($p>0.05$). It is possible that survival of mixed inoculum of A109 and P25 in field soil is influenced more by the indigenous microflora than by the co-inoculant.

A limited spread of inoculum from the point of inoculation may be desirable since in practice, inoculum is not likely to reach all soil or every plant. In a separate field experiment in 1990, P25 was recovered from non-inoculated wheat roots after 80 days at distances of 10 cm and 20 cm from the roots of inoculated wheat plants, despite very dry soils in an abnormally hot summer. However, recovery at 20 cm was low (10 - 100 cfu g^{-1} root; below statistically reliable counts). A109 was not recovered more than 10 cm from inoculated roots. Populations of P25 are stimulated more than A109 by the rhizosphere, which may explain the difference in horizontal spread between A109 and P25.

REFERENCES

Thompson, I. P., Cook, K. A., Lethbridge, G. and Burns, R. G. (1990). Survival of two ecologically distinct bacteria (*Flavobacterium* and *Arthrobacter*) in unplanted and rhizosphere soil: laboratory studies. Soil Biol. & Biochem. 22:1029-1037.
Thompson, I. P., Young, C. S., Cook, K. A., Lethbridge G. and R. G. Burns. (1992). Survival of two ecologically distinct bacteria (*Flavobacterium* and *Arthrobacter*) in unplanted and rhizosphere soil: field studies. Soil Biol. & Biochem. 24:1-14.

IMPLEMENTATION OF PRE-TESTS FOR GEMMO
RELEASE IN THE ENVIRONMENT

P.Bauda*, B.Baleux, P.Lebaron, M.C. Lett, J.C. Hubert,
G.Faurie,Y.Duval-Iflah and D.Prieur

Centre des Sciences de l'Environnement
1 rue des Récollets
57000 Metz
FRANCE

The aim of this work was to design microcosms and to validate their use as standardized systems in order to predict, to some extent, recombinant DNA (rDNA) dissemination in the environment. Five laboratories participated in (i) the construction of model bacterial strains and plasmids and the checking of their genetic stability, (ii) the definition of techniques for recovering and identifying specific microorganisms from the microcosms, (iii) the implementation of microcosms or artificial experimental systems representative of ecological niches; soils, waters, attached bacterial populations, animal digestive tracts.

STRAINS AND PLASMIDS: CHOICE AND ENUMERATION

In a first step, due to its well-known genetics, *E. coli* was chosen as donor and recipient strain to validate the microcosms and the methodologies.Autochtonous populations will be used as recipients in further steps.The characteristics of strains and plasmids designed for this study are presented in Table 1. They were chosen or constructed to allow, from samples taken in the different microcosms, the screening of donors, recipients and possible transconjugants on selective agar plates as a function of the antibiotic resistance profile of the different cells. The rDNA consists in the thymidine kinase gene of the vaccine virus used as gene marker in pCE plasmids. The presence of rDNA in donors and transconjugants was checked by DNA-DNA hybridization with the thymidine kinase gene of vaccine virus as a specific probe.The total number of bacterial cells was monitored, in addition to classic plate counts, by epifluorescence counts using 4',6-diamidino-2-phenylindole as a DNA specific fluorochrome emiting a blue fluorescence when excited at 365nm.The use of the two techniques was necessary, especially in aquatic media where the presence of viable but nonculturable cells is well described (Colwell *et al*, 1985).

CHOICE OF MICROCOSMS

Five different microcosms have been designed to monitor GEMMO behaviour in conditions simple enough to be standardized and reproduced. The characteristics of the five experimental systems used are briefly given below. (1) The soil microcosm was constituted by polypropylene tubes containing 5g of a brown soil defined by the following balanced texture: clays 31.4%, silt 36.4%, sand 32.2%, pH 6.5, organic carbon 2.7%, organic

The Release of Genetically Modified Microorganisms
Edited by D.E.S. Stewart-Tull and M. Sussman, Plenum Press, New York, 1992

149

Table 1. Characteristics of strains and plasmids used in the study

STRAINS	PLASMIDS
E. coli UB 281 : *pro, met,* Nal[r]	pCE325: Amp[r],Tet[s], Cm[r], bom[+],tk,par
" " UB1636 : *his, try, lys,* Sm[r]	pCE328: Amp[r],Tet[s],Cm[r], bom[-],tk,par
" " UB1637 : *his, try, lys,* Sm[r],recA	R388 : Inc W, tra[+],Tp[r]
" " UB1832 : *his, try, lys,* Sm[r], Rif[r]	pUB2380 : Kn[r], mob[+]
	pCL6 : R388::Tn3926

pCE325 is derived from pBR325 by cloning the thymidine kinase gene of vaccine virus (tk) and "par" sequence of pSC101. pCE328 is derived from pBR328 by cloning tk and par sequence of pSC101.pUB2380 is a gift from P. Bennett (University of Bristol,U.K.). It was isolated from a farm isolate of *E. coli* , in the U.K. R388 was isolated from an *E. coli* hospital isolate.pCL6 was derived from R388 by transposition of Tn 3926, found in an environmental strain of *Yersinia enterocolitica.*

nitrogen 0.35%, water holding capacity 40%. Soil sterilization was ensured when necessary by irradiation. (2) The water microcosm was provided by membrane filter chambers in glass pyrex.The chambers containing bacterial suspensions may be immersed in a 200L recipient of natural water homogenised by oxygen bubbling. (3)The mouse gut microcosm consisted of axenic mice in which the desired bacterial population was introduced orally (Duval-Iflah *et al* ,1980). (4)The marine mussel digestive tract microcosm was chosen because it provides a natural means of concentrating sea-water biomass and it allows the promiscuity of marine bacteria enhancing in this way the probability of conjugative genetic exchange. A few hours contact time between mussels and sea-water containing 10^5 cfu/ml ensures mussel contamination at a level of 10^5 cfu/g of digestive tract. (5) Laboratory bacterial biofilm, produced in fixed-bed reactors, was chosen as an "artificial microcosm" representative of natural attached biomass.The experimental device used to grow biofilms was previously described in Bauda *et al.* (1990).

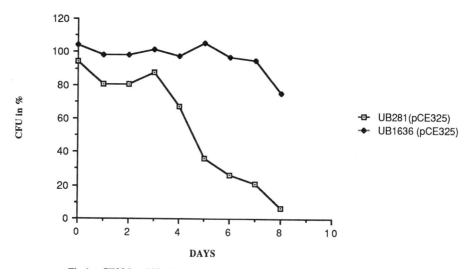

Fig.1 pCE325 stability in strains UB281 and UB 1636 within biofilms.

PLASMID STABILITY

A particular effort was made to choose donor and recipient *E. coli* strains in which the rDNA was stably maintained. The stability study was focused on pCE plasmids whose transfer was investigated in the presence or absence of different helpers (R388, pUB2380, pCL6). Due to its further use as a positive transfer control in the microcosms, R388 stability was also checked.Plasmid stability was investigated in continuously grown laboratory biofilms providing a greater number of generations than batch growth systems. Results, showing the instability of pCE325 in strains UB281 and UB1636 (Figure 1), indicate that these strains and plasmid should not be used in plasmid transfer experiments. This instability, confirmed in the soil microcosm during growth phases, was shown to be directly related to the growth rate of cells.

ACKNOWLEDGEMENTS

Financial support was provided by "Chimie-Ecologie".

REFERENCES

Bauda, P., Menon, P., Block, J.C., Lett, M.C., Roux, B. and Hubert, J.C. (1990) in "Bacterial Genetics in natural Environments" (Fry,J.C., Day,M.J., Eds),pp. 81-88. Chapman and Hall, London.

Duval-Iflah,Y., Raibaud, P., Tancrède, C. and Rousseau, M.(1980) R-plasmid transfer from *Seratia liquefaciens* to*Escherichia coli* in vitro and in vivo in the digestive tract of gnotobiotic mice associated with human fecal flora.*Infect. Immun* . 28: 981-990.

Colwell, R.R., Brayton, P.R., Grimes, D.J., Roszak, D.B., Huq, S.A. and Palmer, L.M. (1985) Viable but non-culturable *V.cholerae* and related pathogens in the environment: implications for the release of genetically-engineered microorganisms.*Biotechnol.* 3: 817-820

PLASMID TRANSFER AND STABILITY OF *PSEUDOMONAS CEPACIA* AND *ENTEROBACTER CLOACAE* IN A CONTINUOUS FLOW CULTURE SYSTEM

L.Sun[1], M.J. Bazin[1] and J.M.Lynch[2]

Biosphere Sciences Division[1]
King's College London
Campden Hill Road
London W8 7AH

Microbiology & Crop Protection Department[2]
Horticulture Research International
Worthing Road
Littlehampton
West Sussex
BN17 6LP, UK

It has been proposed that utilizing bacteria with plasmid-borne genes which are eliminated naturally might be a relatively safe option for releasing genetically-modified microorganisms (GEMMOs) into the environment. The results of our research indicate that this might not be the case. In our flow-through column reactors, donor cells of *Pseudomonas cepacia* containing the transmissible plasmid R388::Tn1721 with genes encoding resistance to trimethoprim and tetracycline, passed these characteristics to recipients. Recipient cells were chromosomally resistant to nalidixic acid so that transconjugants were resistant to all three antibiotics. Plasmid transfer occured at relatively high frequencies but the plasmid, in both donor and transconjugant, disappeared exponentially from the thermodynamically open system. Thus, *P. cepacia* with R388::Tn1721 would appear to have the characteristics favourable for genetic release.

We have found that R388::Tn1721 can be transferred intergenerically from *P. cepacia* to *Enterobacter cloacae*, although at very low frequencies. When *E. cloacae* was grown in a column reactor in an identical fashion to that described above, the plasmid persisted in both donor and transconjugant cells and the populations of both reached stable steady-state densities.

These results provide evidence for the possibility of plasmid-borne traits becoming stabilized in the environment under circumstances in which the released plasmid host disappears. Therefore, it is insufficient to ensure the disappearance of a trait that a plasmid-borne gene disappears from the species in which it was released as the gene may be transfered to an indigenous species in which it is stabilized.

ACKNOWLEDGEMENT

We thank K.C. Wang Education Foundation of Hong Kong, the Committee of Vice-Chancellors and Principals of the Universities of the United Kingdom and Horticulture Research International, Littlehampton, for financial support to one of us (L.S.).

The Release of Genetically Modified Microorganisms
Edited by D.E.S. Stewart-Tull and M. Sussman, Plenum Press, New York, 1992

153

RISK ASSESSMENT IN RELEASES OF NITROGEN-FIXING
ENTEROBACTER INTO SOIL; SURVIVAL AND GENE
TRANSFER, AS INFLUENCED BY AGRICULTURAL
SUBSTRATES

Walter Klingmüller

Department of Genetics
University of Bayreuth
Universitätsstrasse 30
W-8580 Bayreuth, Germany

In strains of nitrogen-fixing *Enterobacter agglomerans*, isolated from the rhizosphere of cereals, the nif-genes are located on large plasmids. Nif-plasmid pEA9 (200 kb) is self transmissible between closely-related strains. To collect data on possible uncontrolled gene spread, for planned releases of such bacteria, this plasmid was labelled with transposon Tn5,

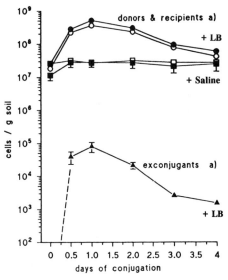

Fig.1. Survival (donor and recipient cells per g soil) and plasmid pEA9 transfer (exconjugants per g soil) in homologous matings of *E.agglomerans* 339 in non-sterilized soil (50 g samples in Erlenmeyer flasks) for increasing mating periods with (a) and without (b) addition of Luria Broth. Mating conditions: 22°C, pH 5.2, humidity 15.5%. The symbols (open symbols: recipients) give the means and 1xS.E.M. from 6 samples each (2 independent experiments, with 3 replicates each). The limit of detection for exconjugants without Luria Broth (b) was 6.4×10^{-8} surviving donors), the number of exconjugants with Luria Broth (a) was zero for zero time conjugation (indicated by the broken line).

The Release of Genetically Modified Microorganisms
Edited by D.E.S. Stewart-Tull and M. Sussman, Plenum Press, New York, 1992

155

Table 1. Plasmid transfer in natural soil. 1 day incubation at 22°C, different bacteria and plasmids, with either complete medium (LB) or sucrose (sucr.) added*).

	Enterobacter (pEA9)***) sucrose[+]		E. coli (RP4) sucrose[-]	
	+LB	+sucr.	+LB	+sucr.
donor**)	2.6×10^8	2.9×10^8	1.6×10^6	6.9×10^6
recipient	1.8×10^8	2.0×10^8	3.4×10^6	7.0×10^5
exconj.	4.0×10^4	1.3×10^5	7.6×10^4	9.0×10^1
transfer frequency	1.5×10^{-4}	4.5×10^{-4}	4.7×10^{-2}	$< 10^{-4}$

*)   no exconjugants without additives
**)   per g soil, for details of method see explanation to Figure 1
***)   Inoculation was with 10^7 cells of the plasmid carrying donor and a homologous plasmid less strain
as recipient

carrying a recombinant insert, and used in mating experiments between homologous *Enterobacter* strains with natural soil as substrate. Comparative experiments were done with *E.coli* (RP4).

Survival and plasmid transfer are described, for non-sterile soil, with or without added Luria broth (Figure 1), with or without added sucrose, and with or without mashed sugar beets.

With washed inoculants of *Enterobacter*, in saline solution, no plasmid transfer could be detected. However, the addition of broth, but also of sucrose (0.6 mg/g soil) elicited strong cell propagation, together with plasmid transfer (Table 1, left part, with sucrose-utilizing *Enterobacter;* optimum for sucrose after incubation for 1 day: 4.5×10^{-5} exconjugants per donor). Cell propagation and plasmid transfer with mashed sugar beets (8 mg/g soil) were similar to that with sucrose. A sucrose non-utilizing *Escherichia coli* strain, containing plasmid RP4 (high transfer rate to a plasmidless *E.coli* as recipient, with broth; Table 1, third column) gave little if any plasmid transfer with sucrose (Table 1, fourth column). The sucrose beet mash effect indicates a risk in bacterial releases, if these go into fields, where sugar beet has been grown before, and if the released bacteria can utilize sucrose. Hence, in general, in bacterial releases, remains from crops grown in the field the year before have to be taken into account. They may offer the risk of boosting plasmid transfer. Avoiding an unfavorable combination of a former crop with bacteria degrading its main contents is a promising principle for biological containment.

REFERENCES

Klingmüller, W., Dally, A., Fentner, C. and Steinlein, M. (1990) Plasmid transfer between soil bacteria. In "Bacterial genetics in Natural environments" (Fry,J.C. and Day,M.J. Eds) pp.133-151.Chapman and Hall;London.

Klingmüller, W. (1991) Plasmid transfer in natural soil: a case by case study with nitrogen-fixing *Enterobacter*. FEMS Microbiology Ecology 85, 107-116.

Klingmüller, W. (1991) Contained experiments and risk assesment for releases of nitrogen-fixing *Enterobacter,* with special focus on possible plasmid spread. In "Biological Monitoring of Genetically Engineered Plants and Microbes" (Mackenzie,D.R. and Henry, S.C. Eds), pp.45-53. Agricultural Research Institute, Bethesda, Maryland, USA.

DIRECT DETECTION OF *E.AGGLOMERANS* DNA SEQUENCES IN SOIL

Sonja Selenska and Walter Klingmüller

University of Bayreuth
Department of Genetics
D 8580 Bayreuth
Germany

We present a non-radioactive detection of Tn5 and nif sequences of a GEMMO *E.agglomerans* 19-1-1 in *Eco*R I digested DNA recovered by a simple direct lysis procedure from inoculated soil samples, where the target bacteria no longer gave colonies on a selective agar medium.

A modification of the DNA recovery procedure allows isolation of total soil DNA and RNA from the same sample.

A 550 bp sequence from the *npt* II gene of Tn5 was amplified in DNA and RNA samples, recovered from soil, inoculated with *E.agglomerans* 19-1-1, with Taq Polymerase (BRL) and rTth DNA polymerase (PERKIN ELMER CETUS), respectively.

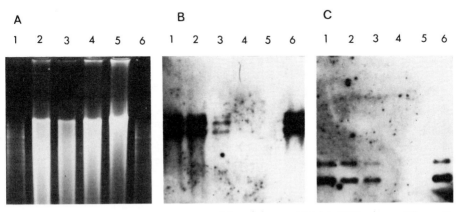

Fig. 1 Detection of *Tn5* and *nif* DNA sequences of *E.agglomerans* 19-1-1 in DNA recovered from soil at different days after inoculation. A) Electrophoresis of *Eco*R I digested DNA recovered: 2) 3 days after inoculation with 2.3 x 10⁷ cells/g soil of *E.agglomerans* 19-1-1; 3) 70 days after inoculation with the same bacteria; 4) 3 days after inoculation with a plasmidless and *Tn5* -free *E.agglomerans* 339; 5) DNA recovered from uninoculated soil; 1) and 6) total DNA of *E.agglomerans* 19-1-1, digested with *Eco*R I endonuclease; B) Southern blot of A) and PhotoGene detection of the Tn5 containing sequences; C) Reprobing of A) with a BioNick labelled 3 kb *nif* fragment of *E.agglomerans nif* plasmid pEA9, containing 1 *Eco*R I site.

The Release of Genetically Modified Microorganisms
Edited by D.E.S. Stewart-Tull and M. Sussman, Plenum Press, New York, 1992

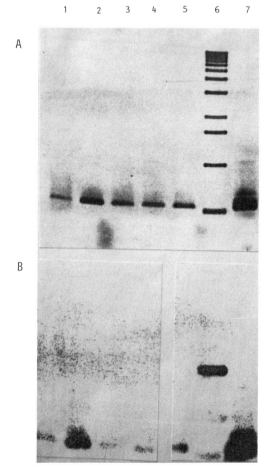

Fig. 2 Amplification of a 550 bp fragment of the *npt* II gene in soil DNA and RNA samples. A) Electrophoresis of PCR-amplified products in 1% agarose gel. 1) reverse rTth PCR-amplification in a soil RNA sample, recovered at the 1st day after inoculation with *E.agglomerans* 19-1-1; 2) reverse rTth PCR-amplification in total RNA of *E.agglomerans;* 3) PCR-amplification in a soil DNA sample recovered 3 days after inoculation with *E.agglomerans* 19-1-1; 4) PCR- amplification in a soil DNA sample recovered 7 days after inoculation: 5) PCR-amplification in a soil DNA sample recovered 35 days after inoculation; 6) DNA-kilobase ladder (BRL); 7) PCR-amplification in total DNA of *E.agglomerans* 19-1-1.

MOLECULAR ANALYSIS OF NUCLEIC ACIDS RECOVERED FROM SOIL

The amounts of soil DNA prepared by our simple procedure (Selenska and Klingmüller, 1991 a,b) were significantly larger than described by others (Steffan *et al.* 1988; Tsai and Olson, 1991). The molecular weight of recovered DNA was about 25 kb. The most important advantage of the procedure, however, was that the resulting DNA, in contrast to that recovered by other procedures, did not inhibit any of the restriction endonucleases tested. Thus it was possible to trace particular DNA sequences of *E.agglomerans* 19-1-1 in soil samples inoculated with these bacteria as described by Klingmüller *et al.* (1990) and Klingmüller, (1991). As shown in Figure 1 Tn5 and *nif* specific sequences were detected in samples analysed 70 days after the introduction of the inoculant bacteria. At that moment conventional plating of samples gave no more colonies. We suggest that the detected DNA could represent inoculated bacteria, which after such a long exposure in soil, no longer express the Km^r gene of Tn5, and/or non-culturable target bacteria, as well as extracellular DNA released from dead bacteria. Among 56 analysed soil DNA samples only one case of a rearrangement of a Tn5 sequence in soil conditions was detected.

Amplification of a 550 bp Fragment of the *npt* II Gene

It was possible to optimize PCR amplification in soil DNA samples, as shown in Figure 2. As a downstream primer the sequence: 5' GCAGCTGTGCTCGACGTTG TCACTGAAG 3' was used, this is situated 177 nucleotides after the beginning of the *ORF* of the *npt* II gene of *Tn5*. As an upstream primer we used the sequence: 5' CAAGAAGGCGATAGAAGGCG 3', which is situated 6 nucleotides before the two stop codons at the end of the gene (Beck *et al*, 1982). For DNA amplification we used Taq Polymerase (BRL). The reaction conditions were as follows: 10 mM Tris.HCl pH 8.3; 50 mM KCl; 3.0 mM MgCl$_2$; 500 mM each dNTP; 0.5 mg of each primer; 5 U Taq Polymerase and 3-5 ml soil DNA. Reaction mixtures were heated for 7 min. at 95° C and then 40 cycles (2 min. at 95° C; 1 min. at 45° C; 3 min. at 72° C) were performed in a Coy TempCycler. After the 40th cycle, the PCR amplification was extended for 15 min. at 72° C. Reverse rTth polymerase amplification in the soil RNA sample was also successful, but only when target bacterial RNA was added.

REFERENCES

Beck, E., Ludwig, G., Auerswald, E.A. and Schaller, H. (1982). Nucleotide sequence and exact localization of the neomycin phosphotransferase gene from *Tn5*. Gene 19, 327-336.

Klingmüller, W. (1991) Plasmid transfer in natural soil: a case by case study with nitrogen-fixing *Enterobacter*. FEMS Microbiology Ecology 85, 107-116.

Klingmüller, W., Dally, A., Fentner, C. and Steinlein, M. (1990). Plasmid transfer between soil bacteria. In "Bacterial Genetics in Natural Environments," (Fry,J.C. and Day,M.J. Eds) pp.133-151. Chapman and Hall, London

Sambrook, J., Fritsch, E.F. and Maniatis, T. (1989). Molecular Cloning: A Laboratory Manual Cold Spring Harbor Laboratory Press, Cold Spring Harbor, New York.

Selenska, S. and Klingmüller, W. (1991). Direct detection of *nif*-gene sequences of *Enterobacter agglomerans* in soil. FEMS Microbiol. Letters 80, 243-246.

Selenska, S. and Klingmüller, W. (1991). DNA recovery and direct detection of Tn5 sequences from soil. Letters in Appl. Microbiol. 13, 21-24.

Steffan, R,J., Coksoyr, J., Bej, A.K. and Atlas, R.M. (1988). Recovery of DNA from soils and sediments. Appl. Environ. Microbiol. 54, 2908-2915.

Tsai, Y. and Olson, B.H. (1991). Rapid method for direct extraction of DNA from soil and sediments. Appl. Environ. Microbiol. 57, 1070-1074.

THE SURVIVAL OF MICROORGANISMS IN THE OPEN AIR

A.M. Bennett, J.M. Watts and J.E. Benbough

Biosafety Unit, Division of Biologics
Public Health Laboratiry Service
Centre for Applied Microbiology and Research
Porton Down, Salisbury, UK

The deliberate or accidental release of genetically-modified organisms into the environment has been a cause for concern of the environmental lobby and is the subject of two EC Directives (Commission of the European Community 1990a; 1990b). Each of these directives stresses the necessity of performing risk assessments of the likely spread and dissemination of the genetically-modified organisms into the environment. Since one of the methods of both accidental and deliberate dissemination of microorganisms will be to the atmosphere it is important to have some knowledge of how microorganisms will survive in the aerosol state in order to carry out this risk assessment.

When microbes are aerosolized many physical, chemical and biological factors affect their decay with time. Strains of microorganisms vary widely in their stability in the aerosol state with some spore-formers being almost indestructible under ordinary environmental conditions. Most microorganisms decay with time and this can vary considerably with conditions such as low relative humidity, sunlight, oxygen and some pollutants. A project is being carried out at PHLS CAMR to find the effect of environmental factors on the aerostability of commonly used industrial microorganisms and genetically-modified organisms.

The survival of microorganisms in air is studied by two novel techniques developed at PHLS CAMR. The Henderson apparatus allows the aerostability of microorganisms to be studied at a variety of relative humidities in an enclosed environment (Druett 1969). In this apparatus microbial aerosols are generated into air of known relative humidity which can be stored for long periods of time (24 hours) in a rotating drum which help prevent gravitational deposition. The microthread technique exposes "captive aerosols" of microorganisms sprayed onto submicron spiders webs to the open air (May and Druett 1968) . The microorganisms suspended on the webs have most of their surface area exposed to the air and are likely to behave in a similar way to aerosols. The Henderson apparatus is used to expose microorganisms to controlled environmental factors while the microthread technique is used to determine the effect of exposure of microorganisms uncontrollable but realistic environmental conditions.

Preliminary studies have been carried out with two strains of Escherichia coli, JM83 and MRE162. The aerostability of the original JM83 strain was compared to that of a JM83 containing a genetically- modified plasmid. It was found that the original strain was approximately three times more stable than the modified strain. No evidence was found of plasmid loss caused by aerosolization. The MRE162 was significantly more stable than the JM83 under every environmental condition tested. Both strains aerostability increased as the relative humidity was increased from 55% to 85%. Their aerostability was lower in the outdoor environment than indoors probably due to the presence of the open air factor, an environmental germicide related to the concentration of industrial pollutants in the atmosphere. Survival was dependent on wind direction and it was found that if the air originated from a potentially

The Release of Genetically Modified Microorganisms
Edited by D.E.S. Stewart-Tull and M. Sussman, Plenum Press, New York, 1992

161

heavily-polluted area, that is, London, the organisms survived an average of three-fold less than if the air came from the direction of the country or the sea.

The survival of aerosolized microorganisms in the atmosphere has been shown to be dependent on many environmental factors. A knowledge of how these factors affect the survival of the microorganisms could be combined with a mathematical model of atmospheric dissemination and dispersal to give a realistic risk assessment of the spread of genetically-modified microorganisms released to the environment. These techniques will be applied to a wide range of industrially important microorganism, in both wild-type and genetically-modified forms.

REFERENCES

Commission of the European Communities (1990a) Directive on the contained use of genetically modified microorganisms. 90/219/EEC, 23.04.90 (OJ L177, 08.05.90). CEC, Brussels.

Commission of the European Communities (1990b) Directive on the deliberate release into the environment of genetically modified organisms. 90/220/EEC, 23.04.90 (OJ L177, 08.05.90). CEC, Brussels.

Druett, H. A. (1969) A mobile form of the Henderson apparatus. Journal of Hygiene 55, 437-448.

May, K. R. and Druett, H. A. (1968) A microthread technique for studying the viability of microbes in a simulated airborne state. Journal of General Microbiology 51, 353-366.

SURVIVAL AND DISPERSAL OF GEMMOs ASSOCIATED WITH ANIMALS

M.J. Bale[1], P.M. Bennett[2], J.E. Beringer[1], J.A. Bale[3] and M.H. Hinton[3]

Department of Botany[1]
University of Bristol
Woodland Road
Bristol BS8 1UG

Department of Pathology and Microbiology[2]
University of Bristol
University Walk
Bristol BS8 1TD

Division of Veterinary Public Health and Food Safety[3]
University of Bristol
Langford House
Langford
Avon BS18 7DU

INTRODUCTION

When considering the release of genetically-modified microorganisms it is important that the potential for interaction with birds and animals is understood, because animals may transport GEMMOs over large distances. We have examined this in a model system with derivatives of a natural *E. coli* strain which efficiently colonizes chicks from surfaces, feed and bedding. The model consists of ground dwelling chicks and flying birds. The environment, feed, faeces and bedding are monitored for the introduced strain. To enhance their detection a series of PCR primers have been synthesized and methods developed to extract DNA from environmental samples.

SURVIVAL OF GENETICALLY-MARKED *ESCHERICHIA COLI*

E. coli ECO80 was marked with the Tn*1721*-derived transposons (Ubben and Schmitt, 1986) Tn*1732* (ECO890) and Tn*1725* (ECO848). Their survival was examined on plastic surfaces, in broiler mash, bedding, tap-water and trough-water. All strains survived well in broiler mash and in bedding but less well in trough-water and on surfaces (results not shown).

COLONIZATION OF CHICKS

Groups of one-day-old chicks were given different doses of the marked *E. coli* strains by a variety of routes. These included feed, contact with contaminated plastic or by housing on

The Release of Genetically Modified Microorganisms
Edited by D.E.S. Stewart-Tull and M. Sussman, Plenum Press, New York, 1992

naturally-contaminated bedding. The dose needed to colonize 50% of the birds (ID_{50}) was calculated by linear regression. The ID_{50} was lowest for feed (9 cfu/g), followed by bedding (7.5×10^3 cfu/g) and surfaces (1.1×10^6 cfu/cm).

EXTRACTION OF DNA FROM ANIMAL SAMPLES

DNA was extracted from environmental samples by blending, and differential centrifugation followed by a Percoll density gradient, lysis and purification (Martin and Macdonald, 1981; Steffan and Atlas, 1988). This rapid method gave good yields of DNA, largely free of humic acid or protein. PCR was done with primers within Tn*1732* and Tn*1725* as well as in the flanking chromosome, which allowed specific amplification of each strain. All samples gave the correct PCR product when spiked with target DNA, indicating that they did not inhibit PCR. The correlation between plate counts and PCR, however, was variable (Table 1). PCR was also done with bacteria from selective media without pre-treatment and proved useful for confirming the identity of the marked *E. coli*.

Table 1. The colonization and dispersal of genetically marked *E. coli* by budgerigars and starlings.

Bacteria in feed cfu/g (length of trial)	Mean count flying bird faeces cfu/g	Proportion of chicks positive	Bacterial count in control litter	Detection by PCR in litter DNA sample
A: Budgerigars				
2.8×10^3 (10 days)	3.9×10^1 (n=8)	0/20	ND	ND
9.8×10^2 (7 days)	4.1×10^1 (n=4)	0/20	ND	ND
3.1×10^4 (14 days)	3.8×10^1 (n=14)	0/20	ND	ND
3.1×10^3 (15 days)	7.9×10^4 (n=8)	10/20	$<2.5 \times 10^1$	1/2
1.1×10^4 (8 days)	1.0×10^2 (n=4)	0/20	ND	ND
B: Starlings				
1.0×10^2 (14 days)	2.9×10^2 (n=8)	Control 0/11 Contam. 9/9	$<2.5 \times 10^1$ 4.5×10^4	0/2 1/2
9.7×10^5 (13 days)	2.3×10^5 (n=7)	Control 6/12 Contam. 11/11	4.5×10^3 4.9×10^6	0/2 1/2

Marked bacteria (ECO890 or ECO848) were incorporated into budgerigar seed or into mash shared by the starlings and one group of chicks. Flying bird faeces were collected from a high shelf. Litter samples were taken and DNA extracted as detailed above. In the starling experiments the litter samples are divided into those chicks given clean mash (Control) and those given contaminated mash (Contam.). ND = not determined

DISPERSAL OF BACTERIA BY FLYING BIRDS

Budgerigars and starlings, were used in the model studies. Budgerigars are seed-eating birds which had minimal contact with the chicks. Conversely, the starlings shared food and water with the chicks.

Budgerigars were given contaminated seed (Table 2a). In most cases this resulted in low levels of colonization and excretion and no chicks were infected. When effective colonization occurred, 50% of the chicks were also colonized, which reflected the higher ID_{50} for contamination *via* the bedding. DNA extracted from mash, bedding, budgerigar faeces and seed showed evidence of ECO890 after PCR.

Starlings which shared contaminated mash with chicks on one side of the room became colonized at levels which reflected the level of mash contamination (Table 1b). The chicks all became colonized, whilst the control chicks were again only infected when the starlings were effectively colonized. Less than half of the DNA samples from bedding known to be contaminated were positive after PCR.

CONCLUSIONS

Genetically-marked derivatives of *E. coli* can survive in, and colonize chicks from, feed, bedding or surfaces. Adult flying birds are more resistant to colonization. Low numbers of bacteria in the faeces of flying birds did not lead to colonization of ground-dwelling chicks; this was only observed when the flying birds were heavily colonized. PCR was useful for identifying bacteria from plates but was less reliable for detecting the bacteria in DNA from environmental samples.

ACKNOWLEDGEMENTS

We would like to thank V. Allen, E. Coombs, A. Cornish, P. Parsons and A. Pilcher for technical assistance. This work was done under contract PECD7/8/95 from the U.K. Department of the Environment.

REFERENCES

Ubben, D. and Schmitt, R. (1986) Tn*1721* derivatives for transposon mutagenesis, restriction mapping and DNA sequence analysis Gene 41, 145-152.

Martin, N. J. and Macdonald, R. M. (1981) Separation of non-filamentous microorganisms from soil by density gradient centrifugation in Percoll. J. Appl. Bacteriol. 51, 243-251.

Steffan, R. J. and Atlas, R. M. (1988) DNA amplification to enhance detection of genetically engineered bacteria in environmental samples. Appl. Environ. Microbiol. 54, 2185-2191.

STUDY OF INTESTINAL COLONIZATION WITH NORFLOXACIN IN VIVO AND IN VITRO

J.Schlundt, E.M.Nielsen, G.Fischer and B.L.Jacobsen

Institute of Toxicology
National Food Agency of Denmark
Mørkhøj Bygade 19
DK 2860, Denmark

INTRODUCTION

The deliberate release of genetically-modified microorganisms (GEMMOs) has highlighted intestinal colonization in relation to risk assessment of GEMMOs.

The intestinal colonizing ability of 'new' bacteria is influenced by the indigenous microbial flora of the animal in question (Freter, 1955), which is why the use of animal models to investigate intestinal colonizing ability in humans might present problems. In this laboratory we aim to establish a human *in vitro* continuous flow (CF) intestinal model. In order to evaluate the system the colonization ability of *E.coli* strains in rat *in vitro* models is compared with the rat *in vivo* situation. The antimicrobial drug norfloxacin in short time treatment (3 days) can reduce colonization resistance of the intestinal microbiota towards *E.coli* without affecting most other intestinal bacterial groups (Nielsen and Schlundt, 1992).

MATERIALS AND METHODS

Bacterial Strains

BJ18 is a rifampicin (*rif*) resistant strain and BJ19 is a nalidixic acid (*nal*) resistant strain derived by spontaneous mutation from the same *E. coli* strain (rough:K⁻:H2) originally isolated from a rat.

Bacteriological Methods

Faecal samples from rats and liquid samples from CF-cultures were diluted and plated on MacConkey agar with rif (100 µg/ml) or nal (40 µg/ml) and incubated at 37°C for 24 h.

Animals

10 Han:Wist. rats, age 4 months, weight 300g. Five were pretreated with 8 mg norfloxacin (Zoroxcin: Merck, Sharp & Dohme) p.o. by gavage for 3 days prior to the experimental period.

The Release of Genetically Modified Microorganisms
Edited by D.E.S. Stewart-Tull and M. Sussman, Plenum Press, New York, 1992

167

In vivo Experimental Procedure

1.0 ml of a mixture containing 1 x 10^9 cfu BJ18 and BJ19 were dosed p.o. by gavage. Faeces was sampled directly from the rectum.

Continuous Flow Cultures

The two-stage CF-culture system consisted of two 1 litre fermenters as described by Mallett *et al.* (1985) and Nielsen and Schlundt (1992). Both fermenters were inoculated with 1.0 g of rat faeces followed by a ten day period of stabilization before addition of test-bacteria. Norfloxacin treatment: 450 mg/l was added to both stages.

RESULTS AND DISCUSSION

Figure 1 presents the results from administration of *E.coli* strains BJ18 and BJ19 to five rats. In all animals BJ18 attained a concentration of 10^4 to 10^6 cfu/g faeces on day 1 but was then washed out. The concentration of BJ19 started at 10^5 to 10^7 cfu/g faeces, but then fell 2 - 5 log units over the next 5 - 7 days. In one animal this fall resulted in total elimination (below detection limit), whereas BJ19 concentration in the remaining four animals increased to a concentration of 10^3 to 10^6 cfu/g faeces. The total *E.coli* concentration was 10^5 to 10^7 cfu/g faeces in all animals rose to a level of 10^3 to 10^6 cfu/g faeces.

In order to reduce colonization resistance of the system, the experiment was repeated with five rats pretreated with norfloxacin for three days. In all animals BJ18 fell below the detection limit at day 4, whereas BJ19 entered all five animals in a concentration of 10^6 to 10^7 cfu/g faeces, which was maintained during all 18 days of this experiment. In all animals the total *E.coli* count equalled the BJ19 concentration.

In two out of three experimental runs of the CF-culture with BJ18 and BJ19 addition both strains remained in stage II of the CF-culture for the total length of the experiment (20 and 60 days). In the third run BJ18 and BJ19 were both eliminated from the CF-culture. The total *E.coli* count fluctuated between 10^4 to 10^7 cfu/ml.

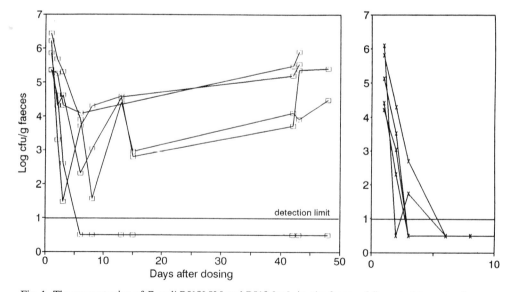

Fig. 1 The concentration of *E. coli* BJ18[X] and BJ19 [□] in the faeces of five rats (shown by five separate lines) dosed p.o. with a mixture of BJ18 and BJ19 at day 0.
In all rats BJ18 is eliminated within 5 days, in four rats BJ19 colonize and in one rat BJ19 is also eliminated.

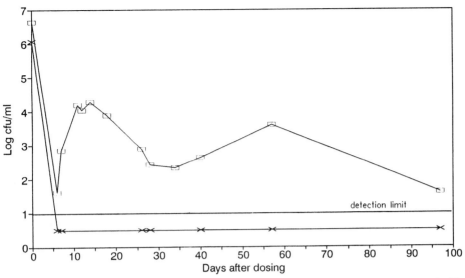

Fig. 2 The concentration of *E. coli* BJ18 [X] and BJ19 [□] in stage II of a CF-culture pretreated with Norfloxacin and dosed with a mixture of BJ18 and BJ19 at day 0. BJ18 is eliminated within 5 days, whereas, BJ19 colonize the system.

This underlined the necessity to standardize the colonization resistance of the CF-culture system as well. In figure 2 the results of a norfloxacin-treated CF-culture with BJ18 and BJ19 addition is shown. Here BJ18 is below the detection limit at day 3, = the first sampling day whereas the BJ19 maintains a concentration of $10^{2,5}$ to 10^4 cfu/ml for the remaining 40 days of the trial, corresponding to the total *E.coli* concentration.

The basis of the studies reported here is two *E.coli* strains with a clear difference in colonization ability *in vivo*, even with presumably small chromosomal differences (nalidixic acid and rifampicin resistance). Thus, these strains would seem reasonable for testing the discriminative potential of the *in vitro* system in relation to colonization ability.

Generally the nalidixic acid-resistant isolate (BJ19) colonized *in vivo* whereas the rifampicin resistant isolate (BJ18) did not. However, in one of five rats BJ19 did not colonize either. Norfloxacin treatment *in vivo* resulted in a more consistent outcome, whereas the norfloxacin treatment of the *in vitro* CF-culture system resulted in colonization of BJ19 and wash-out of BJ18, a result which was not obtained in the conventional CF-culture.

The use of *in vitro* systems to investigate single factors in relation to colonization has gained interest over recent years (Franklin *et al*, 1990). The CF *in vitro* rat gut model described here seems promising, especially when modified with norfloxacin treatment. The success of a similar pig *in vitro* model (Nielsen and Schlundt, 1992) might open up the possibility of establishing a human *in vitro* gut model.

REFERENCES

Franklin, D.P., *et al.* (1990) Growth of *Salmonella typhimurium* SL5319 and *E.coli* F-18 in mouse cecal mucus: role of peptides and iron. FEMS Micr. Ecology 74:229-240.

Freter, R. (1955) The fatal enteric infection in the guinea pig, achieved by inhibition of normal enteric flora. J.Inf.Dis. 97:57-65.

Mallett, A.K., *et al.* (1985) Metabolic adaptation of rat faecal microflora to cyclamate *in vitro*. Fd.Chem.Toxic. 23:1029-1034.

Nielsen, E.M. and Schlundt, J. (1992) The use of norfloxacin to study colonization ability of *E.coli* in *in vivo* and *in vitro* models of the porcine gut. Antimicrobial Agents and Chemotherapy 36: in press.

STUDIES ON GENE FLUX BY FREE BACTERIAL DNA IN SOIL, SEDIMENT AND GROUNDWATER AQUIFER

W. Wackernagel, G. Romanowski and M. G. Lorenz

Genetik
Fachbereich Biologie
Universität Oldenburg
Postfach 2503
W-2900 Oldenburg
Germany

The presence of extracellular high molecular weight DNA in sedimentary and aqueous environments (DeFlaun *et al.*, 1987; Ogram *et al.*, 1987) and the fact that many bacterial species present in these environments can develop natural competence for DNA uptake (Lorenz and Wackernagel, 1988) has raised the question whether horizontal gene transfer by genetic transformation occurs in these habitats. Following several initial studies which suggested a relatively high DNA stability in the evironment (Greaves and Wilson, 1970; Lorenz *et al.*, 1981; Aardema *et al.*, 1983; Lorenz and Wackernagel, 1987; Romanowski *et al.*, 1991) this topic has received increasing attention. Genetic transformation in natural habitats would have profound impacts on biological and ecological aspects of bacterial life including evolution, population dynamics and spread of genetic material not normally part of bacterial genomes. With respect to safety considerations about the deliberate or accidental release of genetically-modified microorganisms a gene flux by bacterial dissemination of free DNA would be of special importance.

STEPS IN GENE TRANSFER BY FREE DNA

We have dissected the possible environmental gene flux by free DNA among bacteria into a series of separate steps: (i) release of functional DNA from donor cells; (ii) distribution of DNA between liquid and solid phases; (iii) protection of DNA against nucleolytic degradation; (iv) development of competence for DNA uptake by cells; (v) transformation of recipients by free DNA and DNA associated with solid surfaces.

We have initiated investigations on each of the five steps by employing experimental model systems: flow-through microcosms with sand as model soil matrix (Lorenz and Wackernagel, 1988), typical soil bacteria with natural competence including *Pseudomonas stutzeri*, *Bacillus subtilis* and *Acinetobacter calcoaceticus* and purified genetic material including chromosomal DNA of these species as well as plasmids replicating in them. The aim of these model system studies is to obtain quantitative data on the processes involved and on biotic and abiotic factors affecting them. This may eventually allow predictions on the efficiency of gene flux in natural

The Release of Genetically Modified Microorganisms
Edited by D.E.S. Stewart-Tull and M. Sussman, Plenum Press, New York, 1992

habitats and an estimation of the turnover of an extracellular bacterial gene pool. The studies are progressing towards systems of increasing complexity (natural soil types, bacterial communities, variable environmental factors).

RESULTS OBTAINED FROM MODEL SYSTEMS

Step (i): Large amounts of high molecular weight DNA (up to 30 µg/ml) including transforming chromosomal fragments and plasmids were released from cells of *A. calcoaceticus* and *B. subtilis* cultured in minimal or rich medium (Lorenz *et al.*, 1991). Most DNA appeared during the death phase when cell lysis was progressing. Even with autologous DNase activity in the culture fluids chromosomal and plasmid markers were detected by transformation 40 h after the cultures reached the stationary phase.

Step (ii): Extracellular linear duplex and plasmid DNA rapidly bound to model soil matrix and to material recovered from a groundwater aquifer (Lorenz and Wackernagel, 1987; Romanowski *et al.*, 1991; unpublished results). The ionic conditions favorable for this process are low concentrations of divalent cations and are met by natural groundwater or interstitial liquid of water-saturated soils.

Step (iii): The protection of DNA adsorbed to purified minerals against nucleolytic degradation was shown previously (see first paragraph). Figure 1 indicates that linear and circular DNA adsorbed to natural groundwater aquifer minerals is 1000-fold less susceptible to nucleolytic degradation than is dissolved DNA. This protective effect of DNA binding to particulates is probably the reason why intact plasmid DNA which was added to non-sterile samples of three soils and re-extracted after various time periods was still detectable by Southern hybridization after 5 days (unpublished).

Step (iv): Competence development and transformation of *P. stutzeri* in nutrient-amended aqueous extracts of four different soils has been shown (Lorenz and Wackernagel, 1991; unpublished). Growth studies indicated that the extracts were limited in carbon/energy and nitrogen and two additionally in phosphorus sources. Interestingly, addition of defined nutrient combinations to the extracts resulting in only phosphorus or nitrogen limitation stimulated transformation 25 to 1000-fold compared to fully supplemented extracts.

Step (v): The chemical compositions of soil environments seem to be suitable for uptake of dissolved DNA (Lorenz and Wackernagel, 1991). Plasmid transformation of competent *A. calcoaceticus* proceeded equally well in non-sterile samples of groundwater and soil extract as

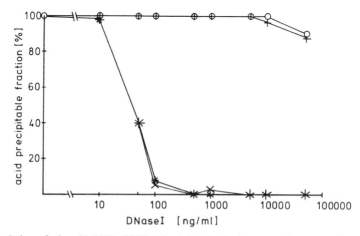

Fig. 1 Degradation of plasmid DNA (22°C, 10 min) adsorbed to groundwater aquifer material (o, supercoiled; +, linearized) or dissolved in sterile groundwater (*, supercoiled; X, linearized) against degradation by DNaseI.

Table 1 Transformation of *A. calcoaceticus* by pKT210 in non-sterile groundwater and aqueous soil extract

Transformation medium	Transformation efficiency $(Cm^R \; cells/\mu g \; pKT210 \cdot 10^9 \; cells)$
Groundwater	$3.9 \cdot 10^5$
Soil extract	$3.9 \cdot 10^5$
Tris-HCl + 0.5 mM Ca^{2+}, pH 8.5	$9.8 \cdot 10^5$

in sterile buffer (Table 1). Other experiments show that mineral-adsorbed homologous DNA can transform *B. subtilis* (Lorenz et al., 1988), *P. stutzeri* (Lorenz and Wackernagel, 1990) and *A. calcoaceticus* (unpublished). Transformation of *P. stutzeri* and *B. subtilis* takes place over a wide range of temperatures (Lorenz and Wackernagel, 1988; 1990).

CONCLUSION

So far these studies have not identified physical, chemical or biological factors which would interrupt environmental gene flux by blocking one of the separate steps required for this process. Our model studies provide the basic knowledge on interactions of cells, DNA and the environment and prepare the ground for the understanding of data from experimental systems of increasing complexity.

REFERENCES

Aardema, B. W., Lorenz, M. G. and Krumbein, W. E. (1983) Protection of sediment-adsorbed transforming DNA against enzymatic inactivation. Appl. Environ. Microbiol. 46, 417-420.

DeFlaun, M. F., Paul, J. H. and Jeffrey, W. H. (1987) Distribution and molecular weight of dissolved DNA in subtropical estuarine and oceanic environments. Mar. Ecol. Progr. Ser. 38, 65-73.

Greaves, M. P. and Wilson, M. J. (1970) The degradation of nucleic acids and montmorillonite-nucleic-acid complexes by soil microorganisms. Soil Biol. Biochem. 2, 257-268.

Lorenz, M. G. and Wackernagel, W. (1987) Adsorption of DNA to sand and variable degradation rates of adsorbed DNA. Appl. Environ. Microbiol. 53, 2948-2952.

Lorenz, M. G. and Wackernagel, W. (1988) Impact of mineral surfaces on gene transfer by transformation in natural bacterial environments, in "Risk Assessment for Deliberate Releases", (Klingmüller, W. Ed.), pp. 110-119. Springer-Verlag, Berlin Heidelberg.

Lorenz, M. G. and Wackernagel, W. (1990) Natural genetic transformation of *Pseudomonas stutzeri* by sand-adsorbed DNA. Arch. Microbiol. 154, 380-385.

Lorenz, M. G. and Wackernagel, W. (1991) High frequency of natural genetic transformation of *Pseudomonas stutzeri* in soil extract supplemented with a carbon/energy and phosphorus source. Appl. Environ. Microbiol. 57, 1246-1251.

Lorenz, M. G., Aardema, B. W. and Krumbein, W. E. (1981) Interaction of marine sediment with DNA and DNA availability to nucleases. Mar. Biol. 64, 225-230.

Lorenz, M. G., Aardema, B. W. and Wackernagel, W. (1988) Highly efficient genetic transformation of *Bacillus subtilis* attached to sand grains. J. Gen. Microbiol. 134, 107-112.

Lorenz, M. G., Gerjets, D. and Wackernagel, W. (1991) Release of transforming plasmid and chromosomal DNA from two cultured soil bacteria. Arch. Microbiol. 156, 319-326.

Ogram, A., Sayler, G. S. and Barkay, T. (1987) The extraction and purification of microbial DNA from sediments. J. Microbiol. Meth. 7, 57-66.

Romanowski, G., Lorenz, M. G. ahd Wackernagel, W. (1991) Adsorption of plasmid DNA to mineral surfaces and protection against DNaseI. Appl. Environ. Microbiol. 57, 1057-1061.

PLASMID TRANSFER BETWEEN *PSEUDOMONAS AERUGINOSA* STRAINS IN NATURAL SOIL

U-A. Temann, R. Hösl and W. Klingmüller

Department of Genetics
University of Bayreuth
Universitätsstr.30
8580 Bayreuth
Germany

Conjugation is one of the mechanisms which could play a role in the transfer of genetic material among bacteria in soil. This must be considered in the assessment of possible risks involved in the release of genetically-modified bacteria and plasmids into the environment.

The conjugative, broad host range plasmid RP4 (Datta et al, 1971) is well suited to study the "worst case" of plasmid transfer in natural soil, not only because of its own high transfer frequencies to various bacterial strains but also in its capacity to mobilize several other plasmids.

The transfer of plasmid RP4 and the mobilization of plasmid RSF1010 (Barth and Grinter,1974) by RP4 between homologous strains of *Pseudomonas aeruginosa* PAO5 in natural soil (sandy loam soil from a test field near Bayreuth University) was studied. Conjugation experiments were carried out in microcosms of 50 g soil /250 ml Erlenmeyer flask. The number of introduced donor and recipient cells was in the range of 10^7 cfu/g soil Standard conditions were 22°C, 15% moisture and 3 days of incubation.

TRANSFER OF PLASMID RP4 IN NATURAL SOIL

It has been shown that transfer of plasmid RP4 occurs between homologous *Pseudomonas* strains, introduced into natural soil (van Elsas et al.1990). In that process the rate of transfer is dependent on various conditions, e.g. ecological and biological factors, soil texture, the availability of nutrients and the number of introduced donor and recipient cells.

The effect of two important ecological factors, temperature and soil moisture, which are subjected to great fluctuation in the natural soil environments, were investigated on the transfer of plasmid RP4.

Transfer of plasmid RP4 between homologous strains of *Pseudomonas aeruginosa* PAO5 is strongly influenced by temperature. In the tested range from 8°C to 43°C the maximum transfer frequency of 8.8 x 10^{-6} T/R corresponds as expected with the optimal growth temperature of 37°C for the tested *Pseudomonas* strain. Below and above this temperature transfer frequencies decrease rapidly. Soil moisture affects also the plasmid transfer in natural soil. In the tested range from 12.5 - 25% water content, the transfer frequency of plasmid RP4 decrease continuously (from 1.1 x 10^{-4} T/R to 2.7 x 10^{-6} T/R) with an increase in soil moisture.

To study the influence of additional nutrients on the transfer of plasmid RP4 we have inoculated natural, non-sterile soil and sterile soil with donor and recipient cells of

The Release of Genetically Modified Microorganisms
Edited by D.E.S. Stewart-Tull and M. Sussman, Plenum Press, New York, 1992

175

Table 1 Transfer frequencies of plasmid RP4 between homologous strains of *Pseudomonas aeruginosa* PAO5 in natural, non-sterile or sterile soil.
For inoculation donor and recipient cells were resuspended either in saline, complete medium (LB) or double concentrated complete medium (2xLB). Transfer frequencies were determined after 4 days of incubation under standard conditions.

soil	Transfer frequencies (T/R) of plasmid RP4		
	saline	LB medium	2xLB medium
soil (nonsterile)	$< 1 \times 10^{-7}$	$7.0 \pm 2.5 \times 10^{-5}$	$2.9 \pm 2.3 \times 10^{-4}$
soil (sterile)	$2.6 \pm 2.4 \times 10^{-6}$	$5.1 \pm 3.0 \times 10^{-4}$	$3.6 \pm 2.4 \times 10^{-4}$

Pseudomonas aeruginosa PAO5, resuspended either in saline (0.85 % NaCl), complete medium (Luria-Bertani-Broth=LB) or double concentrated complete medium (2 x LB).

The addition of nutrients to the soil enhanced plasmid transfer (Table.1). In unsupplemented, natural soil (donor and recipient cells were resuspended in saline for inoculation) transconjugants were detectable but the transfer frequency of plasmid RP4 was very low and just above the detection limit (10 cfu/g soil). Supplementation of natural soil with complete medium increased the transfer. The highest transfer frequency of plasmid RP4 was recorded after the addition of double concentrated complete medium to the soil.

These results for conjugation experiments in natural, non-sterilized soil are applicable to that carried out in sterile soil (Table.1), except that the addition of double concentrated complete medium did not increase further the transfer frequency of plasmid RP4. We have observed higher transfer rates were always greater in sterile soil than in non-sterile, natural soil. The reason for this could be the release of additional nutrients due to the death of naturally resident microorganisms in the soil by sterilization.

MOBILIZATION OF PLASMID RSF1010 BY PLASMID RP4 IN NATURAL SOIL

Mobilization is a mechanism of conjugative plasmid transfer between bacteria depending on the presence of a helper plasmid. We have investigated the mobilization of plasmid RSF1010 which is not self-transmissible but mobilizable by plasmid RP4, in natural soil; first in diparental and later in triparental matings between homologous strains of *Pseudomonas aeruginosa* PAO5.

Mobilization of plasmids can also occur in natural soil. For inoculation, donor and recipient cells were resuspended in complete medium. Three types of transconjugants were isolated from the soil: the first type carried only plasmid RSF1010, the second only plasmid RP4 and the third type of transconjugants carried both plasmids. The resulting transfer frequencies are pesented in Table 2.

There are differences in the transfer rates of plasmids RSF1010 and RP4 alone in diparental and triparental matings. In diparental matings the transfer of plasmid RSF1010 is higher than of plasmid RP4. The reason could be the smaller size of plasmid RSF1010 (RSF1010: 8.7 kb, RP4: 60 kb). In triparental matings these transfer frequencies were opposite; mobilization of plasmid RSF1010 is lower than the transfer of plasmid RP4 because it results from a secondary transfer event. Transfer frequencies for plasmid RSF1010 and RP4 together to one recipient cell are in the same range as for di- and triparental matings.

Table 2 Transfer frequencies of plasmid RSF1010, mobilized by plasmid RP4, and RP4 in di- and triparental matings between homologous strains of *Pseudomonas aeruginosa* PAO5 in natural soil. For inoculation donor and recipient cells were resuspended in complete medium. Transfer frequencies were determined after 3 days of incubation under standard conditions.

Type of mating	Transfer frequencies (T/R)		
	plasmid RSF1010 mobilized by RP4, alone	plasmid RP4, alone	plasmid RSF1010 mobilized by RP4, together with RP4
diparental	$1.1 \pm 0.7 \times 10^{-3}$	$4.7 \pm 4.6 \times 10^{-4}$	$7.6 \pm 4.7 \times 10^{-5}$
triparental	$1.2 \pm 0.5 \times 10^{-4}$	$1.3 \pm 0.5 \times 10^{-3}$	$3.1 \pm 1.0 \times 10^{-5}$

REFERENCES

Barth, P.T. and Grinter, N.J. (1974) Comparison of the deoxyribonucleic acid molecular weights and homologies of plasmids conferring linked resistance to streptomycin and sulfonamides. J. Bacteriol.120: 618-630

Datta, N., Hedges, R.W., Shaw, E., Sykes, R.B. and Richmond, M.H. (1971) Properties of an R Factor from *Pseudomonas aeruginosa*. J. Bacteriol.108: 1244-1249

van Elsas, J.D., Trevors, J.T., Starodub, M.E. and van Overbeek, L.S. (1990) Transfer of plasmid RP4 between Pseudomonads after introduction in soil: influence of spatial and temporal aspects of inoculation. FEMS Microbiol. Ecol.73: 1-12

Moore, R.J. and Krishnapillai, V. (1982) Tn7 and Tn501 Insertions into *Pseudomonas aeruginosa* plasmid R91-5 : Mapping of two transfer regions. J. Bacteriol.149: 276-283

INVESTIGATION OF POSSIBLE GENE TRANSFER FROM GENETICALLY MODIFIED TO INDIGENOUS BACTERIA IN SOIL

Elena Evguenieva and Walter Klingmüller

Department of Genetics
University of Bayreuth
Universitätsstrasse 30
W-8580 Bayreuth
Germany

Bacterial strain: *Enterobacter agglomerans* 339, isolated from fields around Bayreuth.
Plasmid: pEA9, 200 kb, self-transmissible, narrow-host range, with nif-genes and labelled with Tn5:mob.

In these studies an attempt was made to identify heterologous transfer to indigenous soil bacteria.

Mating experiments were made in non-sterile soil in flasks. As a donor a Nal^r, Rif^r strain *E. agglomerans* 339 (pEA9) was used. This strain contains one copy of Tn5 on the plasmid and two copies on the chromosome. As a recipient a Sm^r, Rif^r strain *E. agglomerans* 339-, cured of the plasmid was used.

Donor and recipient cells . 4 x 10^7 /g soil were added. After 5-10 weeks of incubation at 22°C, the cells were harvested and plated out onto LB-agar containing appropriate drugs. Replica-plating was used to distinguish the indigenous bacteria.

Rapid isolation of large plasmids of 400 indigenous soil bacterial clones (no further identified) were made; 15% of them contained large plasmids, which were investigated. Twenty-five percent of the indigenous bacteria, once isolated on agar plates, could not be cultivated for a long time on agar media (after 1-2 passages they died). After several passages on agar media, no plasmids could be isolated from 60% of the culturable clones .

Plasmid transfer to indigenous bacteria was not demonstrated with hybridization with Tn5 probe.

Some of the donor cells detected by plating method were analysed after 42 days of incubation in soil. The total DNA of the clones was digested with *Eco*RI (Fig.1A) and hybridized with Tn5 probe (Fig.1B). In one case (lane 2.) the Tn5 copy on the pEA9 was not detectable, in two cases (lanes 4. and 5.) rearrangements of pEA9 were confirmed after isolation of the plasmids.

The sensitivity of the plating method for recovery of released bacteria from soil was also studied. The soil was inoculated with serial dilutions of overnight cultures of *E. agglomerans*. and after 15 min incubation at 22°C the cells were harvested. About 2-3% of the released *E. agglomerans* 339 Nal^r, Rif^r cells / g soil and about 10% of the released *E. agglomerans* 339 Sm^r, Rif^r cells /g soil were recovered. From non-sterile soil 4-5 times more cells were recovered than from sterilized, by autoclaving, soil. By a modified method of harvesting up to 60% of the released cells /g soil could be recovered.

The Release of Genetically Modified Microorganisms
Edited by D.E.S. Stewart-Tull and M. Sussman, Plenum Press, New York, 1992

179

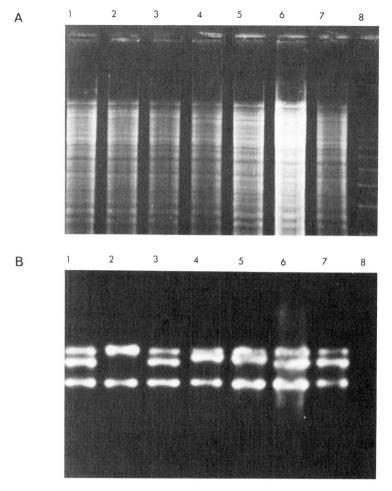

Fig. 1 Detection of Tn5 sequences in total DNA of *E. agglomerans,* digested with *Eco*RI. (A.) Gel electrophoresis of total DNA, digested with *Eco*RI.1. *E. agglomerans* 339 Nal[r], Rif[r] total DNA, digested with *Eco*RI.2.-7. Total DNA of six *E. agglomerans* 339 Nal[r], Rif[r] clones detected by the plating method after 42 days of incubation in soil, digested with *Eco*RI.8. KBL. (B.) Southern blot of A. and Photo-gene detection of Tn5.

REFERENCES

Klingmüller, W., Dally, A., Fentner, C., and Steinlein, M. (1990) Plasmid transfer between soil bacteria. In "Bacterial Genetics in Natural Environments" (Fry,J.C. and Day,M.J. Eds) pp 133-151. Chapman and Hall, London

Selenska, S. and Klingmüller, W. (1991) DNA recovery and direct detection of Tn5 sequences from soil. Letters in Appl. Microbiol. 13, 21-24.

FATE OF PLANT PATHOGENIC PSEUDOMONADS IN BEAN MICROCOSMS

K. Wendt-Potthoff, F. Niepold and H. Backhaus

Institut für Biochemie und Pflanzenvirologie
Biologische Bundesanstalt für Land- und Forstwirtschaft
Messeweg 11
D - 3300 Braunschweig

Pseudomonas syringae pv. *syringae* is a fluorescent pseudomonad colonizing plant surfaces and the causal agent of brown spot disease of beans. For this experiment a mutant with a single insertion of the transposon Tn5 in its chromosome (T12) was used. 4×10^9 cells were spray-inoculated into a microcosm with non-sterile agricultural soil and bush beans (cultivar Red Kidney). The persistence of the bacterial strain and its genetic marker was monitored. Our intention was to test the performance of the microcosm and to develop and compare methods for practical use in monitoring GEMMOs in the environment.

MATERIALS AND METHODS

Stomacher Extraction

The Stomacher lab blender (Seward Medical) is a device with two reciprocating paddles that apply pressure to the sample. The sample plus dilution buffer is placed in a sterile plastic bag before blending. Recovery of bacteria from leaves with stomacher blending is significantly better than with blending, sonication or washing (Donegan *et al.* 1991).

We tested the method for soil samples and compared the recovery of bacteria with that of vortexing and shaking with autoclaved gravel. Samples were treated for 1 minute at high speed (=260 rpm) with 10 mM phosphate (Na_2HPO_4/KH_2PO_4) buffer pH 7,4. Dilutions were plated onto 1/10 Stolp Medium with and without 25 µg/ml kanamycin sulfate and onto *Pseudomonas* Isolation and Enrichment Agar (Krueger & Sheikh 1987). For leaves or roots the liquid was filtered through a Sterivex GS filter unit (Millipore) for DNA preparation.

DNA Preparation via Sterivex GS (Millipore) Filter Unit

This method was developed by Somerville *et al.* (1989) for water samples which are concentrated in the cylindrical filter unit. Cell lysis and proteolysis are carried out within the filter housing. The crude nucleic acid solution is drawn off with a syringe. For leaf wash fluids we made two modifications:
1. The sample was pre-filtered through Miracloth (Calbiochem)
2. The original SET buffer (20% sucrose, 50 mM EDTA, 50 mM Tris Cl pH 7,6) was reduced to 5% sucrose. After ammonium acetate treatment an adsorption to a silica matrix (Geneclean glassmilk, BIO 101) was used for additional purification.

The Release of Genetically Modified Microorganisms
Edited by D.E.S. Stewart-Tull and M. Sussman, Plenum Press, New York, 1992

181

Total DNA Preparation from Soil

We used the small-scale (1 g) preparation method of Porteous and Armstrong (1991) with the following modifications:
1. The Novozym for degrading fungal cell walls was omitted because we were mainly interested in bacterial DNA.
2. A proteinase K concentration of 50 instead of 500 μg/ml in the lysis step was sufficient and cost-saving.

Centrifuge tubes were treated with 1 N HCl between preparations to depurinate any residual DNA which could interfere in the PCR. Aliquots of soil DNA were further purified with the Geneclean II glassmilk kit (BIO 101) for PCR.

Hybridization with Tn5 Probe

The central region of Tn5 (2.7 kb) was labelled with ^{32}P by random primer labelling with ^{32}P dCTP. Hybridization was carried out overnight at 65°C without formamide. Filters were washed at high stringency (0.1 x SSC, 65°C).

Results and Discussion

For bacteria from soil samples recovery efficiency was the same for stomacher blending, vortexing and shaking with autoclaved gravel. The Sterivex preparation could be optimized for leaf wash fluids by prefiltration of the samples and the lower sucrose concentration in the buffer. These modifications prevent clogging of the filters but do not reduce the amount or purity of DNA recovered. About 2.5 μg DNA/g leaf tissue and 5-10 μg DNA/g soil were extracted. Restriction enzymatic cleavage and PCR were possible after glassmilk purification. Bean plants grown from sterilized seeds were efficiently colonized by soil-borne bacteria so that seed inoculation could be more successful in establishing GEMMSO populations on plant surfaces. Until 8 days after inoculation the presence of T12 could be detected by antibody agglutination tests with pathogen specific antibodies (AB). Protein colony blots gave ambiguous results; (cross-reaction of AB with internal proteins of other bacteria and/or media composition modifying bacterial surfaces). Eighty-two days after inoculation the whole plant material of the microcosm was extracted. No Tn5 bearing bacteria could be isolated from these samples. In contrast, bacteria giving positive hybridization signals could be isolated from soil, bean roots and gravel 92 days after inoculation. The characterization of positive colonies is under investigation. Additional information on the presence of Tn5 sequences will be given by PCR of DNA samples from soil, leaves, gravel and old and young bean leaves, stems, pods and *Stellaria* plants.

REFERENCES

Donegan, K., Matyac, C., Seidler, R. and Porteous, A. (1991) Evaluation of Methods for Sampling, Recovery, and Enumeration of Bacteria Applied to the Phylloplane. Appl. Environ. Microbiol. 57, 51-56
Krueger, C.L. and Sheikh, W. (1987) A new selective medium for isolating *Pseudomonas sp.* from water. Appl. Environ. Microbiol. 53 , 895-897
Porteous, L.A. and Armstrong, J.L. (1991) Recovery of bulk DNA from soil by a rapid, small-scale extraction method. Curr. Microbiol. 22, 345-348
Somerville, C.C., Knight, J.T., Straube, W.L. and Colwell, R. (1989) Simple, Rapid Method for Direct Isolation of Nucleic Acids from Aquatic Environments. Appl. Environ. Microbiol. 55, 548-554

DETECTION OF *STREPTOMYCES* MARKER PLASMIDS IN SOIL

V.A. Saunders[1], E.M.H. Wellington[2] and A. Wipat[1]

School of Natural Sciences[1]
Liverpool Polytechnic
Liverpool L3 3AF

Department of Biological Sciences[2]
University of Warwick
Coventry CV4 7AL

Marker plasmids incorporating a thermoregulated *xylE* gene (for catechol 2,3-dioxygenase) were constructed for use in *Streptomyces*. *xylE* was expressed from bacteriophage lambda promoters, λp_L and λp_R, under the control of the temperature-sensitive lambda repressor *c*I857. Marked *Streptomyces lividans* strains retained the XylE phenotype for more than 80 days in soil microcosms and were detectable by DNA hybridization at less than 10 cfu/g dry weight of soil as mycelium and 10^3 cfu/g dry weight of soil as spores, with the *xylE* marker DNA recovered from soil and amplified by the polymerase chain reaction.

STREPTOMYCES MARKER PLASMIDS

Various genetic marker systems have been designed for use in the detection of microorganisms in the environment. To date, however, detection of genetically-modified streptomycetes in soil has depended largely upon the use of antibiotic resistance markers (Rafii and Crawford, 1988; Wellington *et al.*, 1990). While convenient in use such determinants are frequently encountered in natural populations. The functioning of the *xylE* gene in streptomycetes (Ingram *et al.*, 1989; Clayton and Bibb, 1990) has led to its use in the development of a series of *Streptomyces* marker plasmids (pNW1 > pNW5, Figure 1). These plasmids were generated by ligation of broad-host range constructs, derived from the IncQ plasmid pKT230, containing *xylE* from the TOL plasmid pWWO, and lambda promoter *p*L or *p*R (Winstanley *et al.*, 1989) into *Streptomyces* plasmids.

The *xylE* gene in the marker plasmids was expressed in *S.lividans* . Catechol 2,3-dioxygenase (C230) activity increased by up to 10-fold in cultures harboring the regulated constructs, pNW2 (λp_L, *xylE*, *c*I857), pNW3 (λp_R, *xylE c*I857) or pNW5 (λp_R, *xylE*, *c*I857) upon a shift in temperature from 28°C to 37°C. There was no such increase for the unregulated constructs pNW1 (λp_L, *xylE*) and pNW4 (λp_L, *xylE*). Induced C230 activity was, however, lower in *Streptomyces* than in Gram-negative bacteria where similar constructs have been used (Winstanley *et al.*, 1989). Nevertheless λp_L and λp_R proved more efficient than an indigenous promoter , *gal P1* (Ingram *et al.*, 1989) in expressing *xylE* in *S. lividans*. Western blot analysis of polypeptides from marker plasmid-containing strains revealed a band

The Release of Genetically Modified Microorganisms
Edited by D.E.S. Stewart-Tull and M. Sussman, Plenum Press, New York, 1992

183

at ~35,000 daltons corresponding to that of C230. Furthermore C230+ streptomycetes were readily detected as yellow colonies on solid medium after spraying with catechol.

DETECTION AND SURVIVAL OF MARKER PLASMIDS IN SOIL

Nucleic acid hybridization, in conjunction with the polymerase chain reaction (PCR), was employed to detect *xylE* - containing *S.lividans* that had been inoculated into soil microcosms (see Wellington *et al.*, 1990) as spores or mycelium. DNA was recovered from soil samples and a 260 bp region of *xylE* amplified by PCR. Unamplified and amplified DNA were analysed by dot-blot hybridization using a whole *xylE* gene probe. Application of PCR resulted in an increase in detection of *xylE*+ *S.lividans* by four orders of magnitude, mycelium being detectable at 10 cfu/g dry weight of soil and spores at 10^3 cfu/g dry weight of soil (Figure 2). The lower sensitivity of detection for spores may be attributed, at least in part, to the relative efficiency of the spore extraction and lysis procedures.

The *Streptomyces xylE* plasmids survived in soil for longer than 80 days within the spores of the host strain. Strains harboring the regulated (pNW5) or unregulated (pNW4) construct exhibited a similar high degree of maintenance of the C230+ phenotype, although viable spore counts fell below the initial value during this period. Moreover such counts were significantly lower than those for *S.lividans* carrying the parental plasmid (pIJ486). This reduction may be symptomatic of the additional metabolic load imparted by the *xylE* gene in the marker plasmids.

The functioning of *xylE* and of λ promoters in *Streptomyces* opens up the possibility of utilizing the thermoregulated *xylE* marker system in Gram-positive organisms. This system can be detected by a variety of methods and should prove effective in monitoring the survival and spread of streptomycetes in soil and interaction with indigenous species.

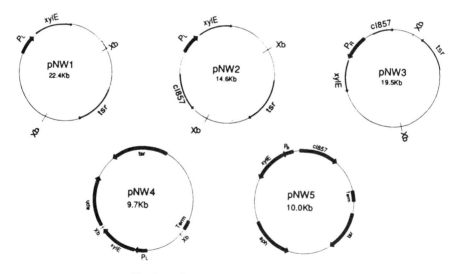

Fig. 1   *Streptomyces* marker plasmids.
pNW1, pNW2, pNW3 derived from *Streptomyces* nonconjugative plasmid pIJ680; pNW4, pNW5 from pIJ486. Xb, site for restriction enzyme *Xba*I; *tsr, aph*, determinants for resistance to thiostrepton, neomycin; *pL*, lambda promoter leftward; *pR*, lambda promoter rightward; *xylE*, gene for catechol 2,3 dioxygenase; Term, transcription terminator from *E. coli* phage fd ; Kb, kilobase pairs.

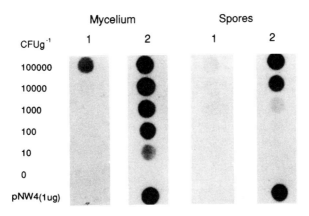

Fig. 2   Dot-blot hybridization analysis for *xylE* in soil DNA using the whole *xylE* gene probe. DNA was isolated from 10g soil microcosms seeded with 0 to 10^5 cfu/g dry weight of soil of either mycelium or spores and probed for the presence of *xylE*. (1) without amplification;   (2) following PCR amplification. 1μg pNW4 DNA was used as a positive control (lane 2 only).

REFERENCES

Clayton, T.M. and Bibb, M.J.(1990) *Streptomyces* promoter-probe plasmids that utilise the *xylE* gene of *Pseudomonas putida*. Nucleic Acid Res. 18, 1077.

Ingram, C., Brawner, M., Youngman, P. and Westpheling, J. (1989) *xylE* functions as an efficient reporter gene in *Streptomyces* spp. : use for the study of *galP1,* a catabolite-controlled promoter. J. Bacteriol. 171, 6617-6624.

Rafii, G. and Crawford, D.L. (1988) Transfer of conjugative plasmids and mobilization of a non-conjugative plasmid between *Streptomyces* strains on agar and in soil. Appl. Environ. Microbiol. 54, 1334-1340.

Wellington, E.M.H., Cresswell, N. and Saunders, V.A. (1990) Growth and survival of streptomycete inoculants and extent of plasmid transfer in sterile and non-sterile soil. Appl. Environ. Microbiol. 56, 1413-1419.

Winstanley, C., Morgan, A.W., Pickup, R.W., Jones, J.G. and Saunders, J.R. (1989) Differential regulation of Lambda p_L and p_R promoters by a *cI* repressor in a broad-host-range thermoregulated plasmid marker system. Appl. Environ. Microbiol. 55, 771-777.

THE REMOVAL AND DISPERSAL OF FOLIAR BACTERIA
BY RAIN SPLASH

Julie Butterworth and Alastair McCartney

Plant Pathology
A.F.R.C.Institute of Arable Crops Research
Rothamsted Experimental Station
Harpenden, Herts. AL5 2JQ

Recent advances in genetic modification have made possible the use of bacteria as novel pest and disease control agents. However, before such agents are released into the environment it is important to know if they are likely to have deleterious effects on non-target organisms or plants. An understanding of the dispersal potential of genetically-manipulated organisms is needed to help in this assessment.

Rain splash as a factor in the spread of disease in crops has long been recognized, and it is vital in the spread of many fungal and bacterial plant pathogens. Little is known, however, of the efficiency of rain in removing bacteria from foliage. The experiments described studied the potential of rain splash for removing and distributing bacteria from the leaves of crop plants.

All experiments were done in the rain tower/wind tunnel facility at Rothamsted Experimental Station. Mono-sized water drops of between 2 and 5mm in diameter, (simulating rain drops), were allowed to fall 11m onto target leaves of *Phaseolus vulgaris* (French bean) cv Prince or *Brassica napus* (oilseed rape) cv Cobra. The leaves were inoculated with populations of *Pseudomonas syringae*, *Klebsiella planticola* or *Bacillus subtilis*, all resistant to 100µg/ml of the antibiotic rifampicin. The leaves were exposed to splash for 30 minutes and splashed droplets were collected on Petri dishes of selective medium or in sterile funnels supported in glass universals at different distances from the target. Samples of water which collected on, and subsequently ran off, the leaves were also collected during the experiments. The number of bacteria present on the leaves before and after each experiment was estimated by washing them in sterile Ringer's solution, plating on selective medium and counting colonies.

During each experiment the two plates closest to the target were changed at 3 minute intervals and the third closest plate was replaced after 15 mins. The splashed drops collected in the funnels and the run-off water from under the target were diluted as appropriate and plated on selective medium. All plates were incubated at 30°C overnight and the colonies counted.

Twenty-five experiments were done for each of the six different combinations of plants and bacteria and a "rain" drop diameter of 2.9mm. A further 24 tests with "rain" drop diameters of 2.4, 2.9 and 4.7mm were done with French bean leaves inoculated with *B. subtilis* as targets.

No statistical differences were found in the proportion of bacteria removed by splash for either bacteria or plant species. For all the experiments the variation in the proportion of the initial population removed was large, but in some cases up to 90% of the bacteria were removed during the experiment.

The Release of Genetically Modified Microorganisms
Edited by D.E.S. Stewart-Tull and M. Sussman, Plenum Press, New York, 1992

187

Table. Proportion of the bacterial population splashed from leaves and the proportion found in the run-off water for different plant and organism species and for different rain drop sizes

Plant/ organism	Drop size (mm)	Proportion of population splashed	Proportion of population in run-off
O/p	2.9	0.001 - 1.07	8.5 - 70
O/k	2.9	0.008 - 0.28	0.5 - 7.6
O/b	2.9	0.004 - 0.04	0.04 - 14
F/p	2.9	0.004 - 0.28	1.6 - 14
F/k	2.9	0.01 - 0.08	1.6 - 4.7
F/b	2.4	0.0 - 0.004	0.004 - 0.8
F/b	2.9	0.001 - 0.02	0.3 - 1.4
F/b	4.7	0.001 - 0.02	0.2 - 1.6

O - oilseed rape. F - french bean.
p - *Pseudomonas syringae*. k - *Klebsiella planticola*.
b - *Bacillus subtilis*.

The proportion of the bacterial population dispersed in splash droplets was generally small (Table) and there was little difference between plant or bacterial species. There were large variations in the proportion splashed in the drop size experiments (Table). However, larger drops tended to remove more cells than smaller ones: the mean values were 0.1, 0.6 and 0.8% for 2.4, 2.9 and 4.7mm rain drop sizes respectively.

In all experiments substantial numbers of bacteria were found in the run off water (Table). There was no significant difference between the proportion of bacteria in the run-off water for the two plant species, however the proportion was usually higher for *P. syringae* and *K. planticola* than for *B. subtilis*. In the second series of experiments, the proportion of bacteria in the run-off water tended to increase with the drop size (Table).

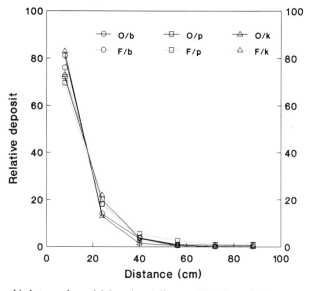

Fig. The relationship between bacterial deposit and distance. The effect of plant and bacterial species.

The numbers of splashed bacteria recovered decreased rapidly with distance from the target leaves. The dispersal patterns were similar for all the experiments with the 2.9mm "rain" drops (Figure). Less than half the bacteria splashed travelled more than 10cm from the target and only about 5% travelled more than 40cm. In the second set of experiments the distances travelled by bacteria increased with "rain" drop size. For example the 4.7mm rain drops splashed half the cells more than 13cm compared with 9 and 7cm for the 2.9 and 2.4mm rain drops. Nevertheless, only a few percent of the bacteria were carried more than 50cm.

For *P. syringae* deposition rates decreased with time by about 20% over the experimental period. For the other two organisms, the rates decreased by about 70% during the experiments. For the 4.7mm drops the deposition rate decrease was similar to that in the first experiments. With the smaller drops deposition rates increased initially then decreased, and at the end of the experiments were about 45% of the maximum value for the 2.9mm drops and about 60% for the 2.4mm drops.

Our experiments show that rain is an efficient means of removing bacteria from leaves, and splash dispersal processes may be similar for a wide range of organisms and plants. The proportions of bacteria dispersed in splash droplets were generally small, as were the distances travelled by most of the bacteria, however large numbers were found in the run-off water collected below the target. These could be an important source for secondary splashing or for travel in ground-water. Large rain drops appear to be more effective in removing and splashing bacteria than small ones. However, bacteria were more rapidly removed by the large drops, suggesting that for longer periods of rainfall the drop size may be less important.

ACKNOWLEDGEMENTS

This work was supported by a grant from the United States Environmental Protection Agency, however it has not been subjected to Agency review and therefore does not necessarily reflect the views of the Agency, and no official endorsement should be inferred.

GROWTH AND SURVIVAL OF GENETICALLY-MODIFIED
PSEUDOMONAS PUTIDA IN SOILS OF DIFFERENT TEXTURE

S.J. Macnaughton, D.A. Rose and A.G. O'Donnell

Department of Agricultural and Environmental Science
University of Newcastle upon Tyne
Newcastle upon Tyne
NE1 7RU, UK

INTRODUCTION

Given current concerns about the risks associated with the use of genetically-modified organisms (GEMMO) in open ecosystems it is important to be able to determine reliably the survival of such organisms and how this survival is influenced by primary environmental factors.

The situation is complicated by the highly variable biotic and abiotic conditions offered within a soil and by soils of different texture.

This study reports on the interaction between soil texture and persistence of a genetically-modified strain of *Pseudomonas putida*.

METHODS

Strain

The recombinant *Pseudomonas putida* PaW8 strain containing the plasmid pLV1016 was described by Winstanley *et al.* (1991). PLV1016 (Inc P) is a cointegrate R68.45 plasmid carrying the *xylE* gene. This is under the control of lambda P_R and its temperature sensitive repressor cI_{857}. The *xylE* gene codes for a catechol-2,3-dioxygenase. The plasmid has a copy number of 2-3 and carries genes for kanamycin resistance. The GEMMO was obtained from Dr. J.R. Saunders, Department of Genetics and Microbiology, The University of Liverpool, Liverpool, L69 3BX, UK.

Soils

Three soils were used in this study; a sandy loam (Rivington series); a clay loam (Hallsworth series); and a peat soil (Wilcocks series). The soils were weighed into pots and remoistened to 50 % (w/w) field capacity with distilled water containing *Ps. putida* PaW8 pLV1016 so that each soil contained GEMMOs at approximately 1×10^7 cfu/g dry weight soil.

Microcosm sampling

Microcosms were sampled at 0, 1, 3, 6, 9, 12, 15, 20, 25, 30, 60 and 90 days. Microorganisms were extracted with a dispersion/differential centrifugation technique (Hopkins

The Release of Genetically Modified Microorganisms
Edited by D.E.S. Stewart-Tull and M. Sussman, Plenum Press, New York, 1992

191

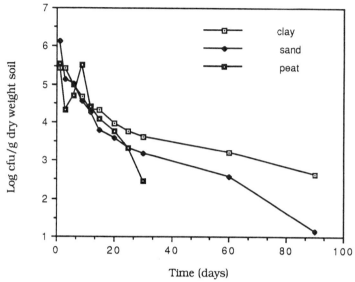

Fig. 1 Survival of *Ps. putida* PaW8 pLV1016 in clay, sand and peat soil over time.

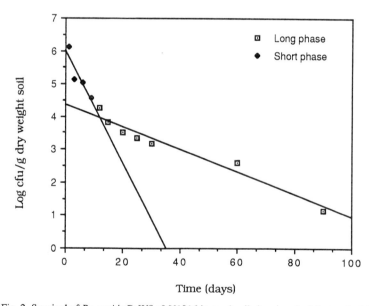

Fig. 2 Survival of *Ps. putida* PaW8 pLV1016 in sand soil showing dual first-order kinetics.

et al. 1991). Serial dilutions of the soil extracts were plated, in triplicate, onto nutrient agar containing cycloheximide (50 µg/ml) for the uninoculated soils, and cycloheximide and kanamycin (50 µg/ml) for the inoculated soils. The plates were incubated for 48 at 28°C, then heated to 42°C for two hours, prior to being sprayed with catechol (1 % w/v). Colonies turning yellow were counted.

RESULTS AND DISCUSSION

Figure 1 shows the effect of soil texture on the survival of the inoculant, with the peat showing a rapid loss of the GEMMO over a period of 30 days.

The sand and the clay soils, however, appear to offer some protection to the inoculant and consequently the GEMMO can be detected over an extended period. These results illustrate the importance of prevailing environmental conditions on the survival of GEMMOs.

A more detailed analysis of the data, however, (Figure 2) has shown that survival is biphasic in both the sand and clay soils but monophasic in the peat. We are currently investigating whether this biphasic mortality is due to the presence of two distinct populations in the inoculant or to an interaction between the inoculant and the indigenous population such as predation.

ACKNOWLEDGEMENTS

We are grateful to the UK Department of the Environment for supporting this work.

REFERENCES

Hopkins, D.W., Macnaughton, S.J. and O'Donnell, A.G. (1991). A dispersion and differential centrifugation technique for representatively sampling microorganisms from soil. *Soil Biology & Biochemistry*. 23(3), 217-225.

Winstanley, C., Morgan, J.A.W, Pickup, R.W. and Saunders, J.R. (1991). Use of a *xylE* marker gene to monitor survival of recombinant *Pseudomoas putida* populations in lake-water by culture on nonselective media. *Applied and Environmental Microbiology*. 57(7), 1905-1913.

SIMPLIFIED SUBTRACTION-HYBRIDIZATION SYSTEM FOR ISOLATION OF STRAIN-SPECIFIC *RHIZOBIUM* DNA PROBES

J. E. Cooper[1,2] and A. J. Bjourson[1]

Department of Agriculture for Northern Ireland[1] and
Department of Food and Agricultural Microbiology[2]
The Queen's University of Belfast
Newforge Lane
Belfast BT9 5PX
Northern Ireland

Subtraction-hybridization is the name given to procedures for removing nucleic acid sequences from the genome of an organism which are shared with related organisms. Sequences remaining are those which are unique to the organism from which they are derived. Scott *et al* (1983) used such a technique to isolate particular genes from transformed mouse cells and Welcher *et al* (1986) extended the concept to bacteria for the isolation of *Neisseria gonorrhoeae* specific clones. This paper describes the development of subtraction systems for generating *Rhizobium* DNA sequences capable of discrimination at inter-strain level when used as probes.

REMOVAL OF COMMON HYBRIDS BY AFFINITY CHROMATOGRAPHY

Bjourson and Cooper (1988) developed a system in which single-stranded, restricted DNA from a potential *R. loti* probe strain was hybridized in solution to excess single-stranded biotinylated and mercurated subtracter DNA from a group of closely-related, cross hybridizing strains. The subtracter DNA and probe strain sequences hybridized to it were removed by two-step affinity chromatography on streptavidin agarose and thiopropyl Sepharose. Sequences eluted from the chromatography column were probe-strain specific, hybridizing with target DNA from the parent strain but not with DNA from any of the strains used as subtracter material. Further developments have resulted in the simplified system described below, which dispenses with the mercury ligand of the subtracter DNA and permits amplification of the small amounts of probe-strain-specific DNA by means of PCR.

SIMPLIFIED SUBTRACTION SYSTEM WITH BIOTINYLATED SUBTRACTER DNA AND PCR AMPLIFICATION OF PROBE-STRAIN-SPECIFIC SEQUENCES

A diagram of the system is shown in Fig 1. Priming of Sau 3A digested probe strain DNA was achieved via ligation to a primer/linker sequence. Subtracter DNA was biotinylated and ligated to a different primer/linker sequence. Amplification of the subtracter DNA by PCR provided the excess amount required for the subtraction hybridization; dUTP was substituted for dTTP in the PCR reaction. A typical subtraction was performed at 65°C for 48 h, after

The Release of Genetically Modified Microorganisms
Edited by D.E.S. Stewart-Tull and M. Sussman, Plenum Press, New York, 1992

195

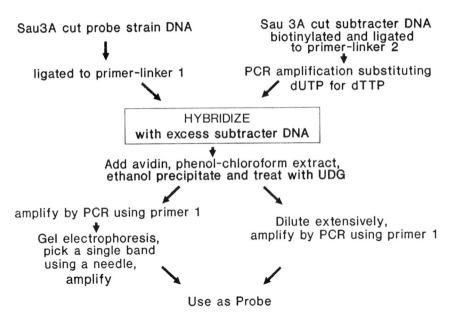

Fig.1 Subtraction system with PCR & Biotinylated DNA.

which any subtracter DNA and subtracter/probe hybrids were removed from the mixture via a phenol- chloroform extraction of an avidin/biotin/DNA complex. Any traces of contaminating subtracter DNA were destroyed by treatment with uracil DNA glycosylase (UDG). After gel electrophoresis or extensive dilution, strain-specific sequences were amplified with primer 1 and used as a probe.

The main advantages of the system are:
1) Probes demonstrate a high level of specificity, permitting inter-strain differentiation in bacteria.
2) Specificity can be further increased by including DNA from more strains in the subtracter matrix.
3) No DNA sequence information is required.
4) When probes are based on amplified DNA from the diluted product they contain the maximum number of base sequences exclusive to a given strain.

REFERENCES

Bjourson, A.J. and Cooper, J.E. (1988) Isolation of *Rhizobium loti* strain-specific DNA sequences by subtraction hybridization. Appl. Environ. Microbiol. 54 : 2852-2855.
Scott, M.R.D., Westphal, K-H, and Rigby, P.W.C. (1983) Activation of mouse genes in transformed cells. Cell 34 : 557-567.
Welcher, A.A., Torres, A.T., and Ward, D.C. (1986) Selective enrichment of specific DNA, cDNA and RNA sequences using biotinylated probes, avidin, and copper-chelate agarose Nucleic Acids Res. 14 : 10027-10044.

PLASMID TRANSFER OF pBR322 DERIVATIVES FROM recA⁻ E. COLI K12 DONOR STRAIN TO VARIOUS NATURAL GRAM-NEGATIVE ISOLATES

Søren Johannes Sørensen

Department of General Microbiology & Institute of Biological Chemistry
Sølvgade 83H, University of Copenhagen
DK1307 Copenhagen K
Denmark

Concern about possible biohazards from accidental or deliberate release of genetically modified microorganisms (GEMMOs) in the environment has been raised during the last years. The fate of recombinant DNA in natural environments probably depends primarily on the nature of its bacterial host and the cloning vector in which it is carried. To minimize the potential transfer of plasmids containing novel genes from a released GEMMO to indigenous bacteria, the novel genes are usually incorporated into presumably safe plasmids - for example pBR 322 and pBR325.

In this study, the transfer of pED6 and pSS1 from *recA⁻ E. coli K12* strain (a MC1000 derivative) to a *E. coli* K12 recipient strain and to various Gram-negative bacterial isolates was investigated. pED6 is a 5.7 kb derivative or pBR322 with a 1.3 kb fragment (the *pyrD* gene from *E. coli*) inserted in the EcoRI site. pSS1 is a 7.8 kb derivative of pBR325 with a 1.8 kb fragment of cDNA of the β-actin gene from chicken, ligated in the PstI site of pBR325. F'lacI^{Q}Z::Tn5 was used as mobilizer plasmid. All mating experiments were done as "broth matings" in 250 ml Erlenmeyer flasks. Before use, overnight cultures were diluted 1:100 into fresh LB-broth (without antibiotics) and grown for 2-3 hours to an optical density at 436 nm of approx 1.0.

"Diparental mating-experiments I-III" were performed with a mixture of 0.4 ml donor-strain culture and 4 ml recipient-strain culture incubated without shaking for 30 min at 37ºC. LB broth (5.0 ml) was added and the mixture incubated for 1 hr with agitation (200 rpm) for phenotypic expression. Following incubation mating mixtures were washed twice by centrifugation through sterile (A+B) medium, serially diluted, and spread on selective MM-plates. The donor strains having a deletion in the *leu*-gene, were not able to grow on this medium.

The F-factor from *E. coli* was able to co-transfer pBR322 during conjugation via a mechanism that involved the transposition of a 5.7 kb DNA fragment, the so called gamma-delta sequence (Tn1000) from F to pBR322. Frequencies of mobilization of pBR322 derivative by F' were approx. 2 x 10⁻⁵ per transconjugant. Mobilization frequency of pSS1 (pBR325 with the chicken β-actin gene), did not differ significantly from the mobilization frequency of pED6.

A "triparental mating-experiment" was performed with a 0.2 ml "donor A" culture (the pED6 containing strain), 2 ml "donor B" culture (the F' containing strain) and 3 ml of the recipient culture (MC4100). The strains were mixed and incubated without shaking at 37ºC.

The Release of Genetically Modified Microorganisms
Edited by D.E.S. Stewart-Tull and M. Sussman, Plenum Press, New York, 1992

197

After 30 min 5 ml LB-broth was added to ensure growth and after 3 hours, further 20 ml LB-broth was added, and the flasks were incubated at 37°C with agitation (200 rpm) for another hour. Mobilization frequency in the triparental mating experiment was 3 x 10^{-6}.

The pBR322 and pBR325 derivatives were readily transferred into a plasmid-free *E. coli* K12 strain. The only major requirement for these transfers was the presence of the mobilizer plasmid, F'. The mobilizer could be present either together with the non-conjugative plasmid ("diparental mating"), or in a second donor ("triparental mating").

"Diparental mating-experiments IV" were performed with 17 different Gram-negative bacterial isolates from seawater as recipient strains and a *recA⁻ E. coli* K12 containing F' and pSS1 as the donor strain, a mixture of 0.4 ml donor-strain culture and 4 ml recipient-strain culture were incubated without shaking for 70 hours at room temperature.

Mobilization of pBR325 from E. coli K12 occured to approx. 30% of marine isolates tested, and *Pseudomonas* sp., *Enterobacter aerogenes*, *Klebsiella oxytoca*, and *Escherichia coli* isolates were recipient-active for pBR325 mobilization by F' conjugation.

SINGLE CELL DETECTION OF BIOLUMINESCENT

PSEUDOMONAS SYRINGAE IN SOIL

D.J. Silcock[1], R.N. Waterhouse[2], J.I. Prosser[2], L.A. Glover[2] and
K. Killham[1]

Departments of Plant and Soil Science[1] and
Molecular and Cell Biology[2]
University of Aberdeen
Scotland

Bioluminescence genes from *Vibrio harveyi* were introduced by plasmid into
Pseudomonas syringae pv. *phaseolicola*. Cells of *Ps. syringae* inoculated into soil slurries,
maintained under microculture conditions, could not be distinguished from soil particles by
conventional light microscopy. However the use of a slow-scan, nitrogen-cooled charge
coupled device with dark-field microscopy enabled detection and visualization of
bioluminescent single cells demonstrating the sensitivity and potential uses of charge coupled
device (CCD) microscopy and bioluminescence for the detection of marked populations in
environmental release experiments.

As greater uses are found for microbial inocula, then so too will increase the potential for
improving or refining these systems through the use of genetic modification. Such
biotechnology promises great benefits through the introduction of genetically modified
microorganisms (GEMMOs) into the environment for purposes such as biological pest control,
improvement of plant resistance to frost and enhanced waste degradation. Investigation of the
fate and impact of genetically modified microbial inocula in the environment necessitates the
development of detection and monitoring techniques under microcosm and field conditions.
Ideally these should have far greater sensitivity than has previously been available from
traditional techniques in microbial ecology.

Monitoring requires rapid, reliable and efficient methods for detection and enumeration.
Bioluminescence-based techniques fulfil these requirements. The *lux* operon of certain *Vibrio*
species has been cloned into a number of microorganisms, and resultant bioluminescence is an
indication of metabolically-active cells. For this work. *lux* A and B genes from *Vibrio harveyi*
were cloned into plasmid pQF70 under the control of a phage promoter, isolated from *Ps.
aeruginosa* pv. *phaseolicola* and the resulting strain expressed high levels of bioluminescence.

Detection and quantification of bioluminescence can be achieved by a number of
techniques, (e.g. luminometry), but it would be convenient for survival studies to have an
indication of the spatial distribution of cells. This requires *in situ* visualization, which can be
achieved using traditional photography, photomultipliers, image intensifiers and charge
coupled devices (CCD's). The last of these is highly sensitive and enables detection of very
low light levels, thus lending itself to use in the soil which acts as a sink for GEMMOs
introduced into the environment. The CCD camera (Wright Instruments Ltd., Enfield, UK)
used for this work was of the nitrogen-cooled, slow-scan type (both features reduce
background noise). It was wall-mounted and encased in a light-tight box, the computer and
monitor for image display situated adjacent to the box. For the purposes of this work the CCD

The Release of Genetically Modified Microorganisms
Edited by D.E.S. Stewart-Tull and M. Sussman, Plenum Press, New York, 1992

199

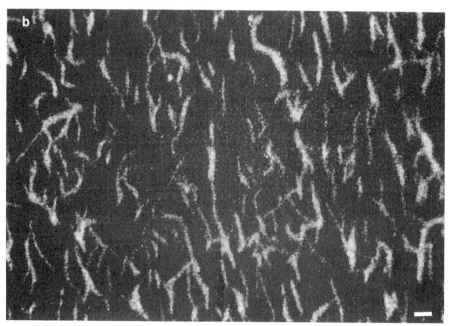

Fig.1 Filaments of *Ps. syringae* cells after 4 h incubation in the absence of soil. Bright field image exposures (Figure 1a) were of 0.02 s duration, whilst dark field exposures (Figure 1b) were of 30 min duration. (Scale bar represents 10 μm).

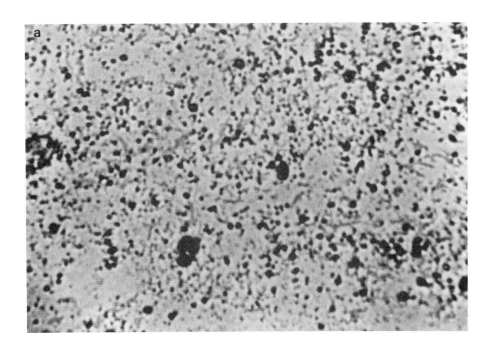

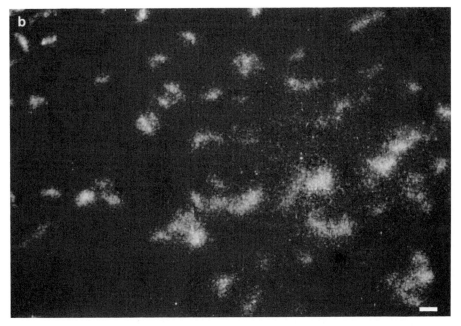

Fig. 2 Bright (Figure 2a) and dark (Figure 2b) field images of *Ps. syringae* cells incubated (2 h) in soil slurry. (Scale bar represents 10 μm).

camera was mounted on a transmitted light microscope (Jenamed Variant, Carl Zeiss, Jena Germany) via a "C" mount.

Our aim as part of the PROSAMO programme has been to assess the potential for CCD detection of single bioluminescent bacteria in soil. Cells of *Ps. syringae* were grown in liquid batch culture and applied to microculture agar slide preparations and incubated at 25°C. Serial dilutions of the cells were added to slurries of non-sterile sandy loam soil. The inoculated slurries were applied to microculture agar slides and also incubated at 25°C. Both sets of microculture slide preparations were imaged under bright and dark-field using CCD-enhanced microscopy following the addition of a small aliquot of n-decyl aldehyde to the slide preparation.

In the case of the *Ps. syringae* incubated in the absence of soil, filaments of cells could be seen under both bright and dark-field (Figures 1a and 1b). The CCD-enhanced microscopy, therefore provided clear imaging of cells where the only illumination was from bioluminescence. In the case of *Ps. syringae* incubated in soil slurries, cells were obscured by soil particles under bright-field. Under dark-field, however, the CCD enabled imaging of the luminescing cells. Comparison of dark-field images for soil slurry and pure culture, therefore, revealed that the presence of soil particles causes the light emitted to be diffused to some extent, but does not impair detection of single cells.

In conclusion, CCD-enhanced microscopy offers a powerful technique for single cell detection of genetically-modified bacteria in soil as part of a package of techniques based on bioluminescence.

ACKNOWLEDGEMENT

This work was undertaken as part of the PROSAMO project.

LUMINESCENCE-BASED DETECTION TO ASSESS SURVIVAL AND ACTIVITY OF A GENETICALLY-MODIFIED INOCULUM IN SOIL

E.A.S. Rattray[1], J.I. Prosser[1], L.A. Glover[1] and K. Killham[2]

Departments of Molecular and Cell Biology[1] and
Plant and Soil Science[2]
University of Aberdeen
Aberdeen, AB9 2UE

This study demonstrated the influence of matric potential on the survival and activity of a genetically modified bacterial inoculum in soil using the recently developed luminescence-based detection techniques. Increasing matric stress had an inhibitory effect on both viable cell concentrations and microbial activity in both glucose-amended and unamended sterile soil. Light output decreased with time, however, while substrate amended respiration remained fairly constant, highlighting important differences between actual and potential microbial activities.

INTRODUCTION

Recent development of molecular-based detection techniques has greatly increased the ability to track microorganisms and their genetic material in the environment. The main impetus for development of such techniques, has been the potential commercial benefits of genetically modified microorganisms (GEMMOs) and the need to monitor them in the environment. Luminescence-based detection techniques involve the introduction of genes for light output, cloned from the marine bacterium *Vibrio fischeri*. The *Escherichia coli* strain used in this study contained plasmid bearing *lux* AB genes to enable expression of luminescence in active cells.

The introduction of microbial inocula into the soil environment necessitates an understanding of the effect of key soil physical and chemical parameters on their survival and activity. One of the most important of these parameters will be water stress, of which the dominant component will be matric potential, except under certain circumstances. The aim of these studies was to investigate the effects of varying matric potential on the viability and activity on *E. coli* (pEMR1) — a genetically-modified strain containing the *lux* genes — in sterile, substrate-amended and non-amended soil using luminescence-based detection.

RESULTS

In the absence of added substrates, growth of *E. coli* was not detected, but following amendment with glucose, significant growth was detected at -5 and -64 kPa. Cells of *E. coli* did not survive beyond 3 days even in the presence of added glucose at -1.5 MPa (Fig.1).

The Release of Genetically Modified Microorganisms
Edited by D.E.S. Stewart-Tull and M. Sussman, Plenum Press, New York, 1992

203

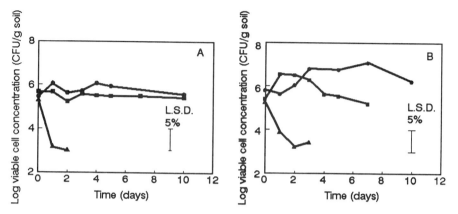

Fig. 1 Viable cell concentration of *E. coli* MM294 (pEMR1) inoculated into sterile soil (a) without and (b) with glucose amendment (50 µg g soil-1) at -5 kPa (●), -64 kPa (■) and -1.5 MPa (▲). Bars indicate least significant differences (P < 0.05).

Activity decreased with increasing matric stress, with light output per cell decreasing to a greater extent than $^{14}CO_2$ production per cell. Both light output per cell and $^{14}CO_2$ production per cell were higher in the presence of glucose than in the absence of glucose at -5 and -64 kPa (Fig. 2).

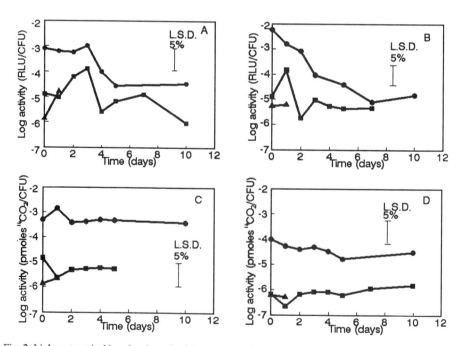

Fig. 2. Light output (a, b) and carbon dioxide production (c, d) per viable cell of *E. coli* MM294(pEMR1) inoculated into sterile soil (a, c) without and (b, d) with glucose amendment at -5 kPa (●), -64 kPa (■) and -1.5 MPa (▲). Bars indicate least significant differences (p < 0.05).

DISCUSSION

This study demonstrated how matric potential influences the survival and activity of a genetically-modified bacterial inoculum in soil, increasing matric stress having an inhibitory effect on both viable cell concentrations and activity in both glucose-amended and unamended, sterile soil

Growth was detected with substrate addition at -5 and -64 kPa. At -5 kPa, growth occured over a 7 day period whereas at the higher matric stress (-64 kPa), growth only occured during the first 24h. Maximum cell concentrations also differed at the two matric potentials. These observations reflect substrate availability due to differing rates of glucose diffusion when water content differs between matric potentials. maintenance of cell concentration also differed between -5 and -64kPa, decreasing sooner at -64 kPa than at -5 kPa. This may reflect the more rapid depletion of substrate from pores of smaller diameter.

Investigation of matric potential on microbial activity must include a consideration of the type of activity that can be measured. This study employed two methods, luminometry and radiorespirometry, measuring actual and potential activity respectively. Light output per cell and $^{14}CO_2$ production per cell exhibited different patterns over time, the former declining more rapidly. The presence of glucose in the soil led to sustained activation of potential activity, whereas actual activity was immediate and short-lived (as demonstrated by light output). Luminometry therefore provided evidence that *E. coli* actively utilized substrates at different matric potentials in the soil, and radiorespirometry highlighted the differences between actual and potential activity. In conclusion, luminometry provided a rapid and sensitive technique contributing to characterization of the fate of microbial inocula in soil. The marked effect of matric stress on the GEMMO suggests that matric potential will be a major determinant of survival of GEMMOs introduced into the soil environment.

THE EFFECT OF CYANOBACTERIAL WATER BLOOM
FORMATION UPON CONJUGATIONAL GENE TRANSFER
BETWEEN ASSOCIATED HETEROTROPHIC BACTERIA

N. Cook

Department of Biological Sciences
Dundee University
Dundee DD1 4HN

When heavy growth of cyanobacteria has occured in freshwaters, very calm conditions can lead to a floating mass or scum of cells, commonly termed a water bloom (Reynolds and Walsby, 1972), which may contain most of the planktonic cyanobacterial population. Through time, cells fall away from the surface scum and sink to the bottom, where they decay (Rother and Fay, 1977). Freshwater cyanobacteria often live in very close association with planktonic heterotrophic bacteria (Paerl, 1982), which can be present in the immediate vicinity of the cyanobacterial cells and filaments in greater numbers than in the surrounding environment (Mikhaylenko and Kulikova, 1973; Caldwell and Caldwell, 1978).

During the formation of a water bloom the relatively large masses of cyanobacterial cells and filaments, moving vertically through the water, might accumulate sufficient densities of heterotrophic bacteria to increase the chances of cell to cell contact necessary for conjugational gene transfer to occur among the heterotrophs. To test this, two *Escherichia coli* strains capable of donating and receiving a plasmid encoding chloramphenicol resistance were mixed with samples from a local reservoir in which a heavy growth of cyanobacteria had developed. The mixture was then poured into a simple experimental water column, constructed from a 1 litre polypropylene measuring cylinder, in which events in a fresh water body subject to the occurence of a cyanobacteria water bloom could be basically reflected, and left at room temperature (approx. 20°C). After appropriate time intervals, samples were taken from various depths, assayed for chlorophyll *a* (Mackinney, 1943) as a measure of cyanobacterial cell density, and plated on LB-agar and LB-agar + chloramphenicol and tetracycline, to determine the number of total *E. coli* and transconjugants, respectively.

Initially, experiments were conducted over 8 h, by which time the cyanobacteria had formed a floating scum at the surface of the column. The majority of the *E. coli* cells detected were in the sample taken from among the floating cyanobacteria, but no transconjugants were obtained from any depth. A control experiment showed that when no cyanobacteria were present in the water column the *E. coli* cells remained evenly distributed throughout the column over the 8 h period; no transconjugants were observed. It thus appeared that many *E. coli* cells had been pulled to the surface by the rising cyanobacteria, thereby increasing their number in this region in comparison to other depths.

The lack of detectable conjugational gene transfer between the *E. coli* strains may have been due to a paucity of nutrients, or to insufficient time in the non-optimum conditions, i.e. low temperature and nutrient availability, of the column. To see whether the greater number of *E. coli* cells among the floating cyanobacteria could lead to more conjugational gene transfer occurring there than at other depths, the experiment was repeated, this time adding 2% L-broth

The Release of Genetically Modified Microorganisms
Edited by D.E.S. Stewart-Tull and M. Sussman, Plenum Press, New York, 1992

207

to the column to stimulate conjugational gene transfer: with this treatment 98% of the transconjugants isolated were obtained from among the floating cyanobacteria.

It was observed, when columns were left for several days, that by 24 h clumps of cyanobacteria, composed of dead or dying cells, had fallen away from the floating scum and sunk to the bottom. After 24 h, the majority of *E. coli* cells detected were isolated from among the settled cyanobacteria at the bottom of the column. Transconjugants were detected, albeit in low numbers, among the floating cyanobacteria, and among the decaying cyanobacteria at the bottom of the column, after 24 h. A control experiment with no cyanobacteria present showed that the *E. coli* cells remained fairly evenly distributed throughout the water column at all times, with no transconjugants being observed. It thus appeared that the clumps of sinking cyanobacteria had pulled a substantial number of *E. coli* cells with them to the bottom of the column, thus increasing the chances of cell to cell contact necessary for conjugational gene transfer to occur between the heterotrophs. When nutrients were added to the column, the effect of the cyanobacteria became more apparent, with greater numbers of transconjugants being obtained among the floating and the settled cyanobacteria than from other regions.

The results of these experiments demonstrated that greater conjugational gene transfer could occur between heterotrophic bacteria among the floating cyanobacteria in a water bloom, and among the settled and decaying cyanobacteria on the bottom, than between heterotrophic bacteria suspended in the water, probably due to the increased number of cells providing more chances of cell to cell contact. Surfaces provided by the cyanobacterial cells and filaments themselves may also have promoted conjugation. Other studies (Cook, N. Ph.D. Thesis, University of Dundee, 1989), showed that transfer of a broad host range of plasmid from *E. coli* to a freshwater *Pseudomonas fluorescens* isolate also occured more frequently in the regions of cyanobacterial accumulation. Thus, the demonstration that cyanobacterial water bloom formation enhanced conjugational gene transfer between the model *Escherichia coli* strains may indicate that genetic exchange between, and from, genetically-modified microorganisms introduced deliberately or inadvertently into freshwater environments might be promoted during such naturally occuring situations.

ACKNOWLEDGEMENT

N. Cook acknowledges the support of the Natural Environment Research Council

REFERENCES

Reynolds, C.S. and Walsby, A.E. (1975) Water blooms. *Biological Reviews*, 50, 437-481.

Paerl, H.W. 1982. Interactions with bacteria. In: *The Biology of Cyanobacteria*. Ed. by N.G.Carr and B.A.Whitton. Blackwell, Oxford, 441-461.

Mikhaylenko, L.Ye. and Kulikova, I.Ya. (1973) Interdependence of bacteria and blue-green algae. *Hydrobiological Journal*, 9, 32-38.

Caldwell, D.E. and Caldwell, S.J. (1978) A *Zoogloea* sp. associated with blooms of *Anabaena flos-aquae*. *Canadian Journal of Microbiology*, 24, 922-931.

Mackinney, J. (1941) Absorption of light by chlorophyll solutions. *Journal of Biological Chemistry*, 140, 315-323.

Rother, J.A. and Fay, P. (1977) Sporulation and development of planktonic blue-green algae in two Salopian meres. *Proceedings of the Royal Society of London* Series B, 196, 317-322

CONSTRUCTION AND DETECTION OF
BIOLUMINESCENT *BACILLUS SUBTILIS* STRAINS

N. Cook[1], D.J. Silcock[2], R.N. Waterhouse[1], D.I. Gray[1]
L.A. Glover[1], K. Killham[1], and J.I. Prosser[1]

Department of Molecular and Cell Biology[1] and
Department of Plant and Soil Science[2]
University of Aberdeen
Aberdeen
AB9 1AS

INTRODUCTION

The introduction of bioluminescence genes (Rattray *et al.*, 1990) enables monitoring of microorganisms in natural environments by several means, including the use of a charged-coupled device (CCD) camera (Hooper and Ansorge, 1990). Here we report the construction of strains of *Bacillus subtilis*, a naturally-occuring soil microorganism, in which the luminescence phenotype is either plasmid or chromosomally encoded. Colonies of these strains can be detected amidst non-bioluminescent microorganisms with a CCD. This technique is potentially useful in survival studies in which it may be necessary to detect low numbers of genetically-modified microorganisms amongst an indigenous microbial population.

MATERIALS AND METHODS

Bacillus subtilis NCTC3610 was obtained from the National Culture Collection of Industrial and Marine Bacteria, Aberdeen. Indigenous soil bacteria were isolated from a sandy loam of the Craibstone series. *B. subtilis* PBC11, in which the bioluminescence phenotype is chromosomally encoded, was constructed by integration of plasmid pRNW111 into the chromosome of *B. subtilis* NCTC3610. Plasmid pOTH951 carries bioluminescence genes on a stable Gram-positive vector plasmid (Janniere *et al.*, 1990).

A nitrogen-cooled slow scan CCD fitted with a 50 mm lens (Wright Instruments, Enfield, U.K.) was used in this study. For detection of bioluminescent *B. subtilis* colonies against a background of indigenous soil bacterial colonies, a 10 g aliquot of non-sterile soil was combined with 100 ml of Ringer's solution and the resultant slurry serially diluted to 10^{-4}. Overnight cultures of *B. subtilis* pOTH951 and *B. subtilis* PBC11 in L-broth were serially diluted to 10^{-8} in Ringer's solution. Each 10^{-2} to 10^{-4} soil slurry dilution was inoculated with 10^{-6} to 10^{-8} dilutions of either *B. subtilis* strain. Plates of LB-agar were spread with 100 μl of each inoculated soil slurry dilution and incubated at 30°C for 24 h. Immediately prior to detection, 2 μl of 100% n-decyl aldehyde was applied to the inverted lid of each Petri dish. For bright field images, plates were exposed to the CCD camera for 0.1 s. For dark-field images, plates were exposed for 0.1 - 30 min, depending on the strain(s) present.

The Release of Genetically Modified Microorganisms
Edited by D.E.S. Stewart-Tull and M. Sussman, Plenum Press, New York, 1992

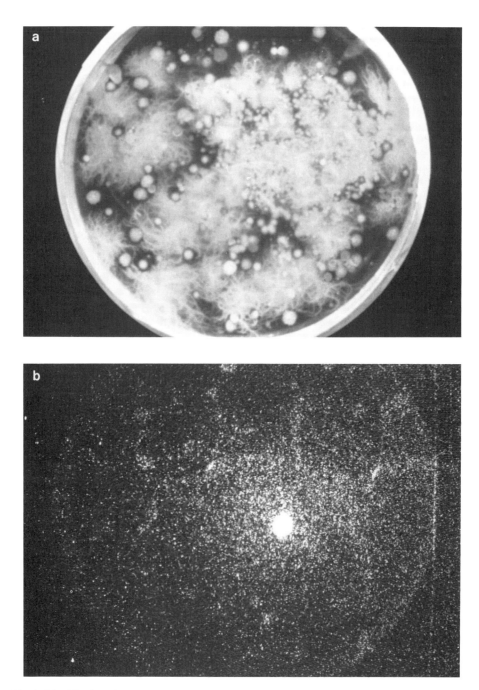

Fig. 1 Identification of *B.subtilis* pOTH951 colonies among soil bacterial colonies by CCD camera. Bright-field image, (a); dark-field image (b).

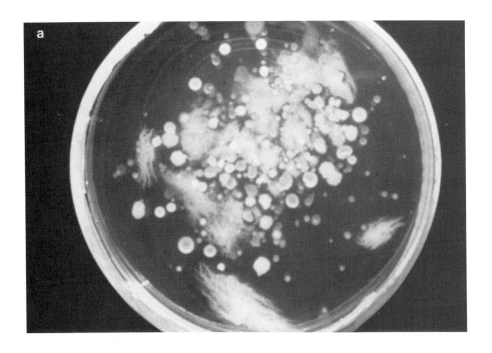

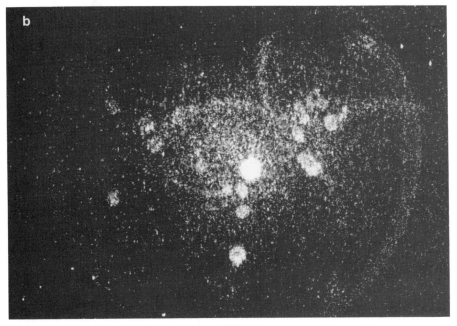

Fig. 2 Identification of *B. subtilis* PBC11 colonies among soil bacterial colonies by CCD camera. Bright-field image, (a); dark-field image (b).

RESULTS

The bioluminescent phenotype of cultures and colonies of *B. subtilis* containing pOTH951 grown under antibiotic selection was detectable visually in a darkened room. The bioluminescent phenotype of *B. subtilis* PBC11, which contains integrated pRNW111, could not be detected by eye.

Figure 1a shows a bright-field image of *B. subtilis* pOTH951 colonies amongst soil bacterial colonies. Overgrowth and crowding make it difficult to identify *B. subtilis* pOTH951 colonies, but they can be identified, through low-level luminescence, after 15 min exposure in dark-field (Figure 1b). Figures 2a and 2b show, respectively, bright and dark field images of colonies of *B. subtilis* PBC11 in the presence of soil bacterial colonies. Following a 30 min exposure, it is possible to distinguish the luminescent PBC11 colonies in a dark-field image.

The small white spots and large central spot on some images result from thermal and electronic noise following a long exposure period. Removal of these is possible by computer image enhancement.

DISCUSSION

Among terrestrial microorganisms, bioluminescence is possessed only by certain species of fungi and a bacterial nematode symbiont (Campbell, 1989). Consequently, there is no significant background amongst soil microorganisms which could mask the detection of bioluminescent GEMMOs, or indigenous microorganisms which may have become bioluminescent following gene transfer. Bioluminescence is thus a useful means of studying survival and transfer of introduced microorganisms and genetic material in soil environments. Under selective conditions, colonies of *B. subtilis* pOTH951 may be visualized by the CCD camera in 10 seconds, demonstrating the potential for rapid analysis of environmental samples, desirable during experiments monitoring survival or gene transfer where large numbers of plates must be examined.

ACKNOWLEDGEMENT

This work was undertaken as part of the PROSAMO project.

REFERENCES

Campbell, A.K. (1989) Living light: biochemistry, functions and biomedical applications. *Essays in Biochemistry,* 24, 41-81.

Hooper, C.E. and Ansorge, R.E. (1990) Quantitative luminescence imaging in the biosciences using the CCD: Analysis of macro and micro samples. *Trends in Analytical Biochemistry,* 9, 269-279.

Janniere, L., Braund, C. and Ehrlich, S.D. (1990) Structurally stable *Bacillus subtilis* cloning vectors. *Gene,* 87, 53-61.

Rattray, E.A.S., Prosser, J.I., Killham, K. and Glover, L.A. (1990) Luminescence-based nonextractive technique for in situ detection of *Escherichia coli* in soil. *Applied and Environmental Microbiology,* 56, 3368-3374.

DETECTION OF LUX-GENE SEQUENCES IN *ESCHERICHIA COLI* MM294 EXTRACTED FROM SOIL USING THE POLYMERASE CHAIN REACTION AND GENE PROBING

D.I. Gray, N. Cook, K. Killham, J.I. Prosser and L.A. Glover

Department of Molecular and Cell Biology
University of Aberdeen
Marischal College
Aberdeen, AB9 1AS, U.K.

INTRODUCTION

The need to detect small numbers of possible transformants in soil microcosms and field experiments is vitally important due to the possibility of foreign genes being introduced from GEMMO's into the indigenous soil bacteria, which unlike many of the introduced strains remain viable in the soil but unculturable in the laboratory (Colwell *et al.*, 1985). The use of simple plate counting techniques to detect soil bacteria requires the efficient removal of cells from the soil, and their subsequent growth on a suitable medium. Our inability to culture many of the naturally occurring soil bacteria limits our ability to monitor the distribution and survival of introduced foreign genes into the soil environment (Trevors and van Elsas, 1989). However with the advent of PCR-technology we have the means to amplify specific target sequence without the prerequisite of cellular growth in culture (Steffan and Atlas, 1988). The only limitation in the application of this technique is the preparation of DNA samples of adequate purity to allow the polymerase chain reaction to occur.

We have developed a two-step isolation procedure which allowed the subsequent amplification of a target sequence contained on a plasmid. The sensitivity of detection of this amplified product was further increased by gene-probe analysis.

MATERIALS AND METHODS

Sterile microcosms consisted of Universal bottles containing 2 g sandy loam soil (Rattray *et al.*, 1990) and were adjusted to a final matric potential of -30KPa. Starved, 1 day old *E.coli* MM294 pBTK5 (Rattray *et al.*, 1990) cells were serially diluted to provide inocula at final concentrations of 10^{10}-10^0 cells/g (air dried wt) soil. Microcosms were incubated for 24 h at 30°C after which cell extractions were performed by the direct addition of sodium deoxycholate (0.1% w/v)/PEG 6000 (0.01% w/v) solution followed by wrist-action shaking at high speed (20 min, 4°C). This suspension was centrifuged at low speed (164 x g) then the resulting supernate was centrifuged at high speed (16300 x g) prior to the final pellet being resuspended in TE buffer (pH 7.4). The pelleted cells were subjected to a standard alkaline-lysis plasmid

The Release of Genetically Modified Microorganisms
Edited by D.E.S. Stewart-Tull and M. Sussman, Plenum Press, New York, 1992

213

isolation procedure (Sambrook *et al.*, 1989). The resulting plasmid DNA was stored at -20°C in distilled water.

The plasmid DNA preparation was purified by passage through a BRL NACS PREPAC Mini-column (Bethesda Research Laboratories) and the first 250 µl column fraction was collected and dialysed against 2000 volumes of sterile TE buffer (pH 7.4), for 24-48 h. The purified DNA was then stored at -20°C.

The general conditions for each 50 µl PCR mixture were as follows: 5 µl of soil-extracted plasmid DNA (positive control 2.0 ng purified pBTK5), 1.0 µM of each primer (5'TGGATCGCTTTGTTCGGCTT3' and 3'AGGCGTTCATGCTGTCTTAC5'), 200 µM of each dNTP, 2.5 U of Taq DNA polymerase (Promega Corporation), 5 µl Promega (1x10) Taq buffer, 26.75 µl sterile water and capped with 50 µl of sterile mineral oil. The PCR was performed for a total of 25 cycles using a PREM programmable restriction enzyme module (Luminar Technology Ltd., Southampton, UK). The initial melting was at 94°C (2 min), further meltings were at 94°C (1 min each), annealings and extensions were done together at 55°C (1 min) and the final primer extension was done at 72°C (5min). The products of the PCR were viewed by running on 2% (w/v) agarose gels stained with ethidium bromide.

Gene-probe analysis was performed after Southern transfer from agarose gels to Hybond-N membranes. A 4 k bp *lux* A-specific gene probe was produced by isolation of a Pst I-Pvu II fragment from pBTK5 which was non-radioactively labeled with digoxigenin-labelled deoxyuridine-triphosphate (Boehringer Mannheim). Hybridizations were performed and the resulting bands detected according to the instructions supplied by Boehringer Mannheim.

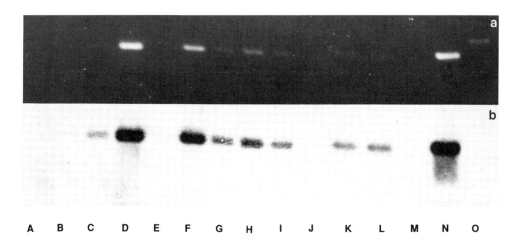

Fig. 1 Amplified products of the *lux* A gene (a) Ethidium bromide stained agarose gel, (b) the corresponding Southern blot of the probed gel. All numbers refer to cells g (air dried) soil.

Lane A, $3x10^0$; **B**, $3x10^2$; **C** $3x10^4$. **Lane D**, positive control reaction - **pBTK5 DNA** extracted from pure cultures. **Lane E**, negative control - uninoculated soil. **Lane F**, $3x10^{10}$; **G** $3x10^9$; **H**, $3x10^8$; **I**, $3x10^7$; **J**, $3x10^6$; **K**, $3x10^{5;}$ **L**, $3x10^4$ · **Lane M**, Negative control reaction - no template present; **N**, positive reaction - as for lane D. **Lane O**, Hind III - digested Lambda DNA size markers.

RESULTS AND DISCUSSION

The amplification conditions for the detection of the lux A gene region were optimized and the product of the amplification region produced on gels was confirmed by Southern blot analysis. The *Vibrio fischeri* specific probes were tested against strains carrying the *lux* A region of *Vibrio harveyi* as well as against other non-*lux* containing Gram-positive and Gram-negative bacteria. All of these control amplifications failed to produce a detectable product as viewed by ethidium bromide stained agarose gels demonstrating their specificity for the target region.

The results of a typical experiment with Boyndie soil which had initially been inoculated with a maximum of 1×10^{10} cells/g (air dried) soil are shown in Fig. 1a, b. The detection level was 1×10^4 cells g (air dried) soil from ethidium bromide stained gels (Fig. 1a) and 1×10^2 using the *lux*-specific probe (Fig. 1b). Attempts to amplify isolated DNA samples prior to the clean-up procedure were never successful while the addition of soil washings to PCR-mixtures containing purified pBTK5 template DNA were similarly unsuccessful, thus demonstrating the requirement for a preliminary DNA clean up procedure.

By this method we have been able to demonstrate the ability of the polymerase chain reaction to detect low cell numbers and the usefulness of *lux*-genes as unique genetic markers in the soil environment. A purification step to clean up the DNA avoids the usual inhibition problems encountered with PCR when attempting to amplify genes from such environmental samples.

ACKNOWLEDGEMENT

This work was undertaken as part of the PROSAMO project.

REFERENCES

Colwell, R.R., Brayton, P.R., Grimes, D.J., Roszak, D.B., Huq, S.A. and Palmer, L.M. (1985) Viable but non-culturable *Vibrio cholerae* and related pathogens in the environment: implications for the release of genetically engineered microorganisms. *Biotechnology* 3, 817-820.

Holburn, W.E., Jansson, J.K., Chelm, B.K. and Tiedje, J.M. (1988) DNA probe method for the detection of specific microorganisms in the soil bacterial community. *Applied and Environmental Microbiology* 54, 703-711.

Rattray, E.A.S., Prosser, J.I., Killham, K. and Glover, L.A. (1990) Luminescence-based nonextractive technique for in situ detection of *Escherichia coli* in soil. *Applied and Environmental Microbiology* 56, 3368-3374.

Sambrook, J., Fritsch, E.F. and Maniatis, T. (1989) *Molecular Cloning: A Laboratory Manual* (Second edition). Cold Spring Harbor Laboratory, Cold Spring Harbor, N.Y.

Steffan, R.J. and Atlas, R.M. (1988) DNA amplification to enhance detection of genetically engineered bacteria in environmental samples. *Applied and Environmental Microbiology* 54, 2185-2191.

Trevors, J.T. and van Elsas, J.D. (1989) A review of selected methods in environmental microbial genetics. *Canadian Journal of Microbiology* 35, 895-902.

MONITORING SURVIVAL OF GENETICALLY-MODIFIED *RHIZOBIUM* IN THE FIELD

P.S. Nicholson, M.J. Jones and P.R. Hirsch

Soil Science Department
AFRC Institute of Arable Crops Research
Rothamsted Experimental Station
Harpenden
Herts AL5 2JQ, UK

In an experiment to investigate the survival of bacterial inoculants in the field and their potential for genetic interaction with native strains, a genetically-modified strain of *Rhizobium* was applied to host and non-host plants (peas and cereals). The field site has been monitored for five years and the inoculant strain shown to persist, although no evidence for gene transfer was found. The soil, which contains the modified strain as an established component of the microflora, offers a unique opportunity for developing improved methods of environmental gene monitoring.

INTRODUCTION

RSM2004, derived from a *Vicia faba* nodule isolate of *Rhizobium leguminosarum* biovar *viciae*, was marked on the chromosome with resistance to rifampicin and streptomycin (spontaneous mutants selected) and on the symbiotic plasmid with the transposon *Tn5*. This enabled detection of plasmid transfer, since *Tn5* encodes neomycin resistance and provides a unique DNA sequence (Hirsch and Spokes, 1988).

To investigate the persistence of microbial inoculants and their potential for genetic interactions with native strains, RSM2004 was introduced into a Rothamsted field. In the laboratory, transfer of the symbiotic plasmid to other rhizobia in sterile soil could be detected only when at least 10^6 of each parent was present. Filter membrane crosses with a range of rhizobia, including field isolates from the release site, ranged from 10^{-3} to less than 10^{-8}, the limit for detection (Hirsch, 1990).

METHODS AND RESULTS

RSM2004 were isolated directly from soil by plating dilutions on TY agar containing the selective antibiotics neomycin and rifampicin (with cycloheximide, a fungicide). At soil dilutions appropriate for detecting RSM2004 populations at or above 100/g soil, there was little background growth. Identity of isolates could be confirmed by examining their plasmid profiles on gels and demonstrating homology of gel or colony blots to *Tn5* DNA probes. Root nodule isolates were screened on selective media: neomycin-resistant colonies, which lacked the rifampicin marker of RSM2004, were putative transconjugants, to be tested further.

The Release of Genetically Modified Microorganisms
Edited by D.E.S. Stewart-Tull and M. Sussman, Plenum Press, New York, 1992

217

Viable cells/g soil

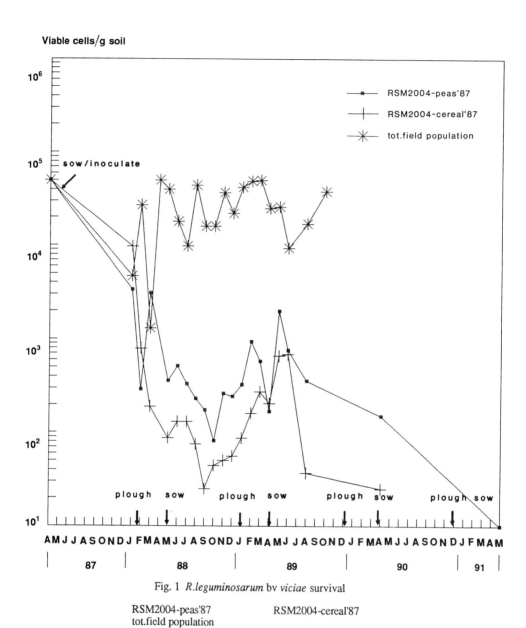

Fig. 1 *R.leguminosarum* bv *viciae* survival

RSM2004-peas'87   RSM2004-cereal'87
tot.field population

RSM2004 estimated from colony counts on selective agar using dilutions of soil where peas or cereals had been inoculated in 1987. Total populations estimated by plant infection tests.

Total populations of rhizobia were estimated by inoculating appropriate host plants with soil dilutions (Vincent, 1970).

The Polymerase Chain Reaction (PCR) is being developed for application to DNA extracted from soil microbial populations. Oligonucleotide primers flanking a central region of Tn5 have been used to generate a PCR product visible on agarose gels. Bands are clearly visible from as little as 0.001 pg RSM2004 DNA but this has yet to be extended to soil extracts.

In 1987 the RSM2004 inoculant was applied to peas and cereals at a rate calculated to equal the native population of $R.$ $leguminosarum$ biovar $viciae$, c. 5.10^4 viable cells/g soil. From over 1000 pea root nodules tested, about 5% contained the inoculant and no transconjugants were detected in nodules (frequency less than 10^{-3}), Figure 1.

No further inoculant was applied but in the second year the plot was sown with peas, phaseolus beans, lucerne and cereals. Fewer than 0.1% of pea nodules contained RSM2004 (i.e. none of the 790 tested) and no transconjugants were detected in any of the legume nodules (frequency less than 6.10^{-4}).

In the third year most of the site was sown with peas to allow more comprehensive sampling: over 2500 nodules were screened and c. 3% contained RSM2004 but none contained transconjugants (frequency less than 4.10^{-4}).

In subsequent years the plot has been sown with cereals and RSM2004 survival has been demonstrated only by direct plating of soil dilutions. This year (1991) the population dropped to the limit of detection of around 10 viable cells/g soil, and background growth by other strains which do not appear to have any Tn5 DNA homology exceeded that of RSM2004.

DISCUSSION

The genetically-marked strain RSM2004 has survived for five years to date, following its introduction into Rothamsted soil. Direct isolation from soil has been facilitated by its antibiotic-resistance markers, although now that the population has declined, the low level of soil bacteria able to grow on the selective media (fewer than 100/g soil) are limiting RSM2004 detection.

No evidence was found for transfer of Tn5 to other strains in the field: if it occurs the frequency is too low for us to detect. This is consistent with observations in laboratory experiments, where transfer was not observed when the parents were present with fewer than 10^6 viable cells/g soil.

PCR should have great potential for monitoring genes in soil biomass extracts. Our experience to date indicates that the equivalent of about 10 bacterial cells/g soil can be detected.

ACKNOWLEDGEMENTS

In addition to funding from AFRC, this work has been supported by the CEC Biotechnology Action Programme, Sector Risk Assessment, and currently by a MAFF Open Contract.

REFERENCES

Hirsch, P.R. (1990) Factors limiting gene transfer in bacteria. In: *Bacterial Genetics in Natural Environments.* Eds J.C. Fry and M.J. Day. Chapman & Hall, London.

Hirsch, P.R. and Spokes, J.R. (1988) *Rhizobium leguminosarum* as a model for investigating gene transfer in soil. In: *Risk Assessment for Deliberate Releases.* Ed. W. KlingmHller. Springer-Verlag, Berlin-Heidelberg.

Vincent, J.M. (1970) *A Manual for the Practical Study of the Root-Nodule Bacteria.* IBP Handbook No.15. Blackwell Scientific Publications, Oxford.

QUANTITATIVE DETECTION OF
GENETICALLY-MODIFIED MICROORGANISMS
BY THE PCR-MPN METHOD

Ginro Endo, Takami Koseki and Eisaku Oikawa

Environmental Engineering Laboratory
Faculty of Engineering
Tohoku-Gakuin University
Tagajo-shi 985, Japan

The Polymerase chain reaction (PCR) method has been used to detect pathogenic microorganisms or the special target microorganisms in environmental water or soil samples(Steffan and Atlas, 1988, Bej *et al.*, 1990, Lampel *et al.*, 1990, Bessesen *et al.*, 1990, Bej *et al.*, 1991). It is considered that PCR offers a highly-sensitive method for the detection of genetically-modified microorganisms (GEMMOs). While some approaches have been made for the quantitative enumeration of GEMMOs, these are rather qualitative than quantitative. To monitor the behavior of released GEMMOs in the environment, quantification of the surviving or persistent population of GEMMOs is necessitated. In this study, the combined method of the polymerase chain reaction (PCR) and the most probable number (MPN) was used, to improve the use of PCR as a quantitative method.

PROCEDURE

The model GEMMO was produced by the transformation of *E. coli* HB101 with the plasmid pIMA2 which possesses *lin* A gene coding the γ-BHC degrading enzyme.This model GEMMO was cultivated and ten-fold dilutions were prepared. Five tubes were prepared for each of three dilutions, each tube was amplified by PCR, and the amplified fragment was detected by agarose gel electrophoresis. The electrophoresis bands were counted and the MPN was determined from the five-tube-MPN table.

RESULT AND DISCUSSION

Figure 1 shows an example of the data of PCR-MPN. In this case, the number of PCR-MPN was estimated as 3.30×10^{12} cells/ml from the MPN table. The relationship between the detection numbers and the optical density values are shown in Figure 2. This result shows that the combined PCR-MPN method is very sensitive and highly quantitative compared to the plate viable count technique. The results showed that the detection sensitivity of PCR-MPN was approximately three orders higher than that of the agar plate counting method. This means a single GEMMO may be monitored by the PCR-MPN method, whereas plate counting cannot achieve a satisfactory result if the target microbial cell population is under 10^3 cells/ml.

The Release of Genetically Modified Microorganisms
Edited by D.E.S. Stewart-Tull and M. Sussman, Plenum Press, New York, 1992

221

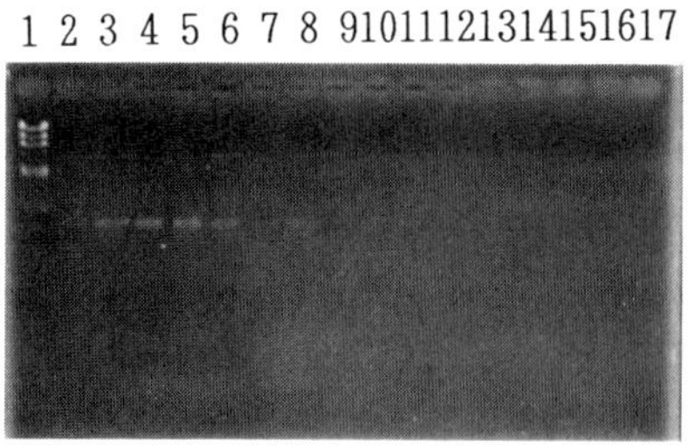

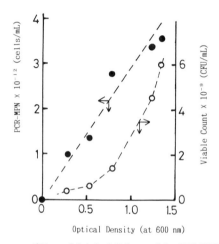

Lane:
1. λ-DNA/Hin d III, size marker
2.
| 10⁻¹¹ diluted sample - PCR amplified
6.
7.
| 10⁻¹² diluted sample - PCR amplified
11.
12.
| 10⁻¹³ diluted sample -PCR amplified
16.
17. Minus (blank) control - PCR amplified

Fig. 1 An Example Data of PCR-MPN. *In this case, PCR-MPN can be determined as 3.3×10^{12} cells/mL.

Fig. 2 Relationships between OD_{600} of Original Culture of the GEMMO and PCR-MPN and CFU

The results obtained from PCR-MPN showed a linear relationship between the MPN and the original microbial density, as measured by optical density. This demonstrates that PCR-MPN method is useful for the quantitative detection of GEMMOs released into the environment, and may be useful for monitoring the consequences after the introduction of GEMMOs.

REFERENCES

Bej, A.K., Steffan, R.J., Haff, L. and Atlas, R.M. (1990) Detection of coliform bacteria in water by polymerase chain reaction and gene probe. Appl. Environment. Microbiol. 56, 307-314.

Bej, A.K., Mahbubani, M.H. and Atlas, R.M. (1991) Detection of viable *Legionella pneumophila* in water by polymerase chain reaction and gene probe methods. Appl. Environment. Microbiol. 57, 597-600.

Bessesen, M.T., Luo, Q., Rotbart, H.A., Blaser, M.J. and Elltson, R.T. (1990) Detection of *Listeria monocytogenes* by using the polymerase chain reaction. Appl. Environment. Microbiol. 56, 2930-2932.

Lampel, K.A., Jagow, J.A., Trucksess, M. and Hill, W. (1990) Polymerase chain for detection of invasive *Shigella flexneri* in food. Appl. Environment. Microbiol. 56, 1536-1540.
Steffan, R.J. and Atlas, R.M. (1988) DNA amplification to enhance detection of genetically engineered bacteria in environmental sample. Appl. Environment. Microbiol. 54, 2185-2191.

BIOLUMINESCENT *RECA* MUTANTS OF *RHIZOBIUM* AS MODEL ORGANISMS IN RISK ASSESSMENT STUDIES

W. Selbitschka, A. Pühler, and R. Simon

Department of Genetics
University of Bielefeld
4800 Bielefeld 1
P.O. Box 8640
Germany

A major concern about the deliberate release of genetically modified microorganisms (GEMMOs) is that the released organism exerts a potential disruptive effect on the biotic community of its target eco-system. Approaches which circumvent such a potential risk involve GEMMOs which are either genetically debilitated (Curtiss, 1988), or carry conditionally lethal genes (Molin *et al.*, 1987; Bej *et al.*, 1988; Contreras *et al.*, 1991; Knudsen and Karlström, 1991). We propose the use of *rec*A mutants of *Rhizobium* as hosts for use in genetic manipulations, since *rec*A mutants show a reduced fitness (Selbitschka *et al.*, 1991). Using newly developed *rec*A integration vectors, bioluminescent *rec*A mutants of *Rhizobium meliloti* and *Rhizobium leguminosarum* biovar *viciae* were constructed. We propose to consider such strains as ideal model organisms in risk assessment studies.

PROPERTIES OF *RECA* INTEGRATION VECTORS

By standard cloning techniques, *rec*A integration vectors for *R.meliloti* (pWS36) and *R.leguminosarum* bv. *viciae* (pWS42) were constructed. The vectors are derivatives of the suicide plasmid pSUP102 (Simon *et al.*, 1983) and therefore, are mobilizable into *Rhizobium* but are unable to replicate there. As a selectable marker, they carry the *aac*C1 gene encoding resistance to gentamicin. Both vectors carry DNA fragments which contain the entire *rec*A coding region of the respective *Rhizobium* species. The *rec*A coding regions are disrupted by the insertion of DNA fragments containing the *npt*II promoter. Multiple cloning sites are situated downstream of the promoter.

THE CONSTRUCTION AND USE OF BIOLUMINESCENT STRAINS

With the described vectors, the firefly *luc* gene which confers bioluminescence, was integrated into the *rec*A coding region of *R.meliloti* and *R.leguminosarum* bv. *viciae*. To achieve this derivatives of vectors pWS36 and pWS42 carrying the cloned *luc* gene were integrated into the rhizobial *rec*A genes via a single crossover. The vector-mediated gentamicin resistance served as the selection marker. Gentamicin-resistant colonies were obtained at a frequency of approximately 10^{-4}. Subsequently, plasmid pAN3, also a pSUP102 derivative and containing the *sac*RB gene (Ried and Collmer, 1987), was integrated via a single crossover into the resulting merodiploid strains. The pAN3-mediated tetracycline resistance was employed to select for the single crossover event. The integration frequency was about

The Release of Genetically Modified Microorganisms
Edited by D.E.S. Stewart-Tull and M. Sussman, Plenum Press, New York, 1992

225

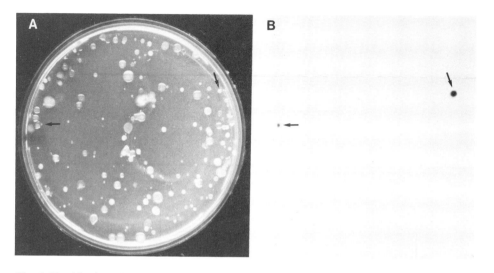

Fig. 1 Identification of single colonies of a bioluminescent *R.meliloti* strain amongst indigenous microorganisms isolated from a non-sterile soil. The arrows indicate identical colonies. (A) Bacteria were grown at 30°C for four days on VMM plates (Vincent, 1970) supplemented with streptomycin (500 mg/l) and cycloheximide (100 mg/l). (B) The same colonies were blotted onto nitrocellulose filters. The filters were wetted with luciferin solution (1mM luciferin in 100 mM sodium citrate, [pH 5.0]).The light emission from *Rhizobium* colonies was detected using a Kodak ortho G-100 film.

3×10^{-5}. Since strains carrying the *sac*RB gene are unable to grow in the presence of 5 % sucrose, it is possible to select for the double crossover. Employing this strategy, bioluminescent clones were selected which carry the *luc* gene integrated into the *rec*A gene. The integration via a double crossover was confirmed by Southern hybridization experiments.

To test whether the *luc* gene could be used as a strain identification marker in microcosm studies, overnight cultures of bioluminescent rhizobial strains were mixed with a suspension of indigenous bacterial cells isolated from a non-sterile soil. Appropriate dilutions were spread onto agar plates. After four days of incubation, the experiment was evaluated. As shown in Fig. 1, single colonies of bioluminescent rhizobia could be detected amongst indigenous bacterial cells of a non-sterile soil.

REFERENCES

Bej, A. K., M. H. Perlin, and R. M. Atlas. (1988) Model suicide vector for containment of genetically engineered microorganisms. Appl. Environ. Microbiol. 54:2472-2477.

Contreras, A., S. Molin, and J-L. Ramos. (1991) Conditional-suicide containment system for bacteria which mineralize aromatics. Appl. Environ. Microbiol. 57:1504-1508.

Curtiss III, R. (1988) Engineering organisms for safety: What is necessary? In: The release of genetically engineered micro-organisms. Sussman, M., C. H. Collins, F. A. S. Skinner, and D. E. Stewart-Tull. (eds.) Academic Press, 6-20.

Knudsen, S. M., and O. H. Karlström. (1991) Development of efficient suicide mechanisms for biological containment of bacteria. Appl. Environ. Microbiol. 57:85-92.

Molin, S., P. Klemm, L. K. Poulsen, H. Biehl, K. Gerdes, and P. Andersson. (1987) Conditional suicide system for containment of bacteria and plasmids. Bio Technology 5:1315-1318.

Ried, J. L., and A. Collmer. (1987) An *npt*II-*sac*B-*sac*R cartridge for constructing directed, unmarked mutations in gram-negative bacteria by marker exchange-eviction mutagenesis. Gene 57:239-246.

Selbitschka, W., W. Arnold, U. B. Priefer, T. Rottschäfer, M. Schmidt, R. Simon and A. Pühler. (1991) Characterization of *recA* genes and *recA* mutants of *Rhizobium meliloti* and *Rhizobium leguminosarum* biovar *viciae*. Mol. Gen. Genet. 229:86-95.

Simon, R., U. Priefer, and A. Pühler. (1983) A broad host range mobilization system for *in vivo* genetic engineering: Transposon mutagenesis in gram negative bacteria. Bio Technology 1:784-791.

Vincent, J. M. (1970) A manual for the practical study of root nodule bacteria. (IBP Handbook 15) Blackwell Scientific Publications, Ltd., Oxford

AN ARTIFICIAL MICROCOSM TO STUDY MICROBIAL
INTERACTIONS IN THE RHIZOSPHERE

D. Pearce[1], M.J. Bazin[1] and J.M. Lynch[2]

Microbial Physiology Research Group, King's College[1]
Campden Hill Road
London W8 7AH

Horticulture Research International[2]
Worthing Road
Littlehampton
West Sussex BN17 6LP

An artificial microcosm for the ectorhizosphere is described where a growth-limiting nutrient is supplied by diffusion at a predetermined rate from a cylindrical ultrafiltration membrane in the form of a hollow fibre. The artificial microcosm is thermodynamically open, reflecting the nature of soil, allowing tighter control of both growth-limiting nutrient and energy fluxes and it incorporates a leaching mechanism. To date the system has been shown to reach steady-state and the control parameters are currently under investigation.

INTRODUCTION

The model system consists of a column reactor packed with a solid substrate surrounding a cylindrical ultrafiltration membrane. Defined nutrient medium is passed down the column and a growth-limiting carbon source in an identical solution is supplied by radial diffusion from the lumen of the ultrafiltration membrane tube.

MATERIALS AND METHODS

The whole artificial microcosm was stable during sterilization in the autoclave at 121°C for 15 minutes. A diffusion gradient was maintained by pumping the nutrient solution through the lumen of the ultrafiltration membrane tube at the rate of 1 ml/h. The leaching rate was set at 1 ml/h. Build-up of nutrient medium was prevented by applying minimal positive air pressure inside the tube (See Figure 1). The effluent of the columns was sampled at timed intervals.

RESULTS

Nutrient solution flowed through the column at an even rate over a two week test period. The mass of effluent was collected at timed intervals and found to range between a minimum of 0.45g and a maximum of 0.55g. Glucose diffused into the nutrient solution percolating through the sand and could be detected in the effluent within 250 minutes. The concentration of glucose in the effluent then increased sharply over 60 minutes to reach a supply rate

The Release of Genetically Modified Microorganisms
Edited by D.E.S. Stewart-Tull and M. Sussman, Plenum Press, New York, 1992

229

equivalent to 0.1% (w/v) glucose in the original nutrient medium. This level was maintained for two weeks.

DISCUSSION

The artificial microcosm produced a steady supply rate of growth-limiting nutrient and provides an experimental system for the study of microbial interactions in the ectorhizosphere. Particularly the influence of:
a) Radial distance from the nutrient source,
b) The effect of increased simulated leaching and
c) The proportion of interactions occurring in the biofilm.

There are a number of advantages employing an artificial microcosm to complement data obtained from soil-root experimental systems:

* Properties of roots and the interactive effects of associated microorganisms are difficult to differentiate,
* Changing parameters that affect microbial activity in order to study their interactions, will also closely affect linked factors that control exudation in a living plant root.
* Microorganisms increase the exudation rate of living plant roots,
* Closer control over the supply rate of growth-limiting nutrient inside the lumen of the ultrafiltration membrane and
* More precise figures for calculating variables can be achieved when the supply rate is accurately known.

The incorporation of a leaching mechanism and the thermodynamically open nature of this artificial ectorhizosphere microcosm, takes the study a step closer to the natural environment when compared with previous experimental models. This is achieved by allowing vertical and radial exchange of energy and materials between the microbial population under investigation and the external substrate.

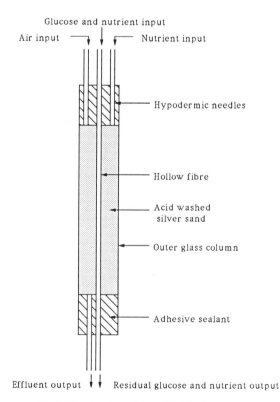

Fig.1 The structure of the artificial microcosm

DETECTION OF BACTERIA FROM NATURAL ENVIRONMENTS BY FLOW CYTOMETRY

Jonathan Porter[1], Clive Edwards[1] and Roger Pickup[2]

Department of Genetics and Microbiology[1]
University of Liverpool
Liverpool
L69 3BX

Institute of Freshwater Ecology[2]
Windermere
Cumbria

Recent interest in microbial ecology has focused upon the detection and analysis of cells without prior culture of the organism. Many approaches that have been suggested use immunofluorescence or nucleic acid analysis. Another method is the use of flow cytometry (FCM) for the rapid analysis of single cells with the potential to measure a wide range of parameters.

ASSESSMENT OF VIABILITY USING FLOW CYTOMETRY

Exponential cells of *Bacillus subtilis* were incubated with three fluorescent dyes, fluorescein diacetate (FDA), 3,3'-dihexyloxacarbocyanine $DiOC_6(3)$ and rhodamine 123 (R123), which only stained cells possessing general cytoplasmic enzyme activity (FDA), or cells capable of generating a membrane potential $DiOC_6(3)$ and R123). Cells treated with dye and gramicidin S (Gram) were used as a control. FDA was the most effective of the three dyes; live cells could also be distinguished when mixed with dead cells (data not shown). $DiOC_6(3)$ also allowed detection of live from dead cells, but the distributions overlapped. R123 was not taken up, or was not retained, by the cells, and was thus ineffective for determining viability of *B. subtilis*.

Vegetative cells of *B.subtilis* were released into sterile lakewater and were successfully detected by FDA (Figure 1(b)). There was some attenuation of the fluorescent signal in non-sterile lakewater, probably due to particulate adherence (Figure 1(c)). The only effective dye that labeled bacteria, from ten-fold concentrated lakewater, was R123, but only after incubation with EDTA, suggesting a predominantly Gram-negative population. No FDA hydrolysis was detected (Figure 1). In all cases, addition of Gram "decreased" the fluorescence.

CONCLUSIONS

Hydrolysis of FDA was an effective indication of cell viability for *B.subtilis* when grown in a nutritionally complex medium and released into filtered sterile water. Non-sterile lakewater caused attenuation of the fluorescence signal, probably due to interference from particulate

The Release of Genetically Modified Microorganisms
Edited by D.E.S. Stewart-Tull and M. Sussman, Plenum Press, New York, 1992

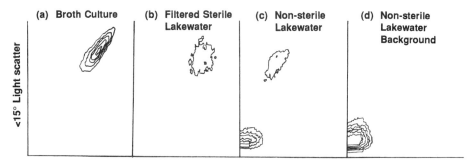

Fig.1 Effectiveness of FDA for detecting viable *B. subtilis* in sterile and non-sterile lakewater.

Exponential phase *B. subtilis* was assessed for viability using FDA in broth [Figure 1(a)], sterile lakewater [Figure 1(b)] and non-sterile lakewater [Figure 1(c)]. No background FDA hydrolysis was detected [Figure 1(d)].

material. The incubation time of cells with FDA was crucial for maximum staining and was found to be four minutes. In a natural lakewater sample, EDTA treatment was needed to ensure uptake of the cationic lipophilic dyes. This means that dyes for measuring bacterial viability are not necessarily universal to all cell types.

ACKNOWLEDGEMENTS

This work was supported by a NERC grant through the TIGER programme.

THE DETECTION OF BACTERIA IN AQUATIC ENVIRONMENTS BY FLOW CYTOMETRY

J.P. Diaper and C. Edwards

Department of Genetics and Microbiology
University of Liverpool
P.O. Box 147
Liverpool L69 3BX UK

Because of the potential risks associated with the accidental or deliberate release of genetically-modified microorganisms (GEMMOs) into natural environments (Fry and Day, 1990) there is an urgent need for methods that rapidly detect and identify target species within complex microbial communities.

The commonest methods used to recover microbes from environmental samples is by growth on selective agar. However, the recovery of microorganisms by plating methods has long been regarded as a problem because non-culturable but viable cells are often present. Therefore plate counts probably yield under-estimations of microbial numbers in the environment (Trevors, 1989). Direct methods of enumeration such as fluorescent antibodies, which do not rely on the ability of an organism to grow on agar, have been extensively used for the detection and identification of bacteria in environmental samples by microscopy (Bohlool and Schmidt, 1980). This can be both laborious and time consuming.

Flow cytometry measures physical or chemical characteristics of individual cells as they pass past optical or electronic sensors (Shapiro, 1990). This technique can rapidly and precisely detect, characterize and identify thousands of cells in a mixed population in a relatively short period of time, compared with the few hundred normally detected by fluorescent microscopy. Thus flow cytometry has the potential, as yet largely untested, to be used as a tool to monitor populations of GEMMOs in the environment.

In these experiments *Staphylococcus aureus* was used as a model organism to evaluate the ability of flow cytometry to detect immunologically bacteria in environmental samples. After staining with FITC-IgG, *S. aureus* cells could clearly be detected by flow cytometry (Fig.1a) whilst other bacteria (i.e. *Pseudomonas fluorescens*) could not be separated from the background fluorescence (Fig. 1b). *S. aureus* could also be detected in lakewater by flow cytometry after staining with FITC-IgG (Fig. 1c) and it was possible to accurately estimate the population size.

These results demonstrate that it is possible to detect immunologically-stained *S.aureus* cells in environmental samples by flow cytometry. The detection limit, although similar to that observed with PCR, in the order of 10^3 cells ml, could be improved by the use of more specific antibodies and by concentrating the samples prior to staining. Flow cytometry also has the additional ability to estimate the numbers of cells present unlike PCR which is more qualitative than quantitative. Flow cytometry therefore appears to have great potential for monitoring microbial populations in the environment as well as for more specific identification techniques required for target species present in low numbers within large diverse populations.

The Release of Genetically Modified Microorganisms
Edited by D.E.S. Stewart-Tull and M. Sussman, Plenum Press, New York, 1992

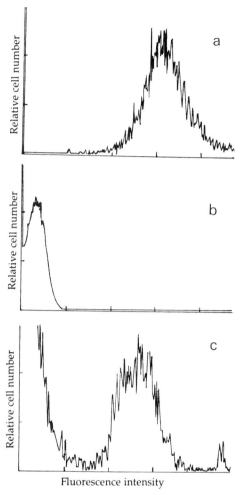

Fig.1 Fluorescent labelling of bacteria with FITC-IgC (a) *S.aureus*; (b) *P.fluorescens*; (c) *S.aureus* in lakewater.

ACKNOWLEDGEMENTS

We are grateful to the NERC for funding this work (Grant No. GR3/7596).

REFERENCES

Bohlool, B, B. and Schmidt, E.L. (1980). The immunofluorescence approach in microbial ecology. Advances in Microbial Ecology 4, 203-241.

Fry, J.C. and Day, M.J. (1990). Bacterial Genetics in Natural Environments. Chapman and Hall: London.

Shapiro, H.M. (1990) Flow cytometry in laboratory microbiology : New directions. ASM News 56, 584-588.

Trevors, J.T. and Van Elsas, J.D. (1989). A review of selective methods in environmental microbial genetics. Canadian Journal of Microbiology 55, 895-901.

BACILLUS THURINGIENSIS AS A MODEL ORGANISM

FOR EVALUATING RISKS ON THE RELEASE OF GEMMOs

Malin Bryne, Susanne Munkberg-Stenfors and Ritva Landén

Department of Microbiology
Stockholm University
S-106 91 Stockholm
Sweden

Bacillus thuringiensis (Bt) is a Gram-positive enthomopathogenic organism. Upon sporulation it produces an inclusion body or crystal, the so called δ-endotoxin which is lethal to susceptible insect species (for review see Aronson *et al*, 1986). Depending on the Bt subspecies used mainly Lepidopteran (moths and butterflies)., Dipteran (flies and mosquitos) or Coleopteran (beetles) insects are affected. The specificity is mainly due to a receptor in the gut of the insect (Van Rie *et al*, 1991). Commercial preparations consisting of a mixture of spores and crystals have been used world-wide for decades as pesticides both in agriculture and forestry. The annual production is currently estimated to be 3000 tonnes and for instance 900 tonnes were spread in China in 1990 (Beer, 1991). One subspecies (*israelensis*) with activity against mosquitos has during the last decade been used for protection of human health against diseases transmitted by insects, such as malaria, arbovirus diseases, filariasis, African trypanomiasis a.s.o. The commercial preparations are considered safe for nontarget insects, predators, plants and animals.

The long-lasting experience of the safe use of this organism makes it suitable for large-scale field tests. The specificity of the δ-endotoxin and the narrow host range minimizes the risks even if the toxin gene should happen to be transfered in the environment. Moreover the organism has a very short survival time in the soil. (Beer, 1991). The genetic transfer systems are reasonably well-studied, both transduction, conjugation and transformation are reported (Carlton and Gonzales, 1985). Thereby a lot of mutants are available for the genetic studies of gene transfer and gen stability test in either a microcosmos or in the environment. A third argument is that companies producing commercial Bt preparations wish to broaden the host range and/or introduce new toxins into existing stains. This is easily done with recombinant techniqucs. The toxin gene has also been introduced into plants (tomato, potato and tobacco) and to microorganisms that grow in other habitats than Bt (*Pseudomonas*, Cyanobacteria).

To use Bt as a model organism in the Swedish environment has an additional advantage since the use of Bt as a pesticide was not approved in Sweden until February 1990. Thereby Swedish soils should not be heavily-contaminated by commercial strains. In order to estimate the natural occurence of this organism in our country, we started to isolate Bt strains from different locations in the southern part of Sweden. As Bt is a spore-former we utilized this in two ways. Bt spores germinate poorly in 0.25 M acetate and by growing samples in such medium and then heating the samples we could enrich the soil samples for Bt (Martin and Travers, 1989). The colonies which appeared were then tested for sensitivity to a five bacteriophages isolated from Swedish soils. A total of 2318 colonies were tested and these resulted in 191 presumptive Bt isolates (8%). After a second screening 51 remained (2%). These were then tested for crystal production (microscope) and pathogenicity for *Trichoplusia ni* larvae. Of the 51 isolates one third produced crystals and 46% were pathogenic for the test

The Release of Genetically Modified Microorganisms
Edited by D.E.S. Stewart-Tull and M. Sussman, Plenum Press, New York, 1992

insect. Our impression is that it is easy to isolate Bt strains by this method, even if the detection level still is very low. However, it seems that Bt is very ubiquitous in Swedish soils, as well as it is world-wide (Martin and Travers, 1989, Smith and Couche, 1991). We think it is necessary to develop more sensitive methods to detect Bt and we are currently investigating if some of the virulence genes that we have cloned could be used as probes in a detection assay.

With the newly-isolated strains as recipients in transduction experiments we have started to investigate gene transfer. The experiments were done under laboratory conditions according to Heierson *et al.* (1987). As the donor a well-characterized Bt var. *kurstaki* Bt2141 genotype *rif, str, spc* was used. The transducing bacteriophage was 64, one of the phages isolated from Swedish soils. As control another *kurstaki* strain, Bt22, was included as recipient. So far only four strains have been tested and for these the transduction frequency ranged from 4×10^{-7} to 0.5×10^{-9}. The frequency for the homologous transduction with Bt22 as recipient was 6×10^{-9}. The conclusion is that under the conditions used the frequency is low but for some isolates considerably higher than for the control.

In summary, Bt fulfils some requirements for a good model organism, but the limited knowledge about the ecology of Bt requires further investigation. The organisms survives poorly in soil and does not multiply in this environment. On the other hand, it has been found in dead insects and on the phylloplane, but the recycling and the role of the organism in the environment is not well-understood and this needs to be examined.

REFERENCES

Aronson, A.I., Beckman, W. and Dunn, P. (1986) *Bacillus thuringiensis* and related insect pathogens. Microbiol. Rev. 50, 1-24.

Beer, A. (1991) Bt - a model agent still waiting to take off? Agrow. 141 22-24.

Carlton, B.C. and Gonzales, J.M. (1985) The genetics and molecular biology of *Bacillus thuringiensis*, in "The molecular biology of bacilli", Vol 2. (Dubanau, D., Ed). pp. 211-249. Academic Press, Orlando, Florida.

Freer, J., Real, M.D., Van Rie, J., Jansens, S. and Peferoen, M. (1991) Resistance to the *Bacillus thuringiensis* bioinsecticide in a field population of *Putella xylostella* is due to a change in a midgut membrane receptor. Proc. Natl. Acad. Sci. USA. 88, 5119-5123.

Heierson, A., Landén, R., Lövgren, A., Dalhammar, G. and Boman, H.G. (1987). Transformation of vegetative cells of *Bacillus thuringiensis* by plasmid DNA. J. Bact. 169, 1147-1152.

Martin, P.A.W., Travers, R.S. (1989) Worldwide abundance and distribution of *Bacillus thuringiensis* isolates. Appl. Environ. Microbiol. 55, 2437-2442.

Smith, R.A., and Couche, G.A. (1991) The phylloplane as a source of *Bacillus thuringiensis* variants. Appl. Environ. Microbiol. 57, 311-315.

Van Rie, J., McGaughey, W.H., Johnson, D.E., Barnett, B.D. and Van Mellaert, H. (1990) Mechanism of insect resistance to the microbial insecticide *Bacillus thuringiensis*. Science 247, 72-74.

PLASMID MEDIATED RESISTANCE TO HEXAVALENT CHROMIUM IN *AGROBACTERIUM*

D.W. Cullen[1], I.M.Packer[1] and D.J.Platt[2]

Department of Biological Sciences[1]
Glasgow Polytechnic
Glasgow, G4 0BA and
Department of Bacteriology[2]
Glasgow Royal Infirmary
Glasgow, G4 0SF

INTRODUCTION

Chromium (Cr) is widespread in the environment as a consequence of mining and industrial activities. Trivalent Cr(III) and hexavalent Cr(VI) are the most stable oxidation states in natural environments. The hexavalent form is the more toxic to bacteria (Nriagu and Nieboer, 1988). Plasmids that confer resistance to Cr(VI) have been reported in several bacterial species (Mergeay, 1991).

The bacteria from a chromium-polluted sediment were investigated and we report here the presence of plasmids that mediate Cr(VI) resistance in *Agrobacterium* species.

METHODS

Bacterial Isolation and Identification

The investigated site was the chromium-polluted Malls Myre Burn in Glasgow. Sediment bacteria were isolated at 25°C on casein peptone starch agar (CPSA; Jones, 1970) that contained Cr(VI) as $K_2Cr_2O_7^{2-}$ (0.48mM). Two isolates, from different sample sites were identified as *Agrobacterium radiobacter* by the API 20 NE system, and were designated M2.10 and M5.4.

Cr(VI) Minimum Inhibitory Concentration (MIC) Determinations

Bacterial suspensions were inoculated onto CPSA media which contained a range of Cr(VI) (as $K_2Cr_2O_7^{2-}$) concentrations. The MIC was taken as the lowest concentration which inhibited growth after 7 days at 25°C.

Plasmid Profiles

Plasmid DNA was prepared by successive treatment of 1.0 ml of culture with Sarcosyl, lysozyme, SDS and a clearing spin to remove chromosomal DNA. Electrophoresis was in a 0.6% horizontal agarose gel for 4 - 5 hours at 100V.

The Release of Genetically Modified Microorganisms
Edited by D.E.S. Stewart-Tull and M. Sussman, Plenum Press, New York, 1992

TABLE 1 MICs of Cr(VI) for various *Agrobacterium* strains

Strain	MIC Cr(VI)(mM)	Comments
M2.10	3.84	Malls Myre isolates
M5.4	3.84	
A. tumefaciens C58	0.18	
A. tumefaciens K749	0.12	
A. radiobacter K84	0.18	Reference Strains
A. radiobacter NCIMB 9042	0.24	
A. radiobacter NCIMB 9161	0.24	
M5.4S1	0.72	Partially susceptible isolates
M5.4S2	1.44	derived from M5.4
M2.10S1	0.96	Partially susceptible isolates
M2.10S2	0.96	derived from M2.10
M2.10S3	0.96	

Isolation of derivatives susceptible to Cr(VI)

Strains of M2.10 and M5.4 with reduced levels of Cr(VI) resistance were obtained by serial subculture, (5 overnight cycles) in unsupplemented broth and identified by replica-plating onto CPSA that contained 1.92 mM Cr(VI).

Transfer of Cr(VI) resistance

Conjugation was done at a ratio of 1:1 donor / recipient, on 0.45 μm filters placed on CPSA for 48 hours. Transconjugants were selected on CPSA that contained both Cr(VI) and a donor counterselective antibiotic.

RESULTS

Levels of Cr(VI) Resistance and Stability of the Resistance Phenotype

The two *Agrobacterium* isolates (M2.10 and M5.4) were more resistant to higher concentrations of Cr(VI) than several *Agrobacterium* reference strains (Table 1). Neither of these strains exhibited resistance to other metals or antibiotics tested.

Partially sensitive derivatives to Cr(VI) were isolated at a frequency of 0.09 - 0.3% by serial subculture, and their levels of resistance, although lower than the parent isolates, were still higher than the reference strains (Table 1).

Plasmid Profiles of Cr(VI) Resistant and Partially Susceptible Isolates

Strain M5.4 (lane 1) harbored 5 plasmids (A-E) and strain M2.10 (lane 5) 6 plasmids (F-K); the plasmid sizes ranged from 7kb to 147 kb (Figure 1). The potential involvement of plasmids that mediated Cr(VI) resistance was supported by plasmid profile analysis of partially sensitive strains. M5.4S1 (lane 2) lacked plasmid C and M2.10S2 (lane 8) lacked plasmid J. Additionally, M5.4S2 (lane 4) displayed a plasmid rearrangement where plasmid C was absent and another larger plasmid (40kb, arrow) was present.

Transfer of Cr(VI) resistance

When a variety of Gram-negative bacteria (enterobacteria, pseudomonads, *Acinetobacter* species) were used as recipients, transfer of Cr(VI) resistance from M2.10 and M5.4 was not detected. Transfer of Cr(VI) resistance did occur from M5.4 to the partially-susceptible derivative M5.4S1. This transfer was detected at a frequency of 4.8×10^{-4} per recipient. The transconjugants when screened exhibited the wild-type resistance phenotype and had inherited plasmid C, thus its role in Cr(VI) resistance was confirmed.

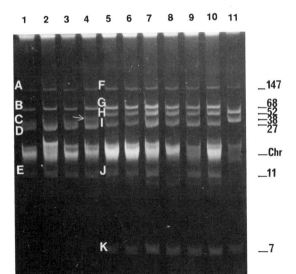

Fig.1 Plasmid profiles of Cr(VI) resistant strains M5.4 and M2.10 and their partially susceptible derivatives.
   Lanes 1 and 3, M5.4; lane 2, M5.4S1 ; lane 4, M5.4S2 ; Lanes 5,7 and 9, M2.10 ; lane 6, M2.10S1 ; lane 8, M2.10S2; lane 10, M2.10S3 ; lane 11, plasmids in *Escherichia coli* 39R861 (147, 64, 36, 7 kb.).
   Approximate sizes of plasmids A to K are indicated to the left of gel (kilobases, kb). Chr, Chromosomal DNA.

Cr(VI) resistance was transfered from both M2.10 and M5.4 to *Agrobacterium tumefaciens* K749 and occured at a frequency of approximately 2.0×10^{-5} per recipient. Transfer to K749 was also detected, though at a reduced frequency (ie 1.25×10^{-8}), from M5.4S1 which lacked a plasmid implicated in Cr(VI) resistance (plasmid C, Figure 1). These observations raised the possibility that more than one plasmid specified Cr(VI) resistance, and this hypothesis was further supported by the plasmid profiles of over 100 K749 transconjugants. Plasmid profile analysis indicated that plasmids A and C from M5.4, plasmid A from M5.4S1, and plasmids H and I from M2.10 were transferred to K749 and each confered Cr(VI) resistance. Conjugational plasmid rearrangements were also observed on the basis of restriction enzyme fingerprinting of transconjugants that harbored plasmids C, H and I. None of the K749 transconjugants exhibited the resistance levels of the donors. The Cr(VI) MIC value for K749 which harbored plasmid A was 0.36mM compared to 0.96mM with plasmids C, H or I, and there did not appear to be any 'additive effect' to the resistance level in transconjugants which had received more than one Cr(VI) resistance plasmid. Preliminary data suggest that each of these four plasmids are not self-transmissable to other derivatives of K749.

DISCUSSION

We have demonstrated that two plasmids (possibly three in the M2.10-plasmid J, Figure 1) determine Cr(VI) resistance in two independent strains of *Agrobacterium*. To our knowledge, this is the first report of plasmid-mediated resistance to Cr(VI) in *Agrobacterium*.
   An investigation into the relationships between these Cr(VI) resistance plasmids and their interaction is in progress. This may highlight potential strategies for their exploitation in the treatment of chromium pollution.

REFERENCES

Jones, J. G. (1970) Studies on freshwater bacteria: effect of medium composition and method on estimates of bacterial population. J Appl Bacteriol 33, 679-689.

Mergeay, M.(1991) Towards an understanding of the genetics of bacterial metal resistance. TIBTECH 9, 17-24.

Nriagu, J. O. and Nieboer, E.(1988) Chromium in the natural and human environments. John Wiley, New York

PLASMID TRANSFER IN STERILE SOIL AS INFLUENCED BY A NON-SIMULTANEOUS INOCULATION OF DONOR AND RECIPIENT

A. Richaume, D. Bernillon and G. Faurie

Laboratoire d'Ecologie Microbienne du Sol
URA CNRS 1450
Université Lyon I
69622 Villeurbanne Cedex
France

Soil is an heterogenous, discontinuous and structured environment composed of particles of various size. The importance of soil structure and physico-chemical properties on bacterial population regulation (Heynen *et al.*, 1988, Postma *et al.*, 1990) and genetic exchange (Stotzky and Krasovsky, 1981, Richaume *et al.*, 1989) have been demonstrated.

In the soil environment, the matrix can be an obstacle to the cell-to-cell contact, which is the obligatory step in the process of conjugative transfer, so it is important to consider soil colonization and bacterial location.

The aim of this work was to study the effect of time lag inoculations of parental strains in different soil microcosms on the occurence of plasmid transfer, and to determine the localization of introduced microorganisms in the soil and the possible relationship between strain location and plasmid transfer.

Transfer of the plasmid R68-45 between *Pseudomonas aeruginosa* PAO25 and *Agrobacterium tumefaciens* GMI9023 was studied in different microcosms of γ-irradiated silt loam soil. In 15ml polypropylene tubes (5g/tubes) 200μl of bacterial suspensions were placed on the surface of the soil and in 100ml glass jars (30g/jar) 1.2ml of bacterial suspensions were thoroughly mixed into the soil with a sterile spatula to ensure a homogeneous inoculation. The soil moisture was adjusted to 40% and the incubation temperature was 28°C. In both microcosms, the recipient was introduced 0 to 29 days before the donor and the experiments stopped 7 days after the donor inoculation.

Bacteria were extracted by blending 5g of soil (equivalent dry weight) in 50 ml of NaCl (0.85%) solution. Donor, recipient and transconjugant colony forming unit (cfu) counts were realized by dilution-plating of soil suspensions on appropriate selective media.

The bacterial localization studies were performed only for the tube microcosms by two different methods: the washing-out of microbial cells (Hattori, 1967) based on repeated washing resulted in two fractions, corresponding to the inner and outer part of the soil, and the grain size fractionation of soil after physical dispersion (Jocteur Monrozier *et al.*, 1991) resulting in 5 fractions according to the particle size (ranging from more than 250μm to less than 2μm).

The Release of Genetically Modified Microorganisms
Edited by D.E.S. Stewart-Tull and M. Sussman, Plenum Press, New York, 1992

241

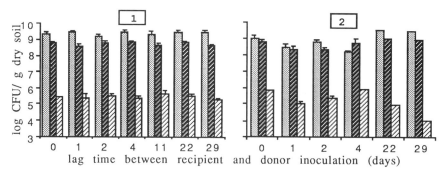

Figure 1 and 2: Transfer of the plasmid R68-45 from PAO25 ▨ to GMI9023 ▨ in microcosms containing 1) 5g of soil and 2) 30g of soil for various lag times of inoculation. (Transconjugants ▨)

Table 1. Weight and CFU percentages in each soil fraction obtained after physical dispersion and grain-size fractionation

Fraction	Weight %	CFU %
<2mm	9	99
20-2mm	29.5	0.06
50-20mm	24	0.35
250-50mm	28.5	0.5
>250mm	9	0.005

The washing out of the soil was realized after 0, 2, 8 and 21 days of inoculation delays. In all cases, the number of bacteria decreased very sharply in successive washings, from more than 21% after the first washing to less than 0.003% after the 15th. The percentages of donor and recipient in the remaining soil for an inoculation lag of 8 or 21 days, determined by the cfu enumerated from the unwashed soil, were 65 and 42% or 44 and 58.5% respectively.

RESULTS

It has been shown that for both microcosms, the numbers of donor and recipient remained quite stable (figures 1 and 2) as did the transconjugant number whatever the inoculation lag time.

The weight percentage and the repartition of the microorganisms in each fraction obtained after the physical dispersion and grain-size fractionation are presented in Table 1. The component of the >250µm and 250-50µ soil fractions were chiefly coarse minerals and vegetal residues, 20-2µm were microaggregates and loamy particles. The <2µm fraction was dispersible clay particles.

DISCUSSION AND CONCLUSIONS

We have shown that plasmid transfer occurred between a donor present for 7 days in the soil, and a recipient previously introduced. The pre-adaptation of the recipient to the environment did not affect its ability to acquire the plasmid. Van Elsas *et al.* (1990) showed that a temporal separation of parental strain introduction into the soil resulted in possible genetic interaction between freshly-introduced and resident organisms. Furthermore, inoculation onto the surface of a small sample gave the same results as thorough mixing of the inocula in a larger sample, indicating that the inoculation procedure has to be adjusted to the size of the soil sample. However, it has to be noted that the bacterial enumeration from part of the 30g (while the totality was used for the 5g microcosms) led to a greater variability in the counts. The behavior of the microorganisms was not affected by the amount of soil in the microcosms. Bacterial localization studies by physical dispersion and grain-size fractionation of the soil which is more drastic than the soil washing method, permits further explanations about the bacterial location in the remaining soil after washings. The percentage of cells decreased rapidly after repeated washings while 40 to 60% were found in the remaining soil. This indicates that the percentage of the population in the soil after washings is about the same after 7 days or longer incubation.

With the soil fractionation, 99% of the population was found in the clay-dispersible fraction after 30 days of incubation. In the non-sterilized soil, the microaggregate fraction (20-2μm) contained the highest microbial biomass (data not shown). This indicates that recently introduced microorganisms in sterile soil cannot extensively colonize pre-existing soil microaggregates and that the energy developed by successive washings was too low to desorb all the adsorbed but accessible microrganisms. The location of the microorganisms, in the dispersible phase or partly adhering onto surfaces of stable soil components, was not an obstacle to cell-to cell contact and genetic exchanges.

From our experiments we can conclude that:

- a genetic interaction between a strain introduced in the soil and a resident micro-organism is possible,
- the inoculation procedure was well-adapted to the size of the microcosms; plasmid transfer and microorganism behavior were not affected by the size of the microcosms,
- there was strong adsorption of bacteria on soil surfaces but no extensive inside colonization of stable microaggregates. In this case soil was not really a barrier to the plasmid transfer.

ACKNOWLEDGEMENTS

We would like to thank Dr. Jocteur-Monrozier for the help with the soil fractionation and J. Viallon for his technical assistance.

REFERENCES

van Elsas, J.D., Trevors, J.T., Starodub, M.E. and van Overbeek, L.S. (1990) Transfer of plasmid RP4 between pseudomonads after introduction into soil; influence of spatial and temporal aspects. FEMS Microbial. Ecol. 73: 1-12.

Hattori,T. (1967) Microorganisms and soil aggregates as their microhabitat. Bull, Inst. Agr. Res. Tohoku Univ. 18: 159-193.

Heynen, C.E., van Elsas, J.D., Kuikman, P.J. and van Veen, J.A. (1988) Dynamics of *Rhizobium leguminosarum* biovar *trifolii* introduced into soil; the effect of bentonite clay on predation by protozoa. Soil Biol. Biochem. 20: 483-488.

Jocteur Monrozier, L., Ladd, J.N., Fitzpatrick, R.W., Foster, R.C. and Raupach, M. (1991) Components and microbial biomass content of size fractions in soils of contrasting aggregation.Geoderma, 49: 37-62.

Postma, J., Hok-A-Hin, C.H., and van Veen, J.A. (1990) The role of microniches in protecting introduced *Rhizobium leguminosarum* biovar *trifolii* against competition and predation in soil. Appl. Environ. Microbiol. 56: 495-502.

Richaume, A., Angle, J.S. and Sadowsky, M.J. (1989) Influence of soil variables on in situ plasmid transfer from *Escherichia coli* to *Rhizobium fredii*. Appl. Environ. Microbial. 55: 1730-1734.

Stotzky, G. and Krasovsky, V.N. (1981) Ecological factors that affect the survival, establishment, growth and genetic recombination of microbes in natural habitats. In "Molecular biology, pathogenicity and ecology of bacterial plasmids". S.B. Levy, R.C. Clowes, E.L. Koenig. Plenum press New York 31-42.

SURVIVAL OF *ENTEROBACTER CLOACAE* ON LEAVES AND IN SOIL DETECTED BY IMMUNOFLUORESCENCE MICROSCOPY IN COMPARISON WITH SELECTIVE PLATING

Jens C. Pedersen and Thomas D. Leser

National Environmental Research Institute
Frederiksborgvej 399
Department of Marine Ecology and Microbiology
P.O. Box 358
DK-4000 Roskilde
Denmark

INTRODUCTION

A major difficulty in studies of the fate of non-indigenous bacteria released into the environment is related to detection. Two common detection methods are selective plating on agar media (SP), and quantitative immunofluorescence microscopy (IF). Both methods have drawbacks; SP techniques rely on the ability of the bacteria to grow on laboratory media, whereas difficulties in separating cells from soil particles may limit the applicability of IF techniques. Furthermore IF does not distinguish between living and dead cells. Comparisons of SP and IF bacterial counts have shown varying results.

As a rhizosphere colonizer *Enterobacter cloacae* is of agricultural interest and various strains can stabilize soil structure or control *Pythium* root disease. This paper presents a comparison between SP and IF microscopy to detect *E. cloacae* released in a soil/plant microcosm and in the field. We investigated the use of IF microscopy to count phylloplane bacteria.

MATERIALS AND METIIODS

Microcosm

Four trays (30x40x30cm) containing plants were housed in each of two 70x80x105 cm growth-chambers.

Field. 2x2 m plots

Soil. Sandy loam

Plants. Dwarf beans (*Phaseolus vulgaris*) and spring barley (*Hordeum vulgare*)

The Release of Genetically Modified Microorganisms
Edited by D.E.S. Stewart-Tull and M. Sussman, Plenum Press, New York, 1992

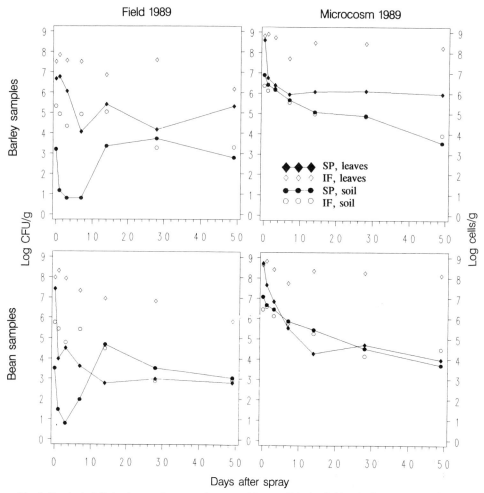

Fig. 1 Survival of *Enterobacter cloacae* on leaves and in topsoil in the field and microcosm experiments.

Bacterial Inoculum. *Enterobacter cloacae* A107 resistant to nalidixic acid was grown overnight at 30°C with shaking in LB medium amended with 500 µg/g nalidixic acid. Cells were washed twice and suspended in Winogradsky's (Wi) salts solution with 0.2% Tween 80 as a wetting agent. This suspension was used (10^8-10^9 cfu/g) as inoculum by spraying plants until runoff.

Samples were taken in triplicate (pooled for IF) from topsoil (upper 2 cm) or randomly selected leaves and blended (soil) or stomacher treated (leaves) with Wi. For IF 45 ml soil homogenates were flocculated with 356 µl 20% $CaCl_2$ and 0.9 g acid-washed PVPP (polyvinylpyrrolidone).

Selective plating (SP) was performed in duplicate on LB amended with 500 µg/g nalidixic acid and 25 µg/g natamycin. Plates were incubated overnight at 30°C. No background was seen.

Immunofluorescence microscopy (IF) was performed on duplicate sub-samples with polyclonal rabbit antiserum raised against intact *E. cloacae* cells treated with formaldehyde, and cross-absorbed with four related Enterobacteriaceae species. Samples (0.1-1 ml homogenate)

were stained on black polycarbonate membrane filters by the following steps (room temperature):
1)   Background blocking (Rhodamine + gelatine + Tween 80)
2)   1% antiserum, 2x1 h
3)   FITC-conjugated swine anti-rabbit serum, 2x30 min.
Stained cells were counted with Zeiss epifluorescence microscope.

RESULTS AND DISCUSSION

IF TECHNIQUE

As the numbers of *E. cloacae* throughout the experiments were above the background levels (max. 10^4 and 5×10^5 cells/g in soil and leaf samples, respectively), *E. cloacae* could be detected with the IF procedure (Fig. 1).

No significant differences were found between bean and barley samples. Different bacterial numbers ($p < 0.0001$), were seen in leaf and soil samples, between experiments, and through time.

Comparison of SP- and IF-results

IF- and SP-results of soil generally coincided. The only exception was the field experiment, Day 0-14, where SP, unlike IF, showed very low numbers of bacteria (<10 cfu/g). The top soil was extremely dry during this period. Later when rain had moistened the soil, SP- and IF results agreed.

While IF-results of barley and bean leaves coincided, SP showed more culturable *E. cloacae* on barley compared to bean leaves.

In future experiments viability of *E. cloacae* will be studied to understand the differences between IF and SP results.

ACKNOWLEDGEMENTS

We thank D. Truelsen, J.L. Rasmussen, B. R. Hansen, H. Nielsen, T. Schyberg and J. Holst for technical assistance.

GENETIC MANIPULATION OF RUMEN BACTERIA

James J. Murray[1], Geoffrey P. Hazlewood[1], and Harry J. Gilbert[2]

Department of Biochemistry[1]
AFRC, IAPGR, Babraham
Cambridge
CB2 4AT

Department of Biological and Nutritional Sciences[2]
University of Newcastle
Newcastle upon Tyne
NE1 7RU

The digestion of plant structural polysaccharides in the rumen is a major, but relatively inefficient, process in ruminant production which depends totally on the activity of microbial enzymes. The rate of polysaccharide digestion could be increased significantly by the genetic modification of rumen bacteria. The species chosen for such a genetic modification should maintain itself at a high population level after introduction into the rumen. Certain strains of *Selenomonas ruminantium*, which have been reintroduced into the rumen, can survive at a constant population level for relatively long periods of time, and are therefore suitable candidates for genetic modification. Cloning vectors will initially be developed from indigenous *S. ruminantium* plasmids.

Plasmid DNA from two strains of S. ruminantium, FB315 and 521C1, was prepared according to the method of Thomson (1990), followed by caesium chloride gradient centrifugation. Full length plasmid DNA from FB315 was cloned and propagated in *E. coli* JM83, and sequenced by the method of Murphy and Kavanagh, (1988). Bal 31 deletions for sequencing were prepared according to Hall and Gilbert, (1988). Sequences were compiled and ordered with the computer programs of Staden, (1980). All other methods were based upon those of Maniatis *et.al.*, (1982).

The plasmid we have cloned is the only high copy number plasmid in FB315, and is about 2.5 kb in size, (Fig.1). The FB315 plasmid is homologous to one of two plasmids present in *S. ruminantium* 521C1, and is slightly smaller than the homologous 521C1 plasmid. It has many unique restriction sites including sites for *Pst* I, *Xmn* I, *Sph* I, *Hinc* II, *Dra* III, *Nsi* I, and *Ava* II. This property, and its low molecular weight, satisfy two criteria for a good cloning vector. The plasmid sequence obtained to date, indicates that the proteins encoded by the plasmid include a replication protein, which is homologous to the replication proteins encoded by other plasmids listed in Fig.2. All these plasmids belong to the family of small Gram-positive plasmids which replicate by the rolling circle mechanism. However, *S. ruminantium* is a Gram-negative organism. Until it has been more fully characterized we cannot be certain of the mechanism of replication of the FB315 plasmid.

The Release of Genetically Modified Microorganisms
Edited by D.E.S. Stewart-Tull and M. Sussman, Plenum Press, New York, 1992

249

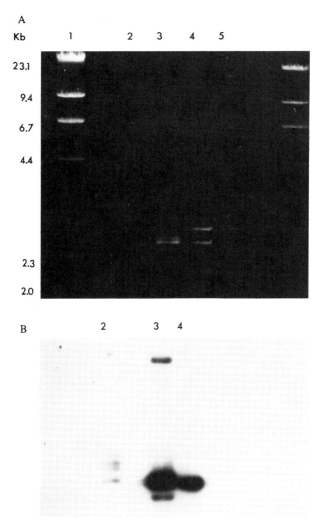

Fig.1 Characterization of FB315 plasmid DNA cloned in pUC 19.
A: Agarose gel electrophoresis of plasmid DNA. B: Southern blot of gel probed with radiolabelled FB315 total plasmid DNA.
Lanes: 1, Lambda DNA, *Hin*d III digest; 2, FB315, uncut; 3, FB315, *Pst* I digest; 4, FB315/pUC19 recombinant plasmid, *Pst* I digest; 5, pUC19, *Pst* I digest.

```
FB315     M R M T I L Q E N E Q D C K L E V I E G T G E V L E D L S G T X R K R P W S E R K S E S V E L L N L F E T
pLAB1000  M S K . . . . . . . . . . . . . . K I L K D V S R N R K E R P W R E R K L E N L Q Y A E Y L R .
pFTB14    M Y S . . . . . . . . . . . S E N D Y S I L E D K T A T G K K R D W R G K K R R A N L M A E H Y E A
pLP1      M S E . . . . . . . . . . . . I F E D K T E N G K V R P W R E R K K I A N V D Y F E L L H .
pTB913    M G V S F N I M C P N S S I Y . S D E K S R V L V D K T K S G K V R P W F R E K K I A N V D Y F E L L H .

FB315     A R K I D E S V I S Q T R L Q A L K D C G S W L T F A Q Q A D . G T R R L A N A N F C R L R L C P L C G W
pLAB1000  I L N F K K A N R V K E S C A E E H L S F K R D P E T G . G R L C L Y Q T W F C K N W R L C P M C A W
pFTB14    L E K R I G A P Y Y G K K A A E R L S E C A E H L S F K R K I G E . G R L K L L Y Q A H F C K V R L C P M C A W
pLP1      I L E F K K R A H D V R G C G E V L R F R K I G E . H L K L Y R V W F C K S R L C P L C N W
pTB913    I L E F K K A E R V K D C A E I L E Y K Q N R E T G E R K K L Y R V W F C K S R L C P M C N W

FB315     R R S L K L F S Q V S R I T D A I L A E K K A . R F I F V T L T V E N V K . G E E L R A T I K R M N E G F
pLAB1000  R R S M G Q S N Q L M Q V L D E A H K Q R K T G . G R F L F L T L T A E N A . S G E R L K Q E V R K M G R A I
pFTB14    R R S M K N S S Q L K Q I I A E A N R Q G K G . W I F L L T L T K N . S A E E L K V S L R A L T K A F
pLP1      R R A M K H G I Q S Q K V V A E V I K Q K P T V R K L F L T L T V K N V Y D G E E L M H V L V C V E P T Y F
pTB913    . . . . . . . . . . . . . . . . . . . . . . . . . . . . . . . . . . . . . . . . . . . . . . . . .

FB315     K C L V Q D K K G M A A S A T F R A D L M G Y M K A I E V T Y N T K R N D F H P H I H C I F E L A P K Y F
pLAB1000  S K L F Q Y K K K . P A K N L L G Y V R S T E I T T I N . K N G T Y H Q H M H V L L F V K P T Y F
pFTB14    R K L T Q Y K K K . V K T S V L G F F R A L E I T T K N H E E D T Y H V L L F V K S S Y F
pLP1      R R M M Q Y K K K . V T K N L V G F M R A T E V T I N K D N S Y N Q H L H V L L F V C V E P T Y F
pTB913    . . . . . . . . . . I N K N L V G F M R A T E V T I N K D N S Y N Q H M H V L V C V E P T Y F

FB315     R G K E G G Y L T H E D W R V M W R S V M K L D Y E P Q V D V R A I K N K X A V A . E V A K Y P V K V S T
pLAB1000  K D S A N . Y I N Q A E W S K L W Q R A M K L D Y Q P I V N V E A V R S N K A K G K N L I A S A Q E T .
pFTB14    . G K N . Y L A Q A E W T S L W K R A M K L D Y T P I V D I R R V K G R V K C R V K K R K G T D S L Q A S A E E T .
pLP1      K N S N N . Y L A Q A E W A K L W Q K A L K V D Y E P V H V Q A V K A N K R K G T D S L Q A S A E E T .
pTB913    K N T E N . Y V N Q K Q W I Q F W K K A M K L D Y D P N V K V Q M I R P . K N K Y K S D I Q S A I D E T .

FB315     D F X K K V K D K E K A A N L D S P R P S K G Q L G P I N Q L N L G P E F N L T L S G P G E F . F R K
pLAB1000  . A K Y Q V K S K D I L T N D Q . E R D L E Q L A G S R Q L I G Y G G I L K K
pFTB14    M E Q K A V L E I S K Y F V K D T D Y F Y L D D A L S A R R L I G Y G G I L K K
pLP1      . A K Y E V K S A D Y M T A D D . E R N L V I K N L E Y A L A G T R Q I S Y G G L L K K
pTB913    . A K Y P P V K D T D F M T D D E . E K N L K R L S D L E E G L H R K R L I S Y G G L L K K

FB315     E G X I K A R L D D I E T G D L V H V E T D . K Q E L N A V A M I L F K Y R A D V G A Y I C .
pLAB1000  E I R K Q L Q L E D V D . A H L I N V D D D K V K I D E V V R E . E V V A K W D Y N K Q N Y F I W .
pFTB14    E I H K K L N N G D A E G G D L V H V G D E D L V K I E E E D D Y T K E Q M E A A A F E V V A Y W H P G I K N Y F I L K .
pLP1      Q I K Q D L K L E D D V H V G D E D Y T K E Q M E A A A K W D F N K Q N Y F I W .
pTB913    E I H K K L N L D D T E E G D L I H T D D D E K A D E D G F S . . . I I A M W N W E R K N Y F I K E
```

Fig.2 The amino acid sequence of the putative replication protein encoded by the FB315 plasmid. Regions of sequence homology between this protein and a number of replication proteins encoded by other plasmids are enclosed in boxes. Identical amino acids and amino acids with similar chemical and physical properties are grouped together as follows: P; A,G; S,T; Q,N; E,D; H; K,R; C; V,L,I,M; F,Y,W. These plasmids originate from the following organisms: pFTB14, from *Bacillus amyloliquefaciens*; pLAB1000, from *Lactobacillus hilgardii* 67; pLP1 from *Lactobacillus plantarum*; pTB913, a deletion derivative of pTB19 from a thermophilic *Bacillus* species. pTB913 arose from pTB19 by *in vivo* recombination in the host bacteria *B. stearothermophilus* CU21 and *B. subtilis* MI113.

REFERENCES

Hall, J. and Gilbert, H.J., (1988). The nucleotide sequence of a carboxymethyl cellulase from *Pseudomonas fluorescens* subsp. *cellulosa*. Mol. Gen. Genet., 213, 112-117.

Maniatis, T., Fritsch, E.F., Sambrook, J., (1982). Molecular Cloning: A Laboratory Manual. Cold Spring Harbor Laboratory, New York.

Murphy, G., Kavanagh, T., (1988). Speeding-up the sequencing of double stranded DNA. Nucleic Acids Res., 16, 5198.

Staden, R., (1980). A new computer method for the storage and manipulation of DNA gel reading data. Nucleic Acids Res., 16, 3673-3694.

Thomson, A.M., (1990). "Gene Transfer in Rumen *Bacteroides* Species," pp. 53-54. Ph.D Thesis, University of Aberdeen.

DETECTING NATURAL TRANSFORMATION OF

ACINETOBACTER CALCOACETICUS, IN SITU,

WITHIN NATURAL EPILITHON OF THE RIVER TAFF

H.G. Williams, M.J. Day and J.C. Fry

School of Pure and Applied Biology
University of Wales College of Cardiff
PO Box 915
Cardiff CF1 3TL
UK

Natural transformation is a process in which competent cells take up and express exogenous DNA. This is a potential mechanism by which recombinant DNA could be spread through natural bacterial populations. The aim of this study was to determine whether genetic information could be transfered by transformation within a natural environment.

Acinetobacter calcoaceticus was chosen as a model organism since it is common in both soil and aquatic habitats (Bauman, 1968). The microencapsulated mutant strain BD413 is competent for natural transformation in laboratory experiments (Juni and Janik, 1969). Untreated whole cell cultures or crude lysates of BD413 (phenotype - His+, Rif^s, Sp^s, Hg^s) prepared by the method of Juni (1972), were used as a source of transforming DNA. An auxotrophic mutant HGW1521(pQM17) (phenotype - His-, Rif^r, Sp^r, Hg^r) derived from BD413 was used as recipient.

The source of DNA (1 ml) and recipient cells (1 ml) were filtered separately onto 0.45 μm cellulose nitrate filters (Whatman), and were placed, filtered side together, under the desired conditions for 24 h. Filters were incubated at 20°C upon standard plate count agar (PCA; Oxoid CM463) as a laboratory standard. Filters were attached to sterile scrubbed stones by rubber bands and incubated in either beaker microcosms (Bale *et al.*, 1987) containing 400 ml of fresh non-sterile river-water at 20°C or *in situ* in the River Taff. They were then collected and vortex mixed in 3 ml B22 salts solution (Bale *et al.*, 1987) containing 100 μg/ml DNase1 (Sigma). Transformants and recipients were enumerated by a drop count technique on B22 minimal media (Bale *et al.*, 1987) + HgCl$_2$ (14 μg/ml) + rifampicin (50 μg/ml; Sigma) + EDTA (400 μg/ml) or PCA + rifampicin (100 μg/ml; Sigma) respectively. The transformation frequency was calculated as the number of transformants per recipient.

The sterility of lysates was confirmed by spreading 100 μl over a PCA plate. The spontaneous mutation frequency of HGW1521(pQM17) to prototrophy in the laboratory was below 10^{-9}. Approximately 10^{-7} colony forming units per recipient were detected when the recipient was incubated *in situ* in the absence of a source of DNA. This could be due to mutation, transfer from the natural population or growth of indigenous microorganisms. The validity of transformants was determined by checking spectinomycin-resistance as a secondary characteristic and the ability to transform auxotrophic mutants of BD413 to prototrophy.

The mean transformation frequencies of HGW1521(pQM17) in the laboratory and *in situ* in the River Taff are shown in Table 1. The transformation frequency with a lysate as the source of transforming DNA incubated in the beaker microcosm, was 1.48×10^{-5} (geometric

The Release of Genetically Modified Microorganisms
Edited by D.E.S. Stewart-Tull and M. Sussman, Plenum Press, New York, 1992

Table 1. Mean transformation frequencies (transformants per recipient) of *Acinetobacter calcoaceticus* HGW1521(pQM17; a histidine auxotroph) to prototrophy, in the laboratory, in microcosms and *in situ*.

Filters Incubated	Source of transforming DNA	Mean Transformation frequency (geometric mean ; n ; standard deviation)
in the laboratory	Lysates	7.08×10^{-4} (-3.15 ; n=3 ; SD=0.17)
upon PCA at 20°C	Whole cells	9.55×10^{-5} (-4.02 ; n=3; SD=0.31)
In situ, in the River	Lysates	5.89×10^{-5} (-4.23 ; n=21 ; SD=0.76)
Taff on a stone surface	Whole cells	1.07×10^{-5} (-4.97 ; n=6 ; SD=0.43)
between 12 and 20°C		

There were significant differences between the mean transformation frequencies (p=0.006). The minimum significant difference (MSD.) between the geometric means was 0.757 (log units). There were significant differences between both the incubation conditions (p=0.003) and source of transforming DNA (p=0.016).

mean, -4.83; n=1). This was not significantly different from the mean frequency observed *in situ*.

The presence of mercury and rifampicin in selective media prevented the growth of the majority of the indigenous population. The use of DNase1 and EDTA was necessary to prevent transformation on the selective media which artificially raised the transformation frequency (unpublished results).

Transformation has been previously demonstrated in the presence of the indigenous population in a sediment microcosm (Stewart and Sinigalliano, 1990). However this appears to be the first reported case of genetic transformation occuring *in situ* within an aquatic environment. These findings support the contention that recombinant DNA from introduced GEMMOs could be transfered to the natural population. Transformation is of extra interest in that surviving DNA may be taken up, incorporated and expressed in a new host after the introduced organism has ceased to be viable.

ACKNOWLEDGEMENT

We should like to gratefully acknowledge the support of the Natural and Environmental Research Council. Attendance costs for REGEM II were paid by the Presidents Fund of the Society for General Microbiology.

REFERENCES

Bale, M.J., Fry, J.C. and Day, M.J. (1987) Plasmid transfer between strains of *Pseudomonas aeruginosa* on membrane filters attached to river stones. J. Gen. Microbiol 133, 3099-3107.

Bauman , P. (1968) Isolation of *Acinetobacter* from soil and water. J. Bacteriol. 96, 39-42.

Juni, E. (1972) Interspecies transformation of *Acinetobacter* : Genetic evidence for a ubiquitous genus. J. Bacteriol. 122, 917-931.

Juni, E and Janik, A. (1969) Transformation of *Acinetobacter calcoaceticus (Bacterium anitratum).* J. Bacteriol. 98, 281-288.

Stewart, G.J. and Sinigalliano, C.D. (1990) Detection of horizontal gene transfer by natural transformation in native and introduced species of bacteria in marine and synthetic sediments. Appl. Environ. Microbiol. 56, 1818-1824.

THE SURVIVAL OF A CHLORAMPHENICOL RESISTANCE PLASMID IN A NATURAL *BACILLUS* POPULATION

I.R. McDonald[1], P.W. Riley[2], A.J. McCarthy[1] and R.J. Sharp[2]

Genetics and Microbiology Department[1]
University of Liverpool
Liverpool
L69 3BX

Division of Biotechnology[2]
C.A.M.R.
Porton Down

Bacillus species are important commercial sources of a variety of industrial enzymes including extracellular proteases and amylases. In addition, the insecticidal properties of *Bacillus thuringiensis* are widely employed for the bio-control of insect pests. Consequently, concerns have been voiced over the deliberate or accidental release of recombinant *Bacillus* strains into the environment, and the potential for gene survival and transfer. In this work a chloramphenicol resistance plasmid, pC194, has been released into a natural *Bacillus* population (in mushroom compost), that has had the levels of antibiotic resistance within the population determined: chloramphenicol; 5.6×10^3 cfu per g of compost.

pC194 has been released into sterile and fresh compost, incubated at 37, 50 and 65°C, and its survival monitored over a 28-day period. The host for these experiments was *B. subtilis* 168 (trp-); this has no selectable marker but can be distinguished from the indigenous population by its different colony morphology.

The results from sterile compost (Fig.1) and fresh compost (Fig.2) release of pC194 would indicate that the plasmid is able to survive in 100% of cfu. Indeed, results from chemostat studies of pC194 in *B. subtilis* confirm its stability in th's host (100% stability after 75 generations). Results from sterile compost would suggest that vegetative cells sporulate after 7-14 days, however in fresh compost the majority of cells remain vegetative even after 13 weeks (Fig.2). Perhaps the greater availability of nutrients in fresh compost, or indeed the release of a toxic substance into the compost during sterilization, cause sporulation. Spores were released into fresh compost to see whether germination would occur; it did not. It was now thought that the difference in numbers was due to an error in determining the spore numbers. However, examination of samples of compost under a scanning electron microscope confirmed that vegetative cells were present in greater numbers than spores, as was also seen in the indigenous population. It would appear from these results that *B. subtilis* pC194 is a suitable host/vector system for monitoring gene transfer in a natural *Bacillus* population.

The Release of Genetically Modified Microorganisms
Edited by D.E.S. Stewart-Tull and M. Sussman, Plenum Press, New York, 1992

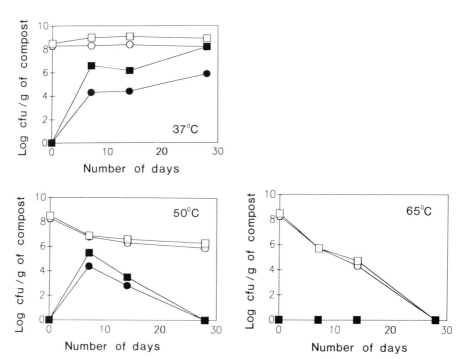

Fig.1 Recovery of *B. subtilis* strain 168 containing plasmid pC 194 from sterile compost incubated at 37, 50 and 65°C.

Nutrient agar (NA): vegetative cells,   ; spores,
NA + chloramphenicol: vegetative cells, O ; spores,
All values are the means of triplicates

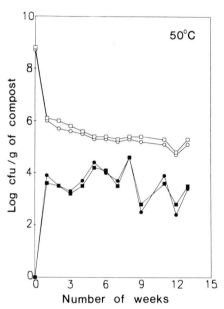

Fig.2 Recovery of *B. subtilis* strain 168 containing plasmid pC 194 from untreated compost incubated at 50°C.

Legend as for Figure 1.

WHAT VALUES IN THE GEMMOs? REFLECTIONS ON REGEM 2

Les Levidow

Centre for Technology Strategy
Open University
Milton Keynes MK7 6AA, UK

The acronym GEMMOs denotes genetically modified microorganisms, which are the basis for a range of new products, including biopesticides, vaccines and bioremediators (degradation of chemical wastes). Although most GEMMO products are intended for contained use, some are now being tested first in contained conditions, in preparation for small-scale releases and then eventual use as living organisms in the environment. Safety considerations go beyond a rerun of the 1970s debate on the contained use of genetic modification. As these newer GEMMOs are designed to persist outside, their release raises wider ecological questions.

Of the GEMMOs which would involve live release, the only one so far approved for commercial use in Britain has been a baker's yeast modified for faster fermentation. As part of the current precautionary approach to risk regulation, the bakery effluent was categorized as a release of GEMMO and was then assessed as entailing no additional hazard. When approval for the use of the yeast was announced in March 1990, the tabloid press featured such headlines as 'A half-baked way to slice up nature'. That response perhaps indicates a residual public unease which could erupt against releases seen as less innocuous or evoking science-fiction scenarios, such as the use of a biopesticide engineered for greater lethality.

As genetic modification is increasingly used to create novel organisms, how might their genetic novelty introduce hazards? And how can we best avoid hazards, before they do actual harm? Answering that question an be particularly difficult in the case of microorganisms, where science is still developing techniques for reliably detecting their survival, persistence and capacity for further genetic exchange. There remains a wide gap between the apparent precision of genetic modification and the potential imprecision of its ultimate ecological effects.

Such uncertainties underlay many presentations at this second international conference on The Release of Genetically Modified Microorganisms. Since the first one, held in 1988, scientists have learned more about GEMMOs' behaviour and have tested more products based on them. That progress has much relevance to risk assessment, especially the step-by-step relaxation of containment and insertion of foreign genes (e.g. OECD, 1986). However, the additional knowledge brings science little closer to any consensus on how to anticipate potential hazards. In itself, better data would not resolve the matter, as risk assessment involves a series of value judgements about the likelihood and acceptability of unintended effects. For a general discussion of value judgements in biotechnology regulation, see Levidow and Tait, 1991; Tait and Levidow, 1992. On implications for a precautionary science, see Levidow, 1992.

SPREADING NOVELTY

An initial quandary is how to detect the fate of the GEMMO and the inserted gene. As some researchers reported to the conference, immunofluorescence techniques have revealed a

The Release of Genetically Modified Microorganisms
Edited by D.E.S. Stewart-Tull and M. Sussman, Plenum Press, New York, 1992

261

far greater survival of microorganisms that was previously suggested by earlier techniques, such as growing them on culture media and then doing plate counts. Rita Colwell, from the Maryland Biotechnology Institute, argued that is is mistaken to assume that a microbe won't persist at all if it doesn't find a niche: the microbe can wait for suitable conditions for resuming growth.

Predictability can become yet more complex in view of potential genetic transfer. John Beringer, Chairman of the Advisory Committee on Releases to the Environment (ACRE), acknowledged that genetic transfer has been a research focus because of both public and scientific interest. He emphasized that conjugal transfer requires a high density of bacteria, in order for them to make physical contact, but noted that such proximity is unnecessary for other modes of horizontal transfer, such as by bacterial viruses or naked DNA. Beringer also cited recent research suggesting that genetic exchange may run much more widely that previously thought, and not simply within 'species' as named by bacteriologists (Smith *et al.*, 1991).

Also, as others noted, cell death does not prevent further transfer. Lysing cells release DNA which retains biological activity, so that even free DNA can remain a viable material for genetic transfer. And recent research has revealed that more free DNA is present than was previously thought. These revelations indicate the great opportunity that has always existed for natural genetic recombination. Thus, when genetic modification crosses whatever might be the natural barriers to that process, the introduced gene in the GEMMO has all the more opportunity to create further genetic novelty among other organisms. Although further research may clarify the extent and likelihood of such transfer, that knowledge will not necessarily clarify implications for potential hazards.

NOVEL NICHES?

Given GEMMOs' capacity for survival, persistence and genetic exchange, what ultimate difference might that make? Might a GEMMO cause a perturbation that will damage the environment? That is how the question was put by Jim Lynch, from the Institute of Horticultural Research, Littlehampton. The answer depends partly upon one's concept of ecological niche, an issue broached by Ron Atlas, from the University of Louisville. Chairing a workshop, he outlined two opposed views for debate: either that a GEMMO would have difficulty finding a niche, or that it would cause an irreversible disturbance.

The conceptual contrast becomes more stark if we adopt a sculpting metaphor from two protagonists in the American debate: either niches are inherently limited by an environment, which would carve the GEMMO back to its original form through natural selection (says Bernard Davis, e.g., 1985); or, genetic novelty can carve out new niches in nature, particularly where a GEMMO has acquired new adaptive traits (says Philip Regal, e.g., 1985, 1988).

On that issue, the strongest view came from John Beringer, who cited a survey of Rhizobium strains isolated from different environments: the survey found only a limited number of isoenzyme combinations. In particular, among variations on homologous genes, there was a non-random distribution of plasmids across chromosome backgrounds (Young and Wexler, 1988). If only a few genetic combinations can confer a selective advantage, then changing a few genes should not make much difference, Beringer said. In a similar vein, Arnold Foudin mentioned USDA's exercise of cataloguing a hundred organisms with characteristics which would make their establishment likely: the field results showed them to be poorly predictable for selective advantage, as the organisms did not create new niches.

However, the authors of the *Rhizobium* study (*Ibid*) do not draw Beringer's inference. Rather, they believe that limited natural recombination could explain the similar mix of genotypes found in different environments. Moreover, they see no evidence for fitness differences (relative selective advantage) among the potential genetic combinations.

PERTURBATIONS AS HAZARDS?

At Regem 2, no one went as far as to argue that genetic novelty can act as an independent variable upon the environment though that may reflect the paucity of ecologists in attendance. At the same time, the notion of novel niches apparently underlay their plea for more basic ecological research. Some ecologists emphasized the need to detect environmental perturbations, which might cause hazards not yet identified.

Jim Lynch welcomed the use of genetic markers to study the impact of introductions on a natural microbial community, measured against its critical baseline factors, especially by monitoring its dominant members. Faced with the daunting task of detecting all microbes in a population, some proposed instead to survey its collective properties. A metabolic profile of a microbial population, for example, could be monitored for changes after introducing a GEM.

Yet some participants considered perturbation in itself to be irrelevant to assessing whether genetic modification could or does make an organism more harmful. Indeed, according to John Beringer, the multiplicity of organisms in an environment probably means that knocking out one probably would not matter. By focussing any anticipation on identifiable hazards, his approach clashed with some ecologists' more open-ended definition of potential harm.

How the problem is conceptualized makes a difference to whatever risk assessment informs implementation of the regulatory rules. For example, European Commission guidelines enquire about the organism's potential interaction with the environment, 'its involvement in environmental processes: primary production, nutrient turnover, decomposition of organic matter, respiration, etc.' (CEC, 1990: 24). In Britain's *Environmental Protection Act* (1990: 117), harm is defined broadly as 'interference with ecological systems'.

ACCEPTABILITY

The prospect of 'knocking out' an organism illustrates how anticipating likely effects can merge into evaluating their acceptability. Question of acceptable risk, and even acceptable damage, arise in judging whether a biopesticidal GEMMO may kill non-target organisms or lead to pesticide resistance.

For example, a baculovirus has been genetically modified to secrete a scorpion toxin into the insect host that it infects. This is the latest stage of a project run by David Bishop in Oxford, at NERC's Institute of Virology and Environmental Microbiology (IVEM). At Regem 2, Robert Possee reported that the novel baculovirus had significantly reduced the time needed for killing the insects and thus could reduce crop damage (Stewart, *et al.*, 1991). IVEM had gathered laboratory evidence to suggest that the virus infects only certain insect pests, and then requested permission to do a small-scale field trial.

However, last summer Advisory Committee on Releases to the Environment (ACRE) raised several concerns about the proposed release. Although the Health and Safety Executive does not disclose the committee's proceedings, at least one committee member (Williamson, 1991) has expressed his concern publicly. In particular, the host range of the virus becomes all the more important it will persist in the environment for several years and poison any infected host. The risk assessment involves a judgement about the type and extent of testing warranted to confirm claims about a narrow host range.

One could also ask whether the host insect population can develop resistance to the virus if used on a large-scale commercial basis. Possee stated that there is no evidence for such a possibility, indeed, that resistance is not possible because all the infected insects die. It is unclear what would count as adequate evidence that resistance could not develop, as is already happening with a naturally-occuring biopesticide, Bt (Tabashnik *et al.*, 1990).

John Beringer took the argument further - from the likelihood of such an effect, to its acceptability. If insects develop resistance to a biopesticide, he said, it would be only a commercial problem, not an environmental problem, unless we suppose that resistant insects have a selective advantage. On another occasion he had responded similarly to concerns about genetically-modified crops going weedy: this would be only a problem for farmers, not the environment, he had said at a conference on '*Risk Assessment in Biotechnology*' June, 1991.

Perhaps unintentionally, such reassurances highlighted the sorts of value judgements which may arise in risk assessment. For example, if insect resistance were to develop against a genetically-modified biopesticide, it might render the wild-type biopesticide less effective; would that effect be environmental harm, at least indirectly? Likewise, if genetic modification were to enhance weedy characteristics of a crop, then how should we regard agriculture in relation to the environment? And considering that most of Britain is a 'disturbed environment', dependent upon human management, rather than a 'natural, balanced ecosystem', how can we anticipate the selective advantage of a novel introduction?

SOLUTIONS OR PROBLEMS?

Those sorts of questions could lead to notional risk-benefit analyses, or even to damage-benefit analyses. As the issue was put more starkly, 'GEMMOs aim to damage the environment a bit more in order to get a better product from agriculture' and thus create more wealth, said Peter Baker, of Britain's Department of Trade and Industry. Apparently he had in mind environment-friendly products such as biopesticides which target pests more effectively and specifically, thus yielding a higher-quality agricultural product. Yet, as suggested earlier, there remain uncertainties about how the genetic design would translate into agronomic practice and overall ecological effect. How might harm be evaluated and then balanced with benefits?

And beyond that question, there lies one of purposes. For example, genetically-modified biopesticides are being designed to overcome pest problems which have been aggravated by intensive monocultural agriculture. Will these GEMMOs be seen as the solution, or as part of the problem?

Rita Colwell expressed a quandary in the face of the commercial pressures. Although risk-assessment programmes for GEMMOs have increased funding for some ecological research, she felt it was difficult for scientists to work out a scheme for scientific investigation to progress, while allowing industry to make wealth. Moreover, she said, 'Jeremy Rifkin has forced our planetary society to look at our science, as we should have done in the 1940s, before nuclear power'. Yet it remains unclear how the present regulatory system could deal with any question other than identifiable hazards. Colwell's concerns were running in the opposite direction than Monsanto's Ken Baker, who had opened the Regem 2 conference by complaining that overly stringent regulation of biotechnology is stifling the European industry.

The final speaker, Ray Kemp, described ways in which regulation can be conducted and presented so as to overcome public distrust about hazards. However, such efforts invariably act as a proxy for assuaging unease about industry's motives and power. The present public sympathy for genetically-modified medicines, usually not themselves GEMMOs, may not so readily extend to products which are seen as solving commercial problems more than health or environmental ones. Potentially controversial is the criterion of industrial efficiency which has been guiding new techniques and their likely agricultural applications, such that unintended effects become difficult to separate from intended ones, particularly when commercial priorities conflict with environmental ones (Levidow, 1991).

As Peter Baker argued, scientists will have to involve themselves in the political issues. Indeed, they may have to persuade the public that the problems they are trying to solve do require genetic modification for the best, safest solution. And that means acknowledging that the very direction of science -- along with its anticipation of potential hazards -- involves choices of values, aims, worldviews. If scientists want to communicate with the public, then they would do well to try acknowledging the value judgements embedded in GEMMOs. Otherwise, tabloid sensationalism may continue its successful appeal to the science-fiction fantasies which express people's unease about the remote powers redesigning our world.

ACKNOWLEDGEMENTS

I would like to thank John Beringer and Peter Baker for helpful editorial suggestions, though the final version remains my responsibility alone. This article arises from a research project, 'Regulating the risks of biotechnology', funded by the Economic and social research Council, project no. R000 23 1611. A similar article appeared in the *Society for General Microbiology Quarterly*, February 1992; and expanded version (Levidow, 1992) takes up implications for a precautionary science.

REFERENCES

Davis, B. (1985). 'Bacterial domestication: underlying assumptions', Science 235: 1329, 1332-5.

CEC (1990). 'Council Directive on the Deliberate Release into the Environment of Genetically Modified Organism', Official Journal of the European Communities, Vol. 33: L117/15-27, 8 May.

Environmental Protection Act 1990. London: HMSO, Part VI, 'Genetically Modified Organisms'.

Levidow, L. (1991).'Biotechnology at the amber crossing', *Project Appraisal*, December, 6(4): 234-38.

Levidow, L. (1992). 'A precautionary science for GEMs?, *Microbial Releases* 1, in press. Springer-Verlag.

Levidow, L. and Tait, J. (1991). 'The greening of biotechnology: GMOs as environment-friendly products', *Science and Public Policy*, October, 18(5): 271-80.

OECD (1986). *Recombinant DNA Safety Considerations*. Paris: Organization for Economic Co-operation and Development.

Regal, P. (1985). 'The ecology of evolution: implications for the individualistic paradigm', in O. Halverson et al., eds, *Engineered Organisms in the Environment: Scientific Issues*, Washington, D.C.: American Society for Microbiology, pp. 11-19.

Regal, P. (1988). 'The adaptive potential of genetically engineered organisms in nature', in J. Hodgson and A.M. Sugden, eds, Planned Release of Genetically Engineered Organisms (TREE/Tibtech), Cambridge: Elsevier, pp. S36-38.

Smith, J.M., Dowson, C.G., Spratt, B.G. (1991). 'Localized sex in bacteria', *Nature* 349: 29-31.

Stewart, L., Hirst, M., Ferber, M., Merryweather, A., Cayley, P., Possee, R. (1991). 'Construction of an improved baculovirus insecticide containing an insect-specific toxin gene', *Nature* 352: 85-88.

Tabashnik B, Cushing N., Finson N., Johnson M. (1990). 'Field development of resistance to *Bacillus thuringiensis* in diamondback moth (Lepidoptera: Plutellidae)', *Jnl Economic Entomology* 83 (5): 1671-76.

Tait, J. and Levidow, L. (1992). 'Proactive and reactive approaches to risk regulation: the case of biotechnology', Futures, April, 24(3): 219-31.

Williamson, M. (1991). 'Biocontrol risks', *Nature* 353: 394.

Young, J.P. W. and Wexler, M. (1988). 'Sym plasmid and chromosomal genotypes are correlated in field populations of *Rhizobium leguminosarum*', *Jnl of General Microbiology* 134: 2731-39.

INDEX

Chromosomal polymorphism, 19
Chymosin,
    from *Kluyveromyces lactis*, 61
    in GEMMOs, 59-61
Clavibacterium xyli, 27
2-chlorobenzoic acid, 35,39
3-chlorobenzoic acid, 35,39
Clostridium butyricum, 125
Colonization,
    intestinal, 167-169
Communities,
    genetic interactions, 15-24
Conjugation, 16
m-chlorophenol, 35
Crown gall, 29
Cyanobacteria,
    water bloom, 207-208

DDT,
    effect of use, 26-27
Detection, 16
    by bioluminescence, 199-205, 209-212
    by flow cytometry, 231-234
    by gene probing, 213-215
    by immunofluorescence, 245-247
    by PCR, 213-215, 221-223
    DNA in soil, 157-159
    GEMMOs in lakes, 141-143
    in river water, 253-255
    of GEMMOs, 119-122
    of streptomycetes, 183-185
Detriment,
    concept of, 93-94
2,4-Dichlorophenol, 127
2,4-dichloro-phenoxyacetate, 41
2,2-dichloropropionate, 8
Dispersal,
    by rain splash, 187-189
    GEMMOs in birds, 163-165
DNA,
    detection in soil, 157-159
    gene flux, 171-174
    persistence of, 117
    probes, 195-196

Emissions,
from fermentation plant, 129-131
Enterobacter, 125
    agglomerans, 120, 179-180
        in soil, 157-159
    cloacae, 116-117, 153
        survival, 245-247
    nitrogen-fixing, 155-156
Enterobacteriaceae, 15
Environmental stress,
    protection against, 28
Environments,
    contaminated, 33-45
Erwinia caratovora, 1, 6

Escherichia coli, 2, 4, 17, 18, 20, 149-151, 155-156
    bioluminescent, 213-215
    bioluminescent in soil, 203-205
    colonization by, 167-169
    enteropathogenic, 117
    in birds, 163-165
    transfer of pBR322, 197-198, 203-205
Exxon Valdez spill, 41,87-88

Fertilizers, oleophilic, 41
Field release,
    baculovirus insecticides, 47-58
Flavobacterium balustinum,
    survival of, 145-147
Flow cytometry, 121-122
    detection by, 231-234
Fluorescent antibody, 121-122
Food enzymes,
    chymosin, 59-61
Food fermentations, 65-66
Food industry,
    GEMMOs in, 59-69
Food poisoning, 17
Frankia, 26
Frost damage,
    protection against, 28
Fusion proteins,
    with prochymosin, 60

GEMMOs,
    abiotic pressures, 3
    as inocula,85-92
    behavioral patterns, 4-7,
    biotic pressures on, 3
    case for, 41-42
    chymosin producing, 59-61
    commercial exploitation, 85-92
    detection, 119-122
    by PCR, 221-224
    environmental pressure on, 1-14
    food fermentations, 62-66
    impact in soil, 137-139
    in agriculture, 25-31
    in birds, 163-165
    in brewing, 64-65
    in soil, 1-14, 179-180, 203-205
    selective advantage, 7
    monitoring of, 39-40, 217-219
    persistence in environment, 115-118
    public confidence, 99-114
    release of, 38-39
    risk perception, 99-114
    safety in food, 66-67
    social implications, 99-114
    survival in environment, 115-118
    value of, 261-265
    wheat-flour bleaching, 64

Opines, 28

Panolis flammea,
    nuclear polyhedrosis virus, 54
PCBs, 38,42,88
Penicillin-binding proteins, 18
Persistence,
    GEMMOs in environment, 115-118
    measurement of, 116-117
    monitoring of, 29-31
    viable non-culturable, 117
Phanerochaete chrysosporium, 41, 42
Phenazine-1-carboxylic acid, 9
Phenol, 35
Phenotype tracking, 29-30
Phosphobacterin, 26
Plasmids, 16
    markers in soil,183-185
    polymorphism, 19
    stability, 151, 153
    survival of, 257-259
    transfer, 153
        Pseudomonas aeruginosa, 175-
        177
        of pBR322, 197-198
Pollutants,
    degradation of, 33
Pollution,
    control, 86-89
    treatment of, 35-38
Polychlorinated biphenyls, 38
Polymerase chain reaction, 30, 120-121,
    213-215, 221-223
Polymorphism,
    chromosomal, 19
    plasmid, 19
Probes,
    DNA, 30, 195-196, 213-215
Protozoa,
    predation by, 3
Pseudomonads,
    decay rates in soil, 5
Pseudomonas, 126-127
    aeruginosa, 5,8, 90
        plasmid transfer, 175-177, 241-
        244
    cepacia, 8, 9, 18, 88, 137-139, 153
    fluorescens, 3, 5, 7, 9, 10, 36, 133-
        135
    containing Tn5, 3
    ice nucleation active, 28
    pathogenic, 181-182
    psychrophilic, 3
    putida, 5, 8, 35-36, 39, 41
        2,4-dichlophenoxyacetate-
        degrading, 6
    in lakes, 141-143
    survival in soils, 191-193
    solanacearum, 1, 126
    stutzeri, 17, 171-174

syringae, 1, 28, 89, 125
    dispersal, 187-189
    in soil, 199-202
Pseudosenescence, 30

Rain splash,
    dispersal by, 187-189
Release,
    baculovirus insecticides, 47-58
Resistance determinants,
    to Co, Zn and Cd, 10
Rhizobia, 1
    interaction in soil,20
Rhizobiaceae, 26
*Rhizobium,*1, 4, 7, 8, 15, 20, 26, 116
    inoculants, 3
    leguminosarum, 6, 20
        bioluminescent, 225-227
        survival of, 217-219
    loti,
        DNA probes, 195-196
    meliloti, 8,18
Rhizosphere,
    interactions in, 229-230
Risk assessment, 93-98, 155-156, 225-
    227, 235-236
    bakers' yeast, 109-112
    input to design, 96-97
    input to operation, 96-97
    modified baculoviruses, 53-54
    pathogenic materials, 97
    radiation, 94
    toxic materials, 97
Risk perception, 99-114
    of GEMMOs, 100
    studies of, 100
Risk,
    communication of, 106-109
    social psychology of, 101-105
Rumen bacteria,
    genetic manipulation, 249-251

Saccharomyces, 30
    cerevisiae,
        glucanase-secreting, 64-65
        lipoxygenase secreting, 64
        starch utilizing, 63-64
        with constitutive maltase, 62-63
Safety,
    of food GEMMOs, 66-67
Salmonella, 18, 20, 117
    aro mutants, 73-75
    choleraesuis, 73
    enterica,
        serotype Derby, 18
        serotype Enteritidis, 17, 72, 75
        serotype Newport, 18
        serotype Typhimurium, 17,18
    flagellin,
        expressing foreign epitopes, 76